Lecture Notes in Networks and Systems 1033

The series "Lecture Notes in Networks and Systems" publishes the latest developments in Networks and Systems—quickly, informally and with high quality. Original research reported in proceedings and post-proceedings represents the core of LNNS.

Volumes published in LNNS embrace all aspects and subfields of, as well as new challenges in, Networks and Systems.

The series contains proceedings and edited volumes in systems and networks, spanning the areas of Cyber-Physical Systems, Autonomous Systems, Sensor Networks, Control Systems, Energy Systems, Automotive Systems, Biological Systems, Vehicular Networking and Connected Vehicles, Aerospace Systems, Automation, Manufacturing, Smart Grids, Nonlinear Systems, Power Systems, Robotics, Social Systems, Economic Systems and other. Of particular value to both the contributors and the readership are the short publication timeframe and the world-wide distribution and exposure which enable both a wide and rapid dissemination of research output.

The series covers the theory, applications, and perspectives on the state of the art and future developments relevant to systems and networks, decision making, control, complex processes and related areas, as embedded in the fields of interdisciplinary and applied sciences, engineering, computer science, physics, economics, social, and life sciences, as well as the paradigms and methodologies behind them.

Indexed by SCOPUS, INSPEC, WTI Frankfurt eG, zbMATH, SCImago.

All books published in the series are submitted for consideration in Web of Science.

For proposals from Asia please contact Aninda Bose (aninda.bose@springer.com).

Alhamzah Alnoor · Mark Camilleri ·
Hadi A. Al-Abrrow · Marco Valeri ·
Gül Erkol Bayram · Yousif Raad Muhsen
Editors

Explainable Artificial Intelligence in the Digital Sustainability Administration

Proceedings of the 2nd International
Conference on Explainable Artificial
Intelligence in the Digital Sustainability
Administration (AIRDS 2024)

 Springer

Editors
Alhamzah Alnoor
Southern Technical University
Basrah, Iraq

Hadi A. Al-Abrrow
University of Basrah
Basrah, Iraq

Gül Erkol Bayram
Sinop University
Sinop, Türkiye

Mark Camilleri
University of Malta
Msida, Malta

Marco Valeri
Niccolò Cusano University
Rome, Italy

Yousif Raad Muhsen
Wasit University
Wasit, Iraq

ISSN 2367-3370   ISSN 2367-3389 (electronic)
Lecture Notes in Networks and Systems
ISBN 978-3-031-63716-2   ISBN 978-3-031-63717-9 (eBook)
https://doi.org/10.1007/978-3-031-63717-9

This Springer imprint is published by the registered company Springer Nature Switzerland AG
The registered company address is: Gewerbestrasse 11, 6330 Cham, Switzerland

If disposing of this product, please recycle the paper.

Preface

Climate change and sustainable development goals are the most pressing global challenges facing society today, with potentially detrimental impacts on individuals, organizations, and societies. The impact of digital technologies on climate change is one of the key research priorities and is considered a researchable spot. Climate-intelligent information systems solutions and environmental, social, and governance intelligence leverage the transformative power of information systems to mitigate adverse environmental impacts. Information systems researchers can use these solutions to create practical impact. Furthermore, digital sustainability activities can advance environmental sustainability goals by creatively deploying technologies that create, use, or transmit electronic source data. The 2nd International Conference (AIRDS 2024) is held to address the theme "Explainable Artificial Intelligence in the Digital Sustainability Administration." The AIRDS 2024 brings together a wide range of researchers from different disciplines. It seeks to call for research contributions to mitigate and adapt to the effects of climate change, as it could cause far-reaching disruptions to communities and the economy worldwide. The main aim of the AIRDS 2024 is to provide a forum for academics, researchers, and developers from academia and industry to share and exchange their latest research contributions and identify practical implications of emerging technologies to advance the wheel of these solutions for global impact. In line with the Fourth Industrial Revolution goals and its impact on sustainable development, AIRDS 2024 is devoted to increasing the understanding and impact of explainable artificial intelligence on individuals, organizations, and societies and how artificial intelligence applications have recently reshaped these entities. In addition to the contribution of explainable artificial intelligence applications to achieve sustainable development goals.

The AIRDS 2024 attracted 89 submissions from different countries worldwide. Out of the 89 submissions, we accepted 26, representing an acceptance rate of 29.21%. The chapters explore the diverse implications of explainable artificial intelligence in various professional and social sectors. It opens with insights from explainable machine learning for real-time payment fraud detection and introduces an advanced machine learning model integrated with explainable artificial intelligence techniques to enhance the detection of payment fraud in real-time scenarios within the digital finance sector. It scrutinizes the potential impacts of robotic process automation on sustainable audit quality, including both opportunities and challenges. This volume also assesses the strategic potential of artificial intelligence in improving customer service and facilitating the retention of organizational human resources. Education continues to be a key theme, with discussions on the factors driving schoolteachers to adopt virtual educational resources to elevate the educational process. Furthermore, the chapters delve into the evolving landscape of digital sustainability within artificial intelligence, considering brand activity and consumer behavior. The book addresses serious considerations of intelligent agriculture decision support tools, artificial intelligence applications in advertising, energy reduction, big data analytics, the Internet of things, classifying solar radiation time series, and a unified technology acceptance model in boosting digital sustainability. The financial sector is not left out, with a case study explaining artificial intelligence's role in sustainable audit quality and financial sustainability. Finally, the volume looks at how artificial intelligence furthers the sustainability of healthcare systems by improving efficiency, disease diagnosis, optimizing resources, and developing in-person and remote care initiatives. Each chapter offers a distinct perspective, providing readers with a well-rounded understanding of the challenges and prospects of artificial intelligence in the digital sustainability administration.

Each submission is reviewed by at least two reviewers, who are considered experts in the related submitted paper. The evaluation criteria include several issues: correctness, originality, technical strength, significance, presentation quality, interest, and relevance to the conference scope. The conference proceedings are published in *Lecture Notes in Networks and Systems Series* by Springer, which has a high SJR impact. We acknowledge all those who contributed to the success of AIRDS 2024. We would also like to thank the reviewers for their valuable feedback and suggestions. Without them, it was impossible to maintain the high quality and success of AIRDS 2024.

Alhamzah Alnoor
Mark Camilleri
Hadi A. Al-Abrrow
Marco Valeri
Gül Erkol Bayram
Yousif Raad Muhsen

Organization

Conference General Chairs

Hadi AL-Abrrow

Department of Business Administration, College of Administration and Economics, University of Basrah, Basrah, Iraq

Camilleri

University of Malta, Msida, Malta

Alhamzah Alnoor

Southern Technical University, Basrah, Iraq

Honorary Conference Chair

Mohanad J. K. Al Asadi

Chancellor of University of Basrah, Basrah, Iraq

Conference Organizing Chair

Abdul Hussain Tawfiq Shibli

Dean of College of Administration and Economics, University of Basrah, Basrah, Iraq

Program Committee Chair

Marco Valeri

Faculty of Economics, Niccolo' Cusano University in Rome, Italy

Publication Committee Chairs

Alhamzah Alnoor

Management Technical College, Southern Technical University, Basrah, Iraq

Mark Camilleri

Faculty of Media & Knowledge Sciences, University of Malta, Msida, Malta

Hadi Al-Abrrow

Department of Business Administration, College of Administration and Economics, University of Basrah, Basrah, Iraq

Marco Valeri

Niccolo' Cusano University, Italy

| Gül Erkol Bayram | School of Tourism and Hospitality Management Department, Sinop University, Sinop, Turkey |
| Yousif Raad Muhsen | Civil Department, College of Engineering, Wasit University, Wasit, Iraq |

Conference Tracks Chairs

Sammar Abbas	Institute of Business Studies, Kohat University of Science and Technology, Pakistan
Marcos Ferasso	Economics and Business Sciences Department, Universidade Autónoma de Lisboa, 1169-023 Lisboa, Portugal
Hussam Al Halbusi	Department of Management at Ahmed Bin Mohammed Military College, Doha, Qatar
Khai Wah Khaw	School of Management, Universiti Sains Malaysia, 11800, Pulau Pinang, Malaysia
Yousif Raad Muhsen	Civil Department, College of Engineering, Wasit University, Wasit, Iraq
Gül Erkol Bayram	School of Tourism and Hospitality Management Department of Tour Guiding, Sinop University, Sinop, Turkey
Gadaf Rexhepi	South East European University, Tetovo, The Republic of Macedonia

Members of Scientific Committee

Muntader F. Saad	Head of the Dept. of Financial Sciences, University of Basrah, Basrah, Iraq
Hadi AL-Abrrow	Department of Business Administration, College of Administration and Economics, University of Basrah, Basrah, Iraq
Khai Wah Khaw	School of Management, Universiti Sains Malaysia, Malaysia
Amjad S. Abdulaali	Dept. of Economics, University of Basrah, Basrah, Iraq
Suhail A. Nasser	Dept. of Accounting, University of Basrah, Basrah, Iraq
Gadaf Rexhepi	Southern European University, The Republic of Macedonia

Wameedh A. Khdair   Department of Business Administration, College
of Administration and Economics, University
of Basrah, Basrah, Iraq

Sammar Abbas   Institute of Business Studies, Kohat University of
Science and Technology, Pakistan

Elham J. Hamid   Dept. of Accounting, University of Basrah,
Basrah, Iraq

Walid M. Rudin   Head of Admin. Info. Systems Dept, University of
Basrah Basrah, Iraq

Rabee K. Thajeel   Dept. of Economics, University of Basrah,
Basrah, Iraq

Baha A. Qasim   Head of Statistics Dept, University of Basrah,
Basrah, Iraq

Gul Erkol Bayram   Sinop University, School of Tourism and
Hospitality Management Department of Tour
Guiding, Turkey

Mohammed J. Mohammed   Dept. of Financial Sciences, University of Basrah,
Basrah, Iraq

Ali N. Hussein   Dept. of Statistics, University of Basrah, Basrah,
Iraq

Faiza Hassan   Dept. of Financial Sciences, University of Basrah,
Basrah, Iraq

Alhamzah Alnoor   Management Technical College, Southern
Technical University, Basrah, Iraq

Hussan Al Halbusi   Department of Management, Ahmed Bin
Mohammed Military College, Qatar

Marcos Ferasso   Economic and Business Sciences Department,
Universidade Autonoma de Lisboa, Portugal

Publicity and Public Relations Committee

Ammar Y. Dhicher   Dean Assistant for Scientific Affairs, University
of Basrah, Basrah, Iraq

Mohanad H. Saleh   Dean Assistant for Students Affairs, University of
Basrah, Basrah, Iraq

Hadi AL-Abrrow   Department of Business Administration, College
of Administration and Economics, University
of Basrah, Basrah, Iraq

Finance Chair

Abdul Hussain Tawfiq Shibli   Dean of College of Administration and
                                      Economics, University of Basrah, Basrah, Iraq

Contents

Explainable Machine Learning for Real-Time Payment Fraud Detection: Building Trustworthy Models to Protect Financial Transactions

Ahmed Abbas Jasim Al-hchaimi[1] [ID], Mohammed F. Alomari[2],
Yousif Raad Muhsen[3,4], Nasri Bin Sulaiman[5]([⊠]), and Sabah Hassan Ali[6]

[1] Thiqar Technical College, Southern Technical University, Basrah, Iraq
`ahmed.alhchaimi@stu.edu.iq`
[2] College of Graduate Studies, Universiti Tenaga National (UNITEN), Kajang, Malaysia
[3] Faculty of Engineering, Universiti Putra Malaysia, Serdang, Selangor, Malaysia
`yousif@uowasit.edu.iq`
[4] Civil Department, College of Engineering, Wasit University, Kut, Wasit, Iraq
[5] Faculty of Engineering, Universiti Putra Malaysia, Serdang, Malaysia
`nasri_sulaiman@upm.edu.my`
[6] Iraqi Ministry of Interior, Baghdad, Al Muthanna, Iraq
`sabah.hassan@mu.edu.iq`

Abstract. In this study, we introduce an advanced machine learning model integrated with explainable AI techniques to enhance the detection of payment fraud in real-time scenarios within the digital finance sector. As online transactions continue to proliferate, so too do the fraudulent activities associate with them. Our approach effectively differentiates between legitimate and fraudulent transactions by meticulously analyzing key features such as transaction amount, type, and the accounts involved. Through a comprehensive evaluation of various machine learning models, the Decision Tree model emerged as the most effective, achieving an accuracy of 95.4048%, precision of 92.9461%, recall of 98.2456%, and an F1-score of 95.5224%. This study not only proposes a robust and explainable machine learning framework but also significantly enhances the transparency of fraud detection decisions. It equips financial institutions with a potent tool to safeguard their customers' assets against fraud, thereby bolstering the reliability and trustworthiness of digital payment systems.

Keywords: Digital Finance Security · Real-time Fraud Analysis · Decision-Tree-Model · Transaction Analysis

1 Introduction

In the swiftly evolving landscape of digital finance, the advent of online transactions has brought unparalleled convenience and efficiency to consumers worldwide. Credit cards, once a luxury, are now a cornerstone of everyday commerce, facilitating billions of transactions across the globe (Al-Enzi et al., 2023; Husin et al., 2023; Atiyah

et al., 2023, b, c; Soltani et al., 2023). The rapid ranging of these digital payments cou-
pled with the increasing number of fraudulent activities has brought great challenges
to financial systems (Alnoor et al., 2024, b, c). Another vulnerable point of financial
institutions is financial fraud (Fig. 1), for instance payment fraud, which damages online
payments security and many customers' faith. Statistically, The Federal Trade Commission represent-ed that more than 3.7 billion dollars was lost to credit card fraud in 2020
(Wright et al., 2020; Ahmed et al., 2024; Avila-Cano et al., 2023).

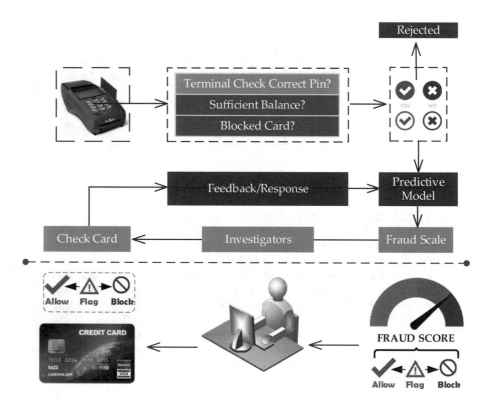

Fig. 1. Typical Online Money Transaction Process.

This demonstrates the need for efficient fraud identification technology. The technology standards that exist today find it hard to meet both the complexities and the
volume of transactions while at the same time assuring seamless user experience associated with digital payment and seamless user experience (Al-Hchaimi et al., 2022, 2023;
Muhsen et al., 2023; Atiyah et al., 2023, b, c; Innan et al., 2024). The fight against the
payment fraud is still complicated by many difficulties. First, the development of fraud
techniques is accelerating rapidly - fraudsters are always outpacing traditional controls,
constantly creating new ways by which fraud could penetrate the screening mechanisms.
Further, the figure of the pecuniary chases the detection process as every transaction has
its particular unevenness. Traditional systems find it difficult to differentiate between

legitimate and fraudulent transactions and make many true-false decisions. This negatively affects the completion of many legitimate transactions and leads to discontent of the customers. One more issue which demands urgent attention is immediate detection of threat. The fraud detection process can take a long time, which is when the fraudulent transaction happens, before the intervention can be made, financial losses occur and logistical nightmares persist in trying to explain them (Muhsen et al., 2023a; Abbas et al., 2023).

Considering these obstacles, the purpose of this project would be the creation of a model that is able to explain fraudulent payments and identifies them before they can be processed. Transactional properties including the amount, the type, and the account numbers are going to be analyzed and a model will be trained using dataset which contains both fraudulent and non-fraudulent transactions (Chen et al., 2019; Ni et al., 2023; Nijwala et al., 2023; Ali et al., 2024; Alnoor et al., 2023; Alnoor et al., 2024, b, c; Cai et al., 2024). The target of it is to provide a highly accurate solution. It is of great importance for financial sector to deliver this accuracy because it helps them to save money, and it is also trusted by their customers which is a vital component of financial markets.

The widespread phenomenon of payment fraud not only becomes more widespread but also, more technologically complex, make it an important research area to address this problem. The financial and psychological consequences that the individual victims who suffered the attack had to experience in addition to the impact of this incident on the overall public trust in online payment systems creating a consistent necessity for advanced detection techniques. This initiative is based on a firm conviction that bringing about honest AI in detecting fraud is capable of changing how financial institutions pick and trip on fraudulent transmissions. With improved transparency and comprehension of these processes, stakeholders will be able to directly participate in the detection of inconsistencies (Ahmadi, 2023; Alnoor et al., 2024, b, c; Pallathadka et al., 2023). This will make electronic payment solutions more relevant to a broader user base, resulting in a higher level of trust in the technology as well as its widespread adoption. The main contribution of this paper can be summarized as follows:

(i) Proposes a novel approach that integrates explainability into the ML model, ensuring that decisions are transparent and justifiable.
(ii) Demonstrates the model's effectiveness in discerning fraudulent transactions as they occur, markedly reducing the potential for financial loss.
(iii) Enhances the model's ability to recognize and learn from complex patterns of fraud.
(iv) Offers a comprehensive evaluation of the model's predictive accuracy, leveraging a diverse dataset to benchmark against existing fraud detection methodologies.
(v) Outlines actionable guidelines for integrating the developed model into existing fraud detection frameworks, ensuring that financial institutions can seamlessly adopt and benefit from this research.
(vi) Aims to empower financial institutions with advanced tools to safeguard against fraudulent transactions, restoring confidence in the security of digital payment platforms.

The reminder of this paper is outlined as: Sect. 2 parents the literature review. Section 3 investigates the research methodology, furthermore, Sect. 5 shows the results and discussion. Finally, Sect. 5 concludes this paper.

2 Literature Review

The prevalence of credit card fraud and its evolving complexity have made it a focal point for numerous studies aiming to harness advanced technologies for more effective detection solutions. Authors of (Cherif et al., 2023) critically evaluated research up to 2021, revealing a considerable exploration gap in deploying deep learning methods for credit card fraud detection. This gap underscores the necessity for novel approaches capable of handling the intricate patterns of fraudulent transactions. For instance, authors of (Asha et al., 2021) demonstrated the superiority of ANN models over traditional SVM and KNN models in accuracy, precision, and recall, yet highlighted the challenges of class imbalance and the need for undersampling strategies. Table 1 lists more details about the transaction fraud and utilized ML approaches for detection, mitigation, and prediction.

This hypothetical table compares various methodologies, from traditional machine learning techniques such as Logistic Regression (LR), Decision Trees (DT), and Support

Table 1. Summary of Prior Studies in the Context of Online Transaction Fraud.

Ref	ML Type	Algorithm(s)	Domain	Dataset	Strengths	Limitations
(Asha et al., 2021)	Traditional	ANN vs. KNN and SVM	Credit card	Public (Kaggle)	High accuracy (0.9992)	Low recall; potential information loss due to under-sampling
(Singh et al., 2023)	Traditional	Shallow NN	Credit card fraud	Public (ULB)	Hybrid swarm intelligence for feature selection	Small dataset; non-deep architecture
(Ni et al., 2023)	Traditional	KNN+LDA+LR	Credit card	Public	High recall across 5 datasets	Low precision
(Jose et al., 2023)	Traditional	DT, RF, ET, GB	Credit card	synthetic (Sparkov)	Effective use of AdaBoost; large dataset	Oversampling may lead to overfitting
(Afriyie et al., 2023)	Traditional	LR, DT, RF	Credit card	synthetic (Sparkov)	Large dataset	Under sampling reduces dataset size; complexity with RF
(Liu et al., 2018)	Deep	GNN	Online payment	Private (collected from Alipay)	Robust model for large datasets	Lack of class imbalance discussion
(Liu et al., 2021)	Deep	GNN	Class imbalance for fraud	Multiple public datasets	Adaptability to various use cases and datasets	Better performance on smaller datasets; less effective for financial fraud

Vector Machines (SVM) to more contemporary approaches employing Artificial Neural Networks (ANN) and hybrid models. Each study's dataset, key strengths, and limitations are delineated, showcasing the evolution of fraud detection strategies over time.

Our paper distinguishes itself by revisiting and refining the use of traditional machine learning models—LR, DT, SVM, and K-Nearest Neighbors (KNN)—infused with the latest advancements in explainable AI to tackle credit card fraud. Unlike previous studies that predominantly leveraged either deep learning approaches or conventional models in isolation, our research synthesizes the strengths of traditional algorithms with the clarity and interpretability afforded by explainable AI. This innovative blend aims to address the critical challenges identified in prior research, including handling large and complex datasets, dealing with class imbalance effectively, and enhancing the precision and recall of fraud detection without sacrificing the model's transparency or interpretability.

Furthermore, our research uniquely combines the reliability and tested nature of traditional machine learning models with the cutting-edge advancements in explainable AI. This dual strategy is designed to not only counteract the limitations identified in previous studies but also to push the frontier in credit card fraud detection towards more transparent, interpretable, and real-time solutions.

3 Methodology

The methodology of our study (see Fig. 2) is designed to develop and evaluate explainable machine learning models for real-time payment fraud detection. This section outlines the systematic approach taken from data preprocessing to the final evaluation of the models.

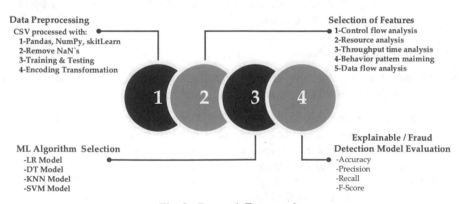

Fig. 2. Research Framework.

3.1 Data Preprocessing

Data preprocessing is the first critical step in our methodology, ensuring the quality and compatibility of our dataset with the machine learning algorithms. The following sub-steps are involved:

(i) Pandas, NumPy, and Scikit-Learn Usage: We employ Pandas for data manipulation, NumPy for numerical operations, and Scikit-Learn for applying pre-processing techniques and machine learning algorithms. These libraries facilitate the handling of CSV files and the execution of complex data transformations.

(ii) NaN Values Removal: We clean the dataset by removing or imputing NaN (Not a Number) values to maintain the integrity of our analysis and ensure that our models are trained on complete and accurate data.

(iii) Training and Testing Split: The dataset is divided into training and testing sets, allowing us to fit our models on one portion of the data and evaluate their performance on another, unseen portion, thus assessing their predictive capabilities.

(iv) Encoding Transformation: The concept of encoding is used to assign a number to all the categories thus making them interpretable to machine learning algorithms.

3.2 Selection of Features

(i) Feature selection aims to identify the most informative and relevant features for predicting fraudulent transactions. This process involves:

(ii) Control Flow Analysis: Analyzing the sequence and operations of transactions to identify suspicious patterns.

(iii) Resource Analysis: Evaluating transaction resources, such as account information and transaction amounts, to pinpoint features indicative of fraud.

(iv) Throughput Time Analysis: Investigating the duration of transaction processes to detect anomalies that may signify fraudulent activities.

(v) Behavior Pattern Mining: Mining user or account behavior over time to discover patterns that differentiate between fraudulent and legitimate transactions.

(vi) Data Flow Analysis: Assessing the movement and destination of transaction data to uncover atypical paths associated with fraud.

3.3 ML Algorithm Selection

We selected a range of machine learning algorithms to construct our fraud detection models, each chosen for its unique strengths in classification tasks:

(i) LR Model: Logistic regression becomes particularly useful for making decisions easy to understand and taking a binary classifications process.

(ii) DT Model: Decision Trees are for their intuitions and ability to model non-linearity without too much need to be undertaken in the manner of data preparation.

(iii) KNN Model: The K-Nearest Neighbors algorithm is selected because of its simplicity of implementation and the classification accuracy it provides with the training set as analogical matrix.

(iv) SVM Model: Support Vector Machines perform exceptionally well as they are famous for their ability to perform in high-dimensional spaces, therefore they are suitable for those datasets that have much number of features.

3.4 Explainable/Fraud Detection Model Evaluation

Our final step involves a thorough evaluation of the developed models using key metrics to measure their performance in detecting fraudulent transactions. These metrics include:

(i) Accuracy: The proportion of total predictions that were correct.

(ii) Precision: The ratio of true positive predictions to the total positive predictions, indicating the quality of positive class predictions.

(iii) Recall: The ratio of true positive predictions to the actual number of positive class samples, reflecting the model's ability to capture positive class instances.

(iv) F-Score: The harmonic mean of precision and recall, providing a balance between the precision and recall metrics.

(v) This comprehensive methodology ensures the development of accurate, reliable, and explainable models for real-time payment fraud detection, addressing both the technical and practical requirements of such systems.

4 Dataset Exploration and Description

The "Online Payments Fraud Detection Dataset" is designed to aid in the identification and analysis of fraudulent transactions in online payment systems. Each record in this dataset encapsulates a transaction's details, allowing for a comprehensive exploration of transaction patterns and potential fraud indicators (Dornadula et al., 2019; Vanini et al., 2023). Below is a breakdown of the dataset's key features:

- step: This attribute signifies a unit of time, with one step equating to one hour. This temporal aspect can be critical in understanding the timing and frequency of transactions, which may help in detecting suspicious patterns often seen in fraudulent activities.
- type: This field describes the type of online transaction executed. Different transaction types may exhibit varying risks and patterns of fraud.
- amount: Represents the monetary value of the transaction. Analysis of transaction amounts can provide insights into typical user behavior and identify outliers that might suggest fraudulent activities.
- nameOrig: Indicates the customer initiating the transaction. By tracking the originator, one can analyze individual behavior and detect potentially fraudulent operations based on historical data.
- oldbalanceOrg: Shows the balance of the originating customer's account before the transaction took place. This can be useful in verifying the authenticity of the transaction amount and the account's typical activity level.
- newbalanceOrig: The balance in the originating customer's account after the transaction. Changes in this balance, when compared with the amount and previous balance, can help in identifying discrepancies that might indicate fraud.
- nameDest: The recipient of the transaction. Monitoring recipients can help in identifying suspicious accounts that frequently receive funds from different sources or are involved in split transactions, a common method in money laundering.
- oldbalanceDest: The initial balance of the recipient's account before receiving the transaction. This feature can aid in understanding the flow of money and whether the credited amounts align with typical account profiles.
- newbalanceDest: Reflects the balance in the recipient's account after the transaction. This is crucial for detecting whether the credited money is being quickly moved out, which is typical in layering stages of money laundering.

- isFraud: A binary indicator where '1' represents a fraudulent transaction and '0' represents a non-fraudulent transaction. This is the target variable used to train models to detect and predict fraudulent transactions.

Understanding the relationships and trends among these attributes can provide crucial insights into typical and atypical transaction behaviors, aiding in the development of robust fraud detection algorithms.

4.1 Exploratory Data Analysis

- **Univariate Analysis** (Caixeta et al., 2023): In this phase, each variable in the dataset is analyzed individually to summarize and find patterns in the data. The first aim is to explore each of the transaction types, their amount and balance distributions. For instance, by looking at the 'type' column we can see that "CASH_OUT" and "PAYMENT" are two of the most frequently used transaction types, and hence, are the prevalent activities inside the dataset. By recognizing these distributions, it is possible to detect those transaction types that have the highest likelihood of being associated with fraud.
- **Bivariate Analysis** (Shi et al., 2023): Bivariate analysis consists of analyzing correlations and interactions between two variables, which helps us determine relationships between data points. It may entail studying the patterns of transaction amounts that vary depending on the type of transactions, or comparing the old and new balances in an account to observe any anomalies that are likely fraudulent behaviors.
- **Multivariate Analysis** (Kadhuim et al., 2023): This analysis is not restricted to two variables but it looks at a multi-factorial environment to unfold the relationships within the dataset in a more complex manner. This maybe realizing how amounts of transactions, types of transaction, and account balances fluctuate over time or different steps. The fact that multivariate analysis helps in finding the patterns and glitches that would not seem obvious when we look at the variables individually cannot be overemphasized.

4.2 Bivariate Analysis: Transaction Type vs. Fraud Occurrence

This part of a bivariate analysis reveals the association between the type and whether it is classified as a fraudulent transaction. Findings that link various types of transactions to the frequency of fraud incidents can lead to complexities for producing fraud prevention measures that aim to specialized goals. Analysis of Transaction Types and Fraud Rates: Analysis of Transaction Types and Fraud Rates:

- CASH_IN: Total: 227130 transactions; Fraudulent: 0 transactions (NaN indicates nothing of this kind happened); Observation: Transactions involving "CASH_IN" do not have any committed fraud recorded. This underscores the fact that digital currency transactions are more naturally safe and tracking exists more clearly.
- CASH_OUT: The figure: 373,063 transactions; Among them: 578 transactions which are fraudulent; Summary: This amount represents the total number of "CASH_OUT" transactions where fraudulent transactions are a small yet remarkable number.
- This type may be more susceptible to fraud due to the nature of withdrawing funds, requiring more stringent monitoring and detection mechanisms.
- DEBIT: Total: 7,178 transactions; Fraudulent: 0 transactions (NaN indicates no occurrences); Analysis: Similar to "CASH_IN", "DEBIT" transactions show no fraudulent activities. This might suggest that direct debit transactions are less likely targets for fraudsters or are well-protected.
- PAYMENT: Total: 353,873 transactions; Fraudulent: 0 transactions (NaN indicates no occurrences); Analysis: Transactions classified under "PAYMENT" do not report any fraudulent instances. This might be due to the types of payments processed, which may generally involve smaller amounts or transactions within controlled environments.
- TRANSFER: Total: 86,189 transactions; Fraudulent: 564 transactions: Analysis: "TRANSFER" transactions, although fewer in total number compared to "CASH_OUT", also exhibit a notable amount of fraud. This type involves moving funds between accounts, which can be particularly attractive to fraudsters attempting to misappropriate large sums.

4.3 Transaction Type Distribution

The distribution of transaction types within the dataset is a critical factor to consider:

- CASH_OUT: 373,641 transactions
- PAYMENT: 353,873 transactions
- CASH_IN: 227,130 transactions
- TRANSFER: 86,753 transactions
- DEBIT: 7,178 transactions

Figure 3 shows that "CASH_OUT" and "PAYMENT" are the most commonly recorded transaction types. Such high frequencies indicate these are routine transactions for many users, potentially making them prime targets for fraudsters due to their frequency and volume. Conversely, "TRANSFER" and "DEBIT" transactions, while less frequent, may involve higher amounts or be linked to more complex transaction networks, potentially increasing their risk for fraudulent activities. Understanding these patterns helps in tailoring fraud detection systems to be more responsive to the nature of each transaction type.

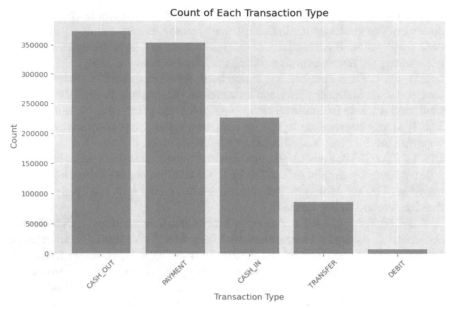

Fig. 3. Transaction Categories per Utilized Dataset.

- The dataset contains five types of transactions: CASH_OUT, PAYMENT, CASH_IN, TRANSFER, and DEBIT
- Among these types, the most common transactions are CASH_OUT and PAYMENT, with counts of 373,641 and 353,873, respectively.
- CASH_IN transactions are also quite common, but less frequent compared to CASH_OUT and PAYMENT, with a count of 227,130.
- TRANSFER and DEBIT transactions are relatively less common, with counts of 86,753 and 7,178, respectively as deployed in Fig. 4.

4.4 Distribution of Fraudulent Transactions

In the exploration of fraudulent transactions within the dataset, it is crucial to understand the proportion of transactions classified as fraud compared to legitimate transactions as shown in Fig. 5. This distribution is a fundamental aspect of the dataset that influences how fraud detection models are developed and evaluated. The dataset shows a significant imbalance in the classification of transactions:

- Not Fraud: 99.9% of the transactions are legitimate. This high percentage reflects the typical nature of transaction datasets where fraudulent activities are rare but potentially very harmful.
- Fraud: Only 0.1% of transactions are fraudulent. Although this represents a small fraction of the total transaction volume, the absolute numbers can still be significant given the large scale of data. This small percentage highlights the challenges in detecting fraud, as the models must identify these rare events without mistakenly classifying legitimate transactions as fraud.

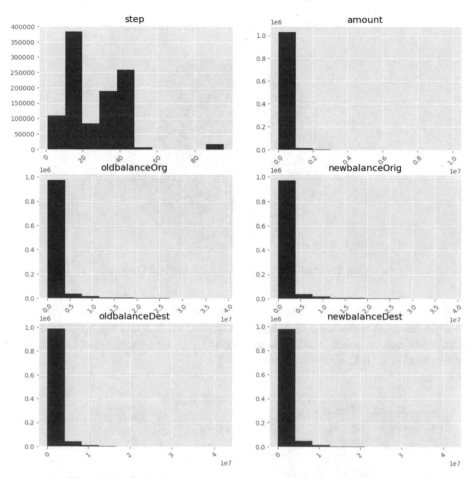

Fig. 4. Distribution of transaction attributes in Online Payment Data.

The stark disparity in the distribution between fraudulent and non-fraudulent transactions poses challenges for data scientists and analysts. Models trained on such data may have a tendency to overwhelmingly predict transactions as non-fraudulent due to the imbalance. This can lead to a high accuracy rate that is misleading because the rare fraudulent cases are the ones of most interest and are more costly when missed. To address this, techniques such as resampling the data, using anomaly detection algorithms, or applying advanced machine learning methods that focus on precision and recall balance, rather than just accuracy, are often employed. Such distribution is the very essence of the tuning process for the sake of ensuring that the model is sensitive enough to catch those critical, fraudulent transactions while causing less redundancy.

4.5 Dataset Analysis Implications

The result of fraud-related analysis implies that "CASH_OUT" and "TRANSFER" exhibit higher fraud prevalence when looked at the data shown on Fig. 6 and Fig. 7. The

Distribution of Fraudulent Transactions

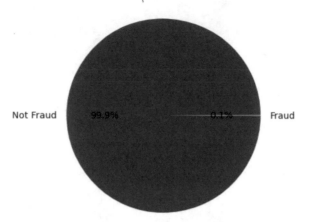

Fig. 5. Transaction Distribution per Utilized Dataset.

dissimilarity in the fraud occurrence types reveals requirement for the transaction-type-specific deterrent forecasting techniques. Institution of financial nature is necessarily required to have efficient fraud detection systems that are properly tuned for the risks categorized in each kind of transaction.

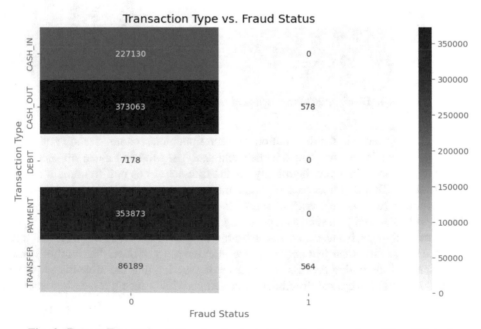

Fig. 6. Dataset Transactions' Mapping Analysis (From the Perspective of Fraud Status).

It can involve the setting of different thresholds of anomaly detections, or machine learning models that learn pattern types of fraud which are distinct from "CASH_OUT" and "TRANSFER". In the given situation, a fraud model should be carefully prepared to ascertain false positives, which could ruin a legitimate transaction, by mining useful information from a large number of legitimate transactions to distinguish between typical and atypical behavior in relatively higher-risk categories.

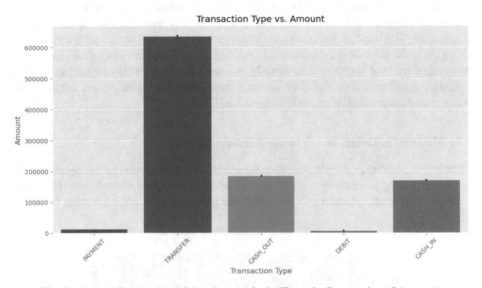

Fig. 7. Dataset Transactions' Mapping Analysis (From the Perspective of Amount).

4.6 Multicollinearity Analysis

Multicollinearity refers to the occurrence of high intercorrelations among independent variables in a dataset, which can lead to statistical issues that affect the performance and interpretability of a regression model (Derraz et al., 2023). In the context of the Online Payments Fraud Detection Dataset, examining multicollinearity helps in understanding how different features relate to each other and whether they might be providing redundant information. For examining features for Multicollinearity, in the provided snapshot of the dataset, the features include:

- step (time of transaction in hours)
- type (type of transaction)
- amount (transaction amount)
- nameOrig (identifier for the customer initiating the transaction)
- oldbalanceOrg (originating account balance before transaction)
- newbalanceOrig (originating account balance after transaction)
- nameDest (identifier for the recipient of the transaction)
- oldbalanceDest (destination account balance before transaction)

- newbalanceDest (destination account balance after transaction)
- isFraud (indicates if the transaction is fraudulent)
- Potential Areas of Multicollinearity

oldbalanceOrg and newbalanceOrig: These two features might be closely related as they both represent the state of the originating account's balance before and after the transaction. High correlation here could be due to the direct impact of the transaction amount on these balances. oldbalanceDest and newbalanceDest: Similar to the originating account balances, these features for the recipient's account might also exhibit high correlation. They represent the state of the account balance before and after the transaction affects the account. in addition, amount, oldbalanceOrg, and newbalanceOrig: The transaction amount is expected to directly influence changes in the originating account balances. Analyzing the correlation between these variables would help determine how independent each is in the context of predicting fraud.

4.7   Methods to Detect and Address Multicollinearity

Correlation Matrix: A simple and effective method to visually inspect potential multicollinearity is by generating a correlation matrix for the numeric variables. High correlation coefficients (close to -1 or 1) indicate potential multicollinearity as shown in Fig. 8.

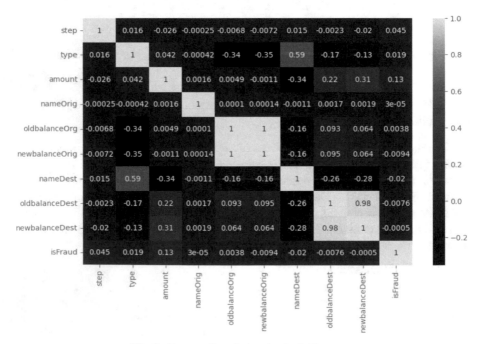

Fig. 8. Dataset Correlation Analysis Heatmap.

The below methods are specifically delas with multicollinearity related issues:

- Variance Inflation Factor (VIF): Calculating the VIF for each variable provides a quantifiable measure of how much the variance of an estimated regression coefficient increases if predictors are correlated. A VIF value greater than 10 is typically considered an indication of multicollinearity.
- Principal Component Analysis (PCA): PCA can be used to reduce dimensionality by transforming the original variables into a new set of variables (principal components) that are linear combinations of the original variables. The new components are orthogonal (independent), effectively removing multicollinearity.

4.8 Analysis of Variance Inflation Factor (VIF) Results

The Variance Inflation Factor (VIF) is utilized to identify the presence and severity of multicollinearity among the independent variables in a dataset, which is crucial for ensuring the reliability of the coefficients in regression models. Below is an interpretation of the VIF results you provided, structured to help understand each variable's impact and possible actions to address any identified issues.

- Step (VIF = 2.877871): This shows moderate multicollinearity. The variance of the regression coefficient is slightly inflated but generally not problematic.
- Type (VIF = 5.055132): The borderline high VIF indicates that this variable has substantial multicollinearity with other variables, suggesting its role might overlap with other features.
- Amount (VIF = 2.126342): Indicates low to moderate multicollinearity, which suggests that the amount of the transaction is fairly independent of the other variables.
- NameOrig (VIF = 2.857206): Similar to 'step', shows moderate multicollinearity and is not a concern.
- OldbalanceOrg (VIF = 709.443407) and NewbalanceOrig (VIF = 716.622550): Both of these variables exhibit extremely high VIF values, indicating significant multicollinearity. This suggests that these variables are providing overlapping information, possibly about the state of the originating account before and after a transaction.
- NameDest (VIF = 3.823784): Displays moderate multicollinearity, which should not typically pose a problem for modeling.
- OldbalanceDest (VIF = 38.440149) and NewbalanceDest (VIF = 41.277884): Both show high VIF values, indicating significant multicollinearity and suggesting overlapping information concerning the recipient's account balances.
- IsFraud (VIF = 1.134756): Shows minimal multicollinearity, indicating that this variable is quite independent of the other variables, which is expected for a dependent variable.

4.9 Handling Imbalanced Data

Given the nature of fraud detection, where fraudulent transactions are much less frequent than non-fraudulent ones, techniques such as Synthetic Minority Over-sampling Technique (SMOTE) can be applied to the training data to balance the dataset. Preparing data for ML involves several essential steps from loading data, handling different data types, splitting the data into training and testing sets, to feature engineering. Each step ensures that the data fed into the model is well-suited for learning patterns and making accurate predictions, crucial for tasks like fraud detection.

4.10 Feature Importance in Fraud Detection Models

Feature importance is a crucial aspect of machine learning model interpretation, helping to identify which variables significantly influence the prediction outcomes. In the context of fraud detection, understanding which features contribute most to the model's decision-making process can provide insights into the nature of fraudulent transactions and inform strategies for prevention and detection. Analysis of Feature Importance (Wadday et al., 2020; Atiyah, 2023; Atiyah et al., 2023, b, c; Husin et al., 2024). Based on the provided importance scores, the features contribute to the model's predictions as follows:

- Amount_Orig (0.40): This feature, representing the amount originally involved in the transaction, has the highest importance score. A high score indicates that the initial transaction amount is a strong predictor of fraud. Large or unusual amounts, especially in the context of the customer's usual transaction patterns, can be a significant red flag for potential fraudulent activity.
- Step (0.20): The time step of the transaction, measured in hours, also plays a substantial role in predicting fraud. This suggests that the timing of a transaction, perhaps in relation to typical customer activity patterns or during unusual hours, might indicate suspicious behavior.
- Type (0.11) and Amount (0.11): Both transaction type and amount have equal importance, indicating that the nature of the transaction and the specific transaction amount are pertinent to detecting fraud. Certain types of transactions might inherently be more susceptible to fraud, or there might be specific transaction amounts that are commonly used by fraudsters to avoid detection.
- Amount_Dest (0.10): The amount received by the destination account is also a significant factor, slightly less so than the type and transaction amount but still notable. This might reflect situations where funds are transferred to accounts that typically do not receive large sums, suggesting a possible redirection of funds for illicit purposes.
- NewbalanceDest (0.07): The new balance of the destination account after the transaction is the least influential but still relevant. Changes in this balance might help identify whether the funds are being quickly moved in a manner consistent with layering stages of money laundering or other fraudulent schemes (Al-Hchaimi et al., 2021; Muhsen et al., 2023b).

in addition, the Implications of Feature Importance.

- Prioritization of Monitoring: The identified feature importance can help financial institutions prioritize certain types of monitoring and controls. For example, transactions involving large amounts or those that occur during abnormal hours should be scrutinized more closely.
- Model Tuning: Categorizing those features that are most opportune to model further refinement helps in phase the model further. Something that has lower effect may probably be the first one that should be eliminated in the basic models. Perhaps the objectivity of accuracy won't change too much.

- Risk Management: Feature importance can become a foundation for risk management strategies, among which are defining transaction limits and requesting extra authentication whenever getting involved in high-risk features like notable originals' amount (amount equal to or above some hundred USD).
- Regulatory Compliance: It is not just the ability to have high-risk accounts highlighted, but understanding which features are indicative of fraud can also help to make sure that regulations, which often force institutions to apply more enhanced due diligence in an area where risk is higher.

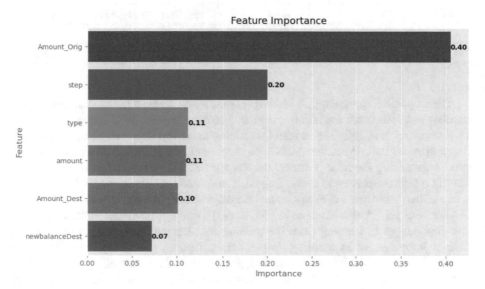

Fig. 9. Dataset Features Importance Rates.

Figure 9 with feature importance analysis is not only beneficial to understand the inner working of the model but also it offers insights for further improvements in fraud detection systems. By concentrating on determining the dominant factors of fraud, financial institutions will be able to prioritize resources, adjust to changing profiles, and incorporate strict safeguards that prevent fraud and protect clients.

5 Results and Discussion

The comprehensive analysis and comparison of machine learning models for fraud detection are segmented into two primary categories: metrics from training and test splits will be used in addition to detailed level evaluations by class. Below you can find the straightening of the categories and each one described according to the presented results.

5.1 Category 1: Training and Testing Model Evaluation

This category focuses on how models perform after being trained on a subset of data and subsequently tested on a separate unseen subset to gauge their ability to generalize.

The metrics used here include Training Accuracy, Testing Accuracy, Precision, Recall, and F1-Score as displayed in Table 2.

Table 2. Category 1 Results: ML Modes Evaluation based on Precision, Recall, and F1-Score Metrics in Training and Testing Aspects.

Model	Training Accuracy	Testing Accuracy	Precision	Recall	F1-Score
LR	87.90%	88.40%	88.89%	87.72%	88.30%
SVM	79.97%	81.40%	96.13%	65.35%	77.81%
K-NN	91.24%	87.53%	87.67%	87.28%	87.47%
DT	100%	95.40%	92.95%	98.25%	95.52%

The Logistic Regression (LR Model): Training Accuracy: 87.90% indicates robust learning but with room for improvement in handling unseen data. Testing Accuracy: 88.40%, which is slightly higher than the training accuracy, suggesting excellent generalization to new data. Precision: At 88.89%, the model is highly accurate when it predicts fraud, minimizing the false positive rate. Recall: 87.72% indicates that the model can identify a high proportion of actual fraudulent transactions. F1-Score: 88.30% shows a balanced harmonic mean of Precision and Recall, affirming the model's effectiveness. In addition, Support Vector Machine (SVM Model) Training Accuracy: 79.97% suggests some challenges in learning from the training set. Testing Accuracy: 81.40% shows a slight improvement in handling unseen data compared to training data. Precision: Extremely high at 96.13%, indicating that almost all positive predictions were correct. Recall: Relatively low at 65.35%, suggesting the model missed a significant number of fraudulent transactions. F1-Score: 77.81%, lower due to the imbalance between Precision and Recall. Moreover, K-Nearest Neighbors (K-NN Model): Training Accuracy: 91.24% shows good learning capability, though slightly overfitting is indicated by the drop in testing accuracy Testing Accuracy: 87.53%, a decent generalization though lower than training accuracy Precision and Recall: Both are nearly equal (around 87.5%), indicating a balanced approach to false positives and negatives F1-Score: 87.47% demonstrates consistent performance between Precision and Recall. Furthermore, Decision Tree (DT Model): Training Accuracy: Perfect at 100%, indicative of overfitting to the training data. Testing Accuracy: Still high at 95.40%, but the drop from training suggests sensitivity to data variations. Precision: 92.95% is excellent, suggesting few false positives. Recall: Very high at 98.25%, showing the model's strength in identifying fraudulent transactions. F1-Score: 95.52%, highlighting exceptional overall performance.

The DT model exhibits the highest performance metrics across all categories, particularly in Recall, making it the most reliable for catching fraudulent transactions. However, its perfect training score points to potential overfitting. In contrast, the Logistic Regression model offers a more balanced performance, making it suitable for scenarios where both types of errors (false positives and false negatives) have significant costs. The choice of model in practical applications should consider these metrics in conjunction with the operational context and the specific costs associated with misclassification.

5.2   Category 2: Data Classifications Model Evaluation

This category assesses model performance based on the ability to correctly classify 'legit' and 'fraud' transactions, focusing on metrics such as precision, recall, and F1-score for each class, as well as overall accuracy as shown in Table 3.

Table 3. Category 2 Results: ML Modes Evaluation based on Precision, Recall, and F1-Score Metrics in Classification Aspect.

Model	Precision (Legit)	Precision (Fraud)	Recall (Legit)	Recall (Fraud)	F1- Score (Legit)	F1-Score (Fraud)	Overall Accuracy
LR	0.879	0.889	0.891	0.877	0.885	0.883	0.884
DT	0.981	0.929	0.926	0.982	0.953	0.955	0.954
SVM	0.738	0.961	0.974	0.654	0.84	0.778	0.814
KNN	0.874	0.877	0.878	0.873	0.876	0.875	0.875

In analyzing the performance of four classification models—Logistic Regression (LR), Decision Tree (DT), Support Vector Machine (SVM), and K-Nearest Neighbors (KNN)—using metrics from confusion matrices illustrated in subplots Fig. 9 a, b, c, and d, a detailed assessment emerges. The Decision Tree model, represented in Fig. 9b, exhibits the highest overall accuracy of 95.4%. This model achieves impressive precision and recall for both classes, particularly excelling in identifying fraudulent transactions with a recall of 0.982 and an F1-Score of 0.955, suggesting a strong alignment between predicted and actual labels. Conversely, the SVM model, shown in Fig. 9c, despite its high precision of 0.961 for fraudulent transactions, suffers from a considerable shortfall in recall for this category at 0.654, resulting in the lowest overall accuracy of 81.4% among the models reviewed. This indicates a higher number of false negatives, where fraudulent transactions are mistakenly classified as legitimate. Both the LR and KNN models, depicted in Figs. 9a and 9d, respectively, demonstrate more balanced metrics across precision, recall, and F1-Scores, but neither reaches the peak performance observed in the Decision Tree model. LR achieves an overall accuracy of 88.4%, while KNN posts 87.5%, showing competent but not exceptional ability to classify transactions accurately. This comprehensive evaluation highlights that while each model offers specific strengths, the Decision Tree model provides the most robust performance for this dataset, though potential issues such as overfitting should be considered for broader application contexts, as shown in Fig. 10.

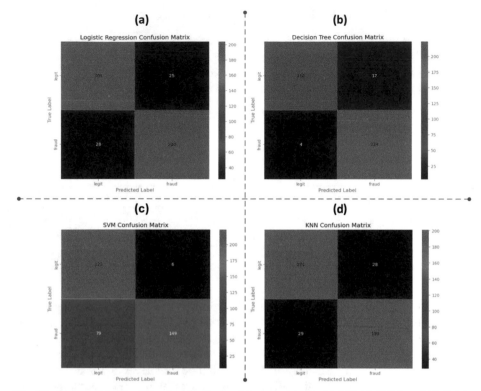

Fig. 10. ML Models Classification Confusion Matrices: (a) LR, (b) DT, (c) SVM, and (d) K-NN.

5.3   Implications of Model Selection in Fraud Detection

The choice of a ML model in fraud detection is crucial due to the significant implications associated with the performance of these models. Errors may result in either costly investigation of innocent people (false positives) or frauds remain − uninvestigated (false negative). Altogether, the costs of errors are too high. Here, we consider why any chosen model together with the implemented approach possesses the feature of fraud detection as a whole.

- DT Model

    Strengths: High Recall and F1-Score: The high recall rate in the Decision Tree's case ensures that an insignificant degree of fraudulent transactions end up being unrecognized, which is particularly important as this risks fraud going undetected. Interpretability: DT aid to highlight decision flowcharts which help regulators to fulfill their task as well as can understand the most influencing features. Risk of Overfitting: The suggestion of the best training accuracy could be explained by the fact that the model is combinatorically complex and hence, it may be possible to steal patterns that are not generalizable from the training data. This may be the reason of the weaker generalization in the real world where no similar data characteristics are provided. Operational Costs: Although the high recall rate reduces the likelihood of frauds not being caught so many missed cases are not expected to be the result of the lower precision level compared to SVM which may however lead to higher operational costs due to more investigating cases as false positives.

- LR Model

Balanced Metrics: By implementing a trade-off which is balanced and decisive, this model avoids the flooding of the system with false positive alarms and with the under-reporting of frauds and scams. Scalability and Simplicity: Logistic Regression models are computationally inexpensive to train and are easily scalable, making them suitable for large datasets. Generalization Capability: This model's performance indicates good generalization, which is essential for consistent performance across diverse transaction scenarios Regulatory Compliance: The transparency and simplicity of logistic regression facilitate explanations of decisions, an important aspect in financial applications subject to strict regulations.

- K-NN Model

Robustness to Noisy Data: K-NN can be very effective if the dataset includes informative labels since it bases classifications on the proximity to other labeled examples. Computational Cost: K-NN involves significant computational expense during the testing phase as it requires calculating the distance from each query instance to all training samples, which can be prohibitive in real-time systems. Scaling Issues: Performance may degrade with very large datasets due to the curse of dimensionality unless dimensionality reduction techniques are employed effectively.

- SVM Model

High Precision: This model's high precision means that when it predicts a transaction as fraudulent, it is very likely to be correct, reducing the burden of unnecessary fraud investigations. Low Recall: The major downside is its lower recall, meaning some fraudulent transactions may not be detected. This can be critical as failing to detect fraud can lead to significant financial losses and damage to customer trust. Model Training and Tuning: SVMs require careful tuning of hyperparameters and can be sensitive to the choice of the kernel, which may require additional resources and expertise.

The selection of a model for fraud detection systems is influenced by several factors, including the acceptable balance between false positives and false negatives, computational resources, the nature of the data, and operational constraints. Each model's characteristics suggest different operational scenarios where they might be the most appropriate. For instance, where interpretability and regulatory compliance are crucial, Logistic Regression and Decision Trees might be preferred, while for high-stake environments where missing a fraud is costlier than investigating a non-fraud, a Decision Tree or even ensemble models combining several approaches could be ideal.

6 Conclusion

The evaluation of various machine learning models for fraud detection elucidates the distinct strengths and weaknesses inherent in each approach. Decision Trees stood out with the highest overall accuracy and recall, making them exceptionally effective for identifying fraudulent transactions. However, their susceptibility to overfitting necessitates cautious deployment. Logistic Regression presented a balanced trade-off between precision and recall, suggesting its suitability for environments where false positives and false negatives have similar repercussions. Conversely, SVMs, despite their high precision, exhibited limitations in recall, potentially overlooking some fraudulent activities. K-NN demonstrated robust performance but may prove impractical for real-time analysis due to its computational demands.

An acknowledged limitation such as: data dependency, static models, feature selection, and class imbalance are potentially affecting real-world applicability of ML approaches in such domains. By addressing these limitations and pursuing the proposed future directions, researchers and practitioners can significantly advance the capabilities of machine learning systems in detecting and preventing fraud, thereby enhancing asset security and fostering trust in financial transactions.

For future research directions in the domain of fraud detection using ML several promising avenues have been identified based on their proven effectiveness in security and network security applications. Dynamic modeling stands out as a crucial area, allowing for adaptations to new fraudulent tactics as they evolve. Additionally, advanced feature engineering is essential for improving the predictive power of models by identifying and utilizing more informative indicators of fraudulent behavior. The use of ensemble techniques, which combine the strengths of multiple models to enhance performance and reliability, also presents a significant opportunity for advancement. Finally, real-time processing capabilities are essential for the immediate detection and mitigation of fraudulent activities, ensuring timely responses to threats. These strategies collectively form a robust framework for enhancing fraud detection systems in increasingly complex security environments.

References

Abbas, S., et al.: Antecedents of trustworthiness of social commerce platforms: a case of rural communities using multi group SEM & MCDM methods. Electron. Commer. Res. Appl. **62**, 101322 (2023)

Afriyie, J.K., et al.: A supervised machine learning algorithm for detecting and predicting fraud in credit card transactions. Decis. Anal. J. **6**, 100163 (2023)

Ahmadi, S.: Open AI and its Impact on Fraud Detection in Financial Industry. J. Knowl. Learn. Sci. Technol. (2023). ISSN 2959-6386

Ahmed, A.D., Salih, M.M., Muhsen, Y.R.: Opinion weight criteria method (OWCM): a new method for weighting criteria with zero inconsistency. IEEE Access (2024)

Al-Enzi, S.H.Z., Abbas, S., Abbood, A.A., Muhsen, Y.R., Al-Hchaimi, A.A.J., Almosawi, Z.: Exploring research trends of metaverse: a bibliometric analysis. In: Al-Emran, M., Ali, J.H., Valeri, M., Alnoor, A., Hussien, Z.A. (eds.) IMDC-IST 2024. LNNS, vol. 895, pp. 21–34. Springer, Cham (2023). https://doi.org/10.1007/978-3-031-51716-7_2

Al-Hchaimi, A.A.J., Flayyih, W.N., Hashim, F., Rusli, M.S., Rokhani, F.Z.: Review of 3D networks-on-chip simulators and plugins. In: 2021 IEEE Asia Pacific Conference on Postgraduate Research in Microelectronics and Electronics (PrimeAsia), pp. 17–20 (2021)

Al-Hchaimi, A.A.J., Sulaiman, N.B., Mustafa, M.A.B., Mohtar, M.N.B., Mohd, S.L.B., Muhsen, Y.R.: Evaluation approach for efficient countermeasure techniques against denial-of-service attack on MPSoC-based IoT using multi-criteria decision-making. IEEE Access (2022)

Al-Hchaimi, A.A.J., Sulaiman, N.B., Mustafa, M.A.B., Mohtar, M.N.B., Mohd Hassan, S.L.B., Muhsen, Y.R: A comprehensive evaluation approach for efficient countermeasure techniques against timing side-channel attack on MPSoC-based IoT using multi-criteria decision-making methods. Egypt. Inform. J. **24**, 351–364 (2023)

Ali, J., Hussain, K.N., Alnoor, A., Muhsen, Y.R., Atiyah, A.G.: Benchmarking methodology of banks based on financial sustainability using CRITIC and RAFSI techniques. Decis. Making: Appl. Manage. Eng. **7**(1), 315–341 (2024)

Alnoor, A., Atiyah, A.G., Abbas, S.: Toward digitalization strategic perspective in the European food industry: non-linear nexuses analysis. Asia-Pac. J. Bus. Adm. (2023)

Alnoor, A., Atiyah, A.G., Abbas, S.: Unveiling the determinants of digital strategy from the perspective of entrepreneurial orientation theory: a two-stage SEM-ANN approach. Glob. J. Flexible Syst. Manage. 1–18 (2024)

Alnoor, A., Chew, X., Khaw, K.W., Muhsen, Y.R., Sadaa, A.M.: Benchmarking of circular economy behaviors for Iraqi energy companies based on engagement modes with green technology and environmental, social, and governance rating. Environ. Sci. Pollut. Res. **31**(4), 5762–5783 (2024)

Alnoor, A., et al.: How positive and negative electronic word of mouth (eWOM) affects customers' intention to use social commerce? A dual-stage multi group-SEM and ANN analysis. Int. J. Hum.-Comput. Interact. **40**(3), 808–837 (2024)

Asha, R.B., Suresh kumar, K.R.: Credit card fraud detection using artificial neural network. Glob. Transit. Proc. **2**, 35–41 (2021)

Atiyah, A.G.: Unveiling the quality perception of productivity from the senses of real-time multisensory social interactions strategies in metaverse. In: Al-Emran, M., Ali, J.H., Valeri, M., Alnoor, A., Hussien, Z.A. (eds.) IMDC-IST 2024. LNNS, vol. 876, pp. 83–93. Springer, Cham (2023). https://doi.org/10.1007/978-3-031-51300-8_6

Atiyah, A.G., Alhasnawi, M., Almasoodi, M.F.: Understanding metaverse adoption strategy from perspective of social presence and support theories: the moderating role of privacy risks. In: Al-Emran, M., Ali, J.H., Valeri, M., Alnoor, A., Hussien, Z.A. (eds.) IMDC-IST 2024. LNNS, vol. 876, pp. 144–158. Springer, Cham (2023). https://doi.org/10.1007/978-3-031-51300-8_10

Atiyah, A.G., All, N.D.A., Zaidan, A.S., Bayram, G.E.: Understating the Social Sustainability of Metaverse by Integrating Adoption Properties with Users' Satisfaction. In: Al-Emran, M., Ali, J.H., Valeri, M., Alnoor, A., Hussien, Z.A. (eds.) IMDC-IST 2024. LNNS, vol 895, pp. 95–107. Springer, Cham (2023). https://doi.org/10.1007/978-3-031-51716-7_7

Atiyah, A.G., Faris, N.N., Rexhepi, G., Qasim, A.J.: Integrating ideal characteristics of chat-GPT mechanisms into the metaverse: knowledge, transparency, and ethics. In: Al-Emran, M., Ali, J.H., Valeri, M., Alnoor, A., Hussien, Z.A. (eds.) IMDC-IST 2024. Lecture Notes in Networks and Systems, vol. 895, pp. 131–141. Springer, Cham (2023). https://doi.org/10.1007/978-3-031-51716-7_9

Avila-Cano, A., Triguero-Ruiz, F.: On the control of competitive balance in the major European football leagues. Manag. Decis. Econ. **44**, 1254–1263 (2023)

Cai, S., Xie, Z.: Explainable fraud detection of financial statement data driven by two-layer knowledge graph. Expert Syst. Appl. **246**, 123126 (2024)

Caixeta, D.C., et al.: Monitoring glucose levels in urine using FTIR spectroscopy combined with univariate and multivariate statistical methods. Spectrochim. Acta Part A Mol. Biomol. Spectrosc. **290**, 122259 (2023)

Chen, Y.J., Liou, W.C., Chen, Y.M., Wu, J.H.: Fraud detection for financial statements of business groups. Int. J. Account. Inf. Syst. **32**, 1–23 (2019)

Cherif, A., Badhib, A., Ammar, H., Alshehri, S., Kalkatawi, M., Imine, A.: Credit card fraud detection in the era of disruptive technologies: a systematic review. J. King Saud Univ.-Comput. Inf. Sci. **35**, 145–174 (2023)

Derraz, R., Muharam, F.M., Nurulhuda, K., Jaafar, N.A., Yap, N.K.: Ensemble and single algorithm models to handle multicollinearity of UAV vegetation indices for predicting rice biomass. Comput. Electron. Agric. **205**, 107621 (2023)

Dornadula, V.N., Geetha, S.: Credit card fraud detection using machine learning algorithms. Procedia Comput. Sci. **165**, 631–641 (2019)

Husin, N.A., Abdulsaeed, A.A., Muhsen, Y.R., Zaidan, A.S., Alnoor, A., Al-mawla, Z.R.: Evaluation of metaverse tools based on privacy model using fuzzy MCDM approach. In: Al-Emran, M., Ali, J.H., Valeri, M., Alnoor, A., Hussien, Z.A. (eds.) IMDC-IST 2024. LNNS, vol 895, pp. 1–20. Springer, Cham (2023). https://doi.org/10.1007/978-3-031-51716-7_1

Husin, N.A., Zolkepli, M.B., Manshor, N., Al-Hchaimi, A.A.J., Albahri, A.S.: Routing techniques in network-on-chip based multiprocessor-system-on-chip for IOT: a systematic review. Iraqi J. Comput. Sci. Math. **5**, 181–204 (2024)

Innan, N., et al.: Financial fraud detection using quantum graph neural networks. Quant. Mach. Intell. **6**, 1–18 (2024)

Jose, S., Devassy, D., Antony, A.M.: Detection of credit card fraud using resampling and boosting technique. In: 2023 Advanced Computing and Communication Technologies for High Performance Applications (ACCTHPA), pp. 1–8. IEEE (2023)

Kadhuim, Z.A., Al-Janabi, S.: Codon-mRNA prediction using deep optimal neurocomputing technique (DLSTM-DSN-WOA) and multivariate analysis. Results Eng. **17**, 100847 (2023)

Liu, Y., et al.: Pick and choose: a GNN-based imbalanced learning approach for fraud detection. In: Proceedings of the Web Conference 2021, pp. 3168–3177 (2021)

Liu, Z., Chen, C., Yang, X., Zhou, J., Li, X., Song, L.: Heterogeneous graph neural networks for malicious account detection. In: Proceedings of the 27th ACM International Conference on Information and Knowledge Management, pp. 2077–2085 (2018)

Muhsen, Y.R., Husin, N.A., Zolkepli, M.B., Manshor, N.: A systematic literature review of fuzzy-weighted zero-inconsistency and fuzzy-decision-by-opinion-score-methods: assessment of the past to inform the future. J. Intell. Fuzzy Syst. **45**(3), 4617–4638 (2023)

Muhsen, Y.R., Husin, N.A., Zolkepli, M.B., Manshor, N., Al-Hchaimi, A.A.J.: Evaluation of the routing algorithms for NoC-based MPSoC: a fuzzy multi-criteria decision-making approach. IEEE Access (2023a)

Muhsen, Y.R., Husin, N.A., Zolkepli, M.B., Manshor, N., Al-Hchaimi, A.A.J., Ridha, H.M.: Enhancing NoC-based MPSoC performance: a predictive approach with ANN and guaranteed convergence arithmetic optimization algorithm. IEEE Access (2023b)

Ni, L., Li, J., Xu, H., Wang, X., Zhang, J.: Fraud feature boosting mechanism and spiral oversampling balancing technique for credit card fraud detection. IEEE Trans. Comput. Soc. Syst. **11**, 1615–1630 (2023)

Nijwala, D.S., Maurya, S., Thapliyal, M.P., Verma, R.: Extreme gradient boost classifier based credit card fraud detection model. In: Proceedings - IEEE International Conference on Device Intelligence, Computing and Communication Technologies, DICCT 2023, pp. 500–504 (2023)

Pallathadka, H., Ramirez-Asis, E.H., Loli-Poma, T.P., Kaliyaperumal, K., Ventayen, R.J.M., Naved, M.: Applications of artificial intelligence in business management, e-commerce and finance. Mater. Today: Proc. **80**, 2610–2613 (2023)

Shi, H., et al.: A comprehensive framework for identifying contributing factors of soil trace metal pollution using geodetector and spatial bivariate analysis. Sci. Total. Environ. **857**, 159636 (2023)

Singh, I., Aditya, N., Srivastava, P., Mittal, S., Mittal, T., Surin, N.V.: Credit card fraud detection using neural embeddings and radial basis network with a novel hybrid fruitfly-fireworks algorithm. In: 2023 3rd International Conference on Intelligent Technologies (CONIT), pp. 1–7. IEEE (2023)

Soltani, M., Kythreotis, A., Roshanpoor, A.: Two decades of financial statement fraud detection literature review; combination of bibliometric analysis and topic modeling approach. J. Financ. Crime **30**, 1367–1388 (2023)

Vanini, P., Rossi, S., Zvizdic, E., Domenig, T.: Online payment fraud: from anomaly detection to risk management. Financ. Innov. **9**, 66 (2023)

Wadday, A.G., Al-hchaimi, A.A.J., Ibrahim, A.J.: IOT Energy consumption based on pso-shortest path techniques. Recent Adv. Electr. Electron. Eng. (Formerly Recent Patents Electr. Electron. Eng.) **13**, 993–1000 (2020)

Wright, J.D., Krzepicki, A.: What is an independent agency to do? The trump administration's executive order on preventing online censorship and the federal trade commission. Admin. L. Rev. Accord. **6**, 29 (2020)

Applying Robotic Process Automation (RPA) in Sustainable Audit Quality: A Literature Review Survey

Aymen Raheem Abdulaali Almayyahi[1,2(✉)], Samir Aboul Fotouh Saleh[2], and Ahmed Kamal Metawee[2]

[1] Management Technical College, Southern Technical University, Basrah, Iraq
Aymen.abdulaali@std.mans.edu.eg
[2] Faculty of Commerce, Mansoura University, Mansoura, Egypt
Aymen.abdulali@stu.edu.iq, {Prof_samir,Metawee68}@mans.edu.eg

Abstract. Audit constitutes one of the most important functions of any organization. Auditors are responsible for the evaluation of accounting activities to ensure compliance of the national and international accounting standards. The technological emergence and industry 4.0 revolution has reformed the conduct of the operations of businesses, although most of the audit operations are still performed by humans. This study aims to investigate the possibilities of audit automation by applying robotics processes for improvements in audit performance and reduction in fraud. We reorganize the findings of the previous studies by exploring the challenges that could hinder the automation of audit activities. The present study relies on the relevant literature to obtain the relevant data. Furthermore, the results of this study highlight the importance of artificial intelligence tools to boost the audit function in different sectors in the context of the era of digitalization transformation.

Keywords: Robotic process automation · Audit · Enterprise resource planning · Artificial intelligence

1 Introduction

The advent of RPA and its integration with artificial intelligence has the transformation accounting and auditing activities across contemporary organizations. As the financial industry embraces automation and data-driven insights, the role of accountants and auditors is evolving with more focus on digital tools, data analytics, and intelligent financial management (Liu & Ishak, 2023; Ali et al., 2024), RPA is a computer program designed to automatically perform a predefined set of operations, transactions, activities, and tasks. It works based on predefined business rules and can integrate with one or more unrelated software systems. The main goal of RPA is to provide the required result while allowing the managing of human exceptions. RPA uses robots to automate tasks performed by humans and thus eliminating the loss of time resulting from work stops, RPA is a low-cost technology that can significantly reduce labor costs as well

© The Author(s), under exclusive license to Springer Nature Switzerland AG 2024
A. Alnoor et al. (Eds.): AIRDS 2024, LNNS 1033, pp. 26–35, 2024.
https://doi.org/10.1007/978-3-031-63717-9_2

(Huang & Vasarhelyi, 2019). RPA uses software applications to perform tasks (Kokina & Blanchette, 2019).

Occasionally, organizations may have an inappropriate usage of RPA. Similarly, some complex tasks may be too difficult to be performed through RPA. In that situation, intelligent process automation (IPA), which is a hybrid of RPA and AI, is used to perform complex tasks. More than one device with human interventions can be used to complete complex tasks (Zhang, 2019; Atiyah et al., 2023).

Generally, audit activities take more human time and effort, therefore, it is imperative to automate them for efficiency and effectiveness. A lot of previous literature has suggested the use of automation in auditing (Huang & Vasarhelyi, 2019; Alnoor et al., 2023). Recently many audit programs and electronic spreadsheets like Excel are in use across many organizations. Although the auditing operations are efficient and improved, the Human auditors are still primarily responsible for the integration of various systems and applications. For example, the external audit operation still requires intensive of worker and relies on manual effort (Srinivasan, 2016; Abbas et al., 2023). Therefore, it is inevitable to use RPA. RPA performs routine operations by automating processes where people interact with multiple applications and analyses through the user interface and following pre-defined steps to make decisions (Huang & Vasarhelyi, 2019; Ahmed et al., 2024). This study is different from previous studies as it proposes a framework based on previous studies for the application of RPA in audits. In doing so risks of implementation and independent evaluation will be taken into account, and therefore it is expected that the proposed framework will be helpful to enhance the performance of audit activities.

2   Literature Review

2.1   Robotic Process Automation in Audit

Audit is one of the most important departments of any organization. Organizations rely more on the audit department for various financial decisions. In organizations, where audit activities are performed manually, there are always chances of errors, delays, and fraud. Because of this, contemporary organizations always prefer automation of the audit department. For successful implementation of RPA, it is important to decide if the audit department is an independent unit or it is merged with the accounting department. Generally, auditing, management accounting, bookkeeping, personnel management, payroll, taxes, etc. software interfaces certainly need to be compatible with each other to share the relevant information (Hyvönen et al., 2008; Alnoor et al., 2024). This inter-dependency of the various functions requires a sophisticated ERP for a successful digital transformation. An ERP system connects various functions to accomplish multifunctional operations, where the central control system allows the chief financial officer to fully control and manage the design and implementation of the system as a whole. (Gotthardt et al., 2020).

One of the challenges associated with clients using RPA is that users do not have complete knowledge of what robots do or how they work. In addition, there is resistance against the use of RPA technology due to their fear that automation will cost employees their jobs. Additionally, safety and security concerns may increase the hesitation of some customers to adopt new technologies, in addition to protecting business processes and the flow of information in an opaque manner, this can increase fears and limit the use of technology, as shown in Fig. 1 (Gotthardt et al., 2020).

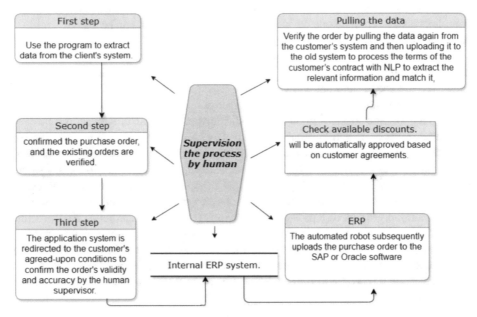

Fig. 1. RPA Protocol.

The above flow chart diagram shows the robotic process automation works process in three different steps. The first step is to use the program to extract data from the client's system. Next, the purchase order is confirmed, and the existing orders are verified. Finally, the auditor supervisor will confirm the order's validity as agreed terms and conditions. The automated robot subsequently uploads the purchase order to SAP or Oracle software, which is the internal enterprise resource planning system. Then if there are any discounts it will be automatically approved based on customer agreements. After that, the orders will be verified by pulling the data again from the customer's system and then uploading it to the old system to process the terms of the customer's contract with NLP to extract the relevant information and match it, the role of the human in this process will be supervision only. The most important feature of the process is reducing costs and increasing efficiency through training and automated education Many auditors prefer manual processing more than decision-making aids or knowledge-based systems (Baldwin et al., 2006).

2.2 Previous Studies

The study of (Farinha et al., 2024) aims to identify the main criteria that organizations should consider when determining which processes to automate. The study identified 32 criteria that organizations and decision-makers should consider before choosing which processes to automate. The most weighted criteria were feasibility, process description, and input and output data. The specified criteria were evaluated by 18 processes in six organizations, and the results were evaluated. A systematic review of the literature was conducted to collect ideas on the standards for Process Automation. The Delphi study was conducted with RPA specialists to improve the parameters obtained from the literature review. The Delphi study included multiple rounds of surveys with open and closed-ended questions to collect expert opinions. The Likert scale was used to evaluate each criterion, and the Fuzzy Delphi method was used to calculate consensus. Meanwhile, the purpose of (Candratio et al., 2023) study is to investigate the application of automated process automation (RPA) in the field of auditing and to understand the factors influencing the adoption of advanced data analytics in auditing. The study aims to provide a framework for the application of RPA technology in auditing, explore its potential benefits in improving audits, as well as explain the reasons behind the adoption or non-adoption of advanced data analytics using a process theory approach. The study also proposes a framework for the use of RPA technology in audit tasks while highlighting its potential impact on Contemporary Accounting Research. Additionally, the study examines the role of Information Technology in the audit context and explores how RPA technology is transforming accounting and auditing practices. It also provides insights from industry reports, such as Deloitte and Ernst and Young, on the readiness of organizations to adopt RPA technology and the potential impact of RPA technology in various sectors. Thus, the study contributes to contemporary accounting research by highlighting the potential impact of RPA technology on audit tasks and processes. In general, the study provides a comprehensive understanding of the application of RPA in auditing and the factors influencing the adoption of advanced data analytics, contributing to the advancement of this field.

Previous studies highlighted the impact of information systems tools for mitigating risk and fostering internal control. Hence, (Plant et al., 2022) study was intended to guide organizations that use DevOps in designing their internal control environment and creating correct audit trails. The study offers a framework that helps organizations mitigate and manage risks in IT teams and departments that use the DevOps model. It also validates the control framework by involving experts with backgrounds in DevOps. It sought to demonstrate that the traditional areas of its control and its core risks remain relevant in the context of DevOps. The framework suggests appropriate risk mitigation practices for organizations using DevOps. The study found that companies in practice often use a combination of manual and automated control. The framework was validated by experts with backgrounds in DevOps, IT risk, and IT audit, as well as by showing it to the study's participants. Besides, (Rinta-Kahila et al., 2023) aimed to understand the dynamics underlying the erosion of skills caused by cognitive automation in organizations. The study explored the interaction between humans' reliance on automation, complacency, and conscious connectivity in the context of skills erosion. The study intended to shed light on the phenomenon of skills erosion and its implications for theory and practice, as well as to identify directions for future research. The most important results were obtained using causal loop modeling based on the dynamics of the system. It explained the dynamic model resulting from the erosion of skills through the interplay of human dependence on automation, complacency, and conscious communication. It shows how the increasing dependence on automation promotes self-satisfaction both at the individual and organizational levels, weakening the awareness of workers across three aspects of work tasks (awareness of activity, maintaining efficiency, and evaluating output), which causes skills erosion. This erosion of skills may remain peculiar, and neither workers nor managers recognize it.

The study of (Shidaganti et al., 2023) aims to understand the importance of human creativity in driving innovation and the growth of companies, especially in the context of technological progress and digitization. The purpose of the study is to highlight the role of creativity in generating new ideas and innovations that can lead to market adaptation and institutional expansion. It also aims to emphasize the need for regular adoption of new technologies to obtain maximum value from them. The study suggests that human creativity is crucial for creativity and innovation because it allows new ideas to be generated. The study recommends that companies and individuals adopt and adapt to technological disruptions to maintain economic success. The study suggests that organizations should focus on the gradual adoption of new technologies to drive innovation and creativity. In addition, it emphasizes the importance of developing innovative digital skills and integrating information systems to keep up with the changing job market. While (Liu & Ishak, 2023) examine the practical use of RPA in higher professional accounting education in China and to explore the integration of RPA into the curriculum. The study also aims to address the current focus on accounting and financial programs in Chinese accounting education, which ignores vital skills in operating a financial robot and data processing. It also aims to highlight the need to strengthen students' organizational and communicative abilities in accounting education. The study emphasizes the importance of developing teachers and adapting curricula in vocational colleges to nurture students with enhanced comprehensive skills and prepare them for the dynamic requirements of the financial sector. Based on insights from academic literature and qualitative analysis, the study aims to provide practical recommendations to promote the integration of RPA into higher professional accounting education in China. The most important finding of the study was that the integration of RPA technology in accounting education in higher professional institutions is relatively limited, especially in China. The current curriculum in higher professional education focuses mainly on basic accounting knowledge and neglects the development of data applications and manipulation skills, leaving students ill-equipped to handle complex financial data analysis and RPA applications. Higher professional institutions should restructure their curricula to prioritize practical skills and use modern tools such as financial robots to align with industry requirements. Teachers in the field of financial robotics are not adequately prepared, there is a need for teacher training and interdisciplinary cooperation to improve their understanding and teaching of cutting-edge technologies. The study emphasizes the importance of nurturing students comprehensive skills, including technical dexterity, soft skills, and practical operational skills, to meet the requirements of the modern financial sector. Integrating RPA technology and practical tools such as financial robots into the classroom can enhance student engagement, understanding of complex concepts, and learning efficiency.

The primary purpose of the (Zhang et al., 2023) was to examine the implementation of RPA in the accounting function and to demonstrate its impact and associated challenges. The study aims to contribute to the literature on accounting and new technologies by providing a comprehensive understanding of the implementation of ARP in accounting. The researchers relied on an exploratory research methodology to understand the implementation of RPA. The study identified five axes related to the adoption of RPA technology in accounting functions which are manpower, IT governance, privacy, and security, system sustainability, and measuring the success of RPA technology. The results of the study have a significant role in the effective implementation of RPA technology for accounting functions. In addition, (Da Silva Costa et al., 2022) aims to provide a summary of the general implementation approaches, observed benefits, challenges, process suitability, and research gaps in the adoption of RPA. In addition to answering specific research questions related to the implementation of RPA, the study also highlighted the benefits, challenges, and characteristics of the implementation process of RPA. The purpose of the study is to provide organizations with good practices for the adoption of RPA and to promote further research on this topic by complementing existing knowledge and proposing new research paths. The study conducted a systematic literature review (SLR) and identified the most general implementation approaches for successful RPA adoption cases, the observed benefits and challenges commonly faced by organizations, the characteristics that make processes more suitable for RPA, and the research gaps in the existing literature. The analysis of 47 research papers resulted in several key insights, including an overview of the RPA technology adoption process across several companies, an analysis of the effects and benefits of RPA technology in organizations, awareness of the implementation challenges, and a comprehensive summary of the characteristics of tasks suitable for RPA. The study also highlighted the need for further research on the implementation of RPA for small organizations.

Despite the automation of audit functions, human intervention[s] cannot be avoided because many of the audit tasks require expert opinion and judgment by the auditors. Thus, RPA requires human supervision for smooth function. In an earlier study the design science research (DSR) methodology was used, and the Attended Process Automation (APA) process automation framework was proposed, which guides the implementation of human-supervised automation in audit processes. The study also discusses the APA framework by applying it to the planning process for individual audits, which is an external audit usually performed by the government audit department. The APA framework also emphasizes the important role of auditors in automated auditing by providing professional judgments that cannot be replaced by the automation process (Zhang et al., 2022). Also, (Griffiths & Pretorius, 2021) address the challenges posed by fraud, especially professional fraud, and to explore the use of RPA as a precautionary measure to reduce the risks associated with fraud. The study aims to explore how automation can reduce the risk of fraud by reducing frequent human interaction with computer systems. The study suggests that automation can reduce the risk of fraud by reducing interaction with computer systems. This can be achieved by automating tasks and simplifying organizational business processes. (Denagama Vitharanage et al., 2020) aims to provide an empirically supported visualization of the benefits of RPA. The purpose of the study is to provide a framework for the expected and unexpected benefits of RPA, classified into operational, managerial, strategic, and organizational topics, across the RPA implementation schedule. The study aimed to better design business cases and plan efforts to realize the benefits of RPA for RPA practitioners. It also aimed to provide a solid foundation for future RPA research by linking case studies and literature findings and suggesting future research directions to better understand the benefits of RPA. The study provided the first empirically supported framework for the benefits of RPA, mapping the expected and unexpected benefits at the introductory, and operational stages of the RPA life cycle. The framework identified seven expected benefits and seven unexpected benefits, providing insights into the potential benefits of implementing RPA. The study identified improvement in accuracy as the most discussed expected benefit and improvement in customer service and satisfaction as the least discussed expected benefit. The study also highlighted the interrelationships between different RPA benefits and proposed future research directions to further investigate these relationships.

3 Methodology

Our research mainly focuses on previous studies in the field to understand the implementation of RPA for enhancing the effectiveness of internal audits. The following terms have been used to search for the relevant literature across different databases as mentioned below. "Robotic process automation, audit, artificial intelligence, benefits and challenges". Our literature search was focused on five main research databases including Google Scholar, Scopus, ResearchGate, ScienceDirect, and ProQuest. A total of 180 articles were initially selected across these five databases and tabulated in terms of titles, abstracts, journals, and the year of publication. Based on the results of the articles, the duplicated ones were removed from the list. In doing the article search, we mainly relied on 22 articles.

4 Discussion

4.1 Data Standardization

To obtain proper performance, high-quality RPA data must be available (Kokina & Blanchette, 2019). The format of the associated data, their source, and compatibility should be considered when considering an automated process automation solution (Moffitt et al., 2018). There are usually two ways to standardize data. First, organizations that apply RPA technology need to consider the need for multifunctional organizational data and associated controls. Second, organizations must ensure that the required data is at the required level of security and quality (Kokina & Blanchette, 2019). RPA technology can cause problems if it is implemented with poor design and poor data. The standard of audit data in the second phase is the standardization of data. The data and the method of audit should be determined and followed by the audit function on the use of the RPA program by the auditor to obtain reliable results (Cohen & Rozario, 2019). In conclusion, the successful use of RPA technology in organizations requires fraud reduction, enhanced audit, and serious standardization of data.

4.2   Challenges and Concerns of Implementing RPA

Like other technologies, RPA is also not error-free. Many challenges and concerns arise while implementing RPA.. The first and foremost challenge is the financial cost, as the use of RPA technology by organizations can be expensive. In addition, it takes a long time to fully develop RPA technology, which negatively affects the audit process in general. Technical ability is the second limit, where auditors must have the knowledge and ability to operate and develop the RPA program (EYARC, 2023). The third challenge is the organizational mindset, as it is difficult to convince managers and decision-makers to accept RPA technology. Some previous studies have shown that auditors have fears about using modern technology. These fears mainly relate to the perception that robots will replace them in the job therefore they refuse to adopt RPA technology in auditing. The knowledge gap within the organization may be a reason for some to worry about the use of robotic process automation technology, especially in small ones. (Liu, 2022).

4.3   Define and Implement a Framework

To ensure the successful implementation of RPA Robotic Process Automation in auditing for enhancing effectiveness and combating fraud, it includes two stages. The first is the stage of defining the process, the second, is designing and building the appropriate technology (RPA technology) through the availability of technical skills, and finally evaluating the results provided by the program and comparing them with manual results. The conditions are set in advance of the audit process to reduce and eliminate fraud. The conditions are represented in both the audit environment and the operational environment. As in the audit environment, the auditor is responsible for these conditions and the process environment. The process identification involves finding the most appropriate processes for the automation process to achieve organizational goals, which are usually characterized as organized and simple tasks with important outputs (Huang & Vasarhelyi, 2019).

The construction and design of the framework for the RPA process is part of the automation environment, where implementing RPA, the current strategy of Information Technology and intelligent automation should be taken into account. The framework considers that the evaluation of the results is the use of the output of the RPA system and the identification of the fraud and the ways to reduce their re-occurrence (Zhang, 2019; Muhsen et al., 2023). The identification of automation processes requires experience in operations and auditing in the context of the specific process environment. This is one of the most important requirements of the first stage, which is to identify the process of automation. The second stage, which is the design and construction of RPA, requires experience in developing RPA technology regardless of the automation environment. Finally, the evaluation of the results of the process can be used in audit analyses within the scope of the audit tasks. This can lead to changes in the process implemented to reduce fraud in the work environment, organizations should construct a new set of skills required in case they do not exist in the business environment (Madakam et al., 2019).

5   Avenues for Future Directions

Analyzing the previous studies highlighted several hotspot points for future research. First, many studies used linear methods to reveal the relationships between antecedences and consequences in terms of RPA studies. However, there is rarely has been reported in previous studies of using nonlinear tools such as Artificial Neural Networks (ANN). By implementing the ANN method academics and practitioners would determine the most contributor factors to achieving high efficiency of using RPA. Several studies utilized and adopted cross-sectional designs to reveal causal relationships but they did not use longitudinal methods. Furthermore, longitudinal design can explore many challenges of using RPA in the auditing aspect. Finally, RPA is considered as a hotspot research area. Thus, using robust statistical methods such as Multi-Criteria decision-making (MCDM) can assign weight to the criteria of RPA and identify the best and worst firms in many sectors based on RPA adoption. Moreover, defining the best and worst company based on RPA would reveal the challenges of using such technology and identify the leading company using RPA to overcome the constraints of adopting this tool in terms of auditing aspects.

6   Conclusion

Automation of robotic processes represents a technology with many benefits when applied to audits. It has a major role in enhancing audit effectiveness, increasing efficiency and quality of audits, and combating corruption and fraud. In addition, reducing the burden on auditors thus increasing their focus on tasks that require human judgment, which increases the possibility of expanding the skills of auditors by making the most of the use of new technologies. Automation has a major role in reducing the time spent by auditors in completing routine repetitive tasks. It is important to identify which activities may be automated. These instructions are typically routine, repetitive, and labor-intensive. They are more suited for RPA technologies to operate, especially when it comes to rules and structured data. Once a robot has been selected, it is important to consider the maintenance that the robots will require in the future to ensure that they can function properly. Finally, a cost-benefit analysis of the technology's use will allow the user to decide whether or not to use RPA technology. The most important threats of RPA to the governance environment in are that the processes are organized or semi-organized while determining the purpose of applying RPA to them. The data must be seriously unified to obtain the correct application of RPA technologies and to enhance the efficiency and effectiveness of auditing. In addition to choosing the appropriate technology that is in line with the project, especially in a large number of vendors of RPA technology, the vendor with the necessary experience should be selected.

References

Abbas, S., et al.: Antecedents of trustworthiness of social commerce platforms: a case of rural communities using multi group SEM & MCDM methods. Electron. Commer. Res. Appl. **62**, 101322 (2023)

Ahmed, A.D., Salih, M.M., Muhsen, Y.R.: Opinion weight criteria method (OWCM): a new method for weighting criteria with zero inconsistency. IEEE Access (2024)

Ali, J., Hussain, K.N., Alnoor, A., Muhsen, Y.R., Atiyah, A.G.: Benchmarking methodology of banks based on financial sustainability using CRITIC and RAFSI techniques. Decis. Making: Appl. Manage. Eng. 7(1), 315–341 (2024)

Alnoor, A., Atiyah, A.G., Abbas, S.: Toward digitalization strategic perspective in the European food industry: non-linear nexuses analysis. Asia-Pac. J. Bus. Adm. (2023)

Alnoor, A., Atiyah, A.G., Abbas, S.: Unveiling the determinants of digital strategy from the perspective of entrepreneurial orientation theory: a two-stage SEM-ANN approach. Glob. J. Flexible Syst. Manage. 1–18 (2024)

Atiyah, A.G., Faris, N.N., Rexhepi, G., Qasim, A.J.: Integrating ideal characteristics of chat-GPT mechanisms into the metaverse: knowledge, transparency, and ethics. In: Al-Emran, M., Ali, J.H., Valeri, M., Alnoor, A., Hussien, Z.A. (eds.) IMDC-IST 2024. LNNS, vol. 895, pp. 131–141. Springer, Cham (2023). https://doi.org/10.1007/978-3-031-51716-7_9

Baldwin, A.A., Brown, C.E., Trinkle, B.S.: Opportunities for artificial intelligence development in the accounting domain: the case for auditing. Intell. Syst. Account. Financ. Manage.: Int. J. 14(3), 77–86 (2006)

Cohen, M., Rozario, A.: Exploring the use of robotic process automation (RPA) in substantive audit procedures. CPA J. 89(7), 49–53 (2019)

Gotthardt, M., Koivulaakso, D., Paksoy, O., Saramo, C., Martikainen, M., Lehner, O.: Current state and challenges in the implementation of smart robotic process automation in accounting and auditing. ACRN J. Financ. Risk Perspect. (2020)

Huang, F., Vasarhelyi, M.A.: Applying robotic process automation (RPA) in auditing: a framework. Int. J. Account. Inf. Syst. 35, 100433 (2019)

Hyvönen, T., Järvinen, J., Pellinen, J.: A virtual integration—the management control system in a multinational enterprise. Manag. Account. Res. 19(1), 45–61 (2008)

Kokina, J., Blanchette, S.: Early evidence of digital labor in accounting: Innovation with Robotic Process Automation. Int. J. Account. Inf. Syst. 35, 100431 (2019)

Liu, S.: Robotic process automation (RPA) in auditing: a commentary. Int. J. Comput. Audit. 4(1), 23–28 (2022)

Liu, X., Ishak, N.N.B.M.: Research on the application and development of RPA in accounting higher vocational education: a Chinese perspective. Int. J. Educ. Humanit. 10(2), 178–182 (2023)

Madakam, S., Holmukhe, R.M., Jaiswal, D.K.: The future digital work force: robotic process automation (RPA). JISTEM-J. Inf. Syst. Technol. Manage. 16, e201916001 (2019)

Moffitt, K.C., Rozario, A.M., Vasarhelyi, M.A.: Robotic process automation for auditing. J. Emerg. Technol. Account. 15(1), 1–10 (2018)

Muhsen, Y.R., Husin, N.A., Zolkepli, M.B., Manshor, N.: A systematic literature review of fuzzy-weighted zero-inconsistency and fuzzy-decision-by-opinion-score-methods: assessment of the past to inform the future. J. Intell. Fuzzy Syst. 45(3), 4617–4638 (2023)

Srinivasan, V.: The Intelligent Enterprise in the Era of Big Data. Wiley, Hoboken (2016)

Zhang, C.: Intelligent process automation in audit. J. Emerg. Technol. Account. 16(2), 69–88 (2019)

Advancing Sustainable Learning by Boosting Student Self-regulated Learning and Feedback Through AI-Driven Personalized in EFL Education

Muthmainnah Muthmainnah[1], Luis Cardoso[2],
Yasir Ahmed Mohammed Ridha Alsbbagh[3], Ahmad Al Yakin[1], and Eka Apriani[1,4(✉)]

[1] Universitas Al Asyariah Mandar, Polewali, Sulawesi Barat, Indonesia
[2] Polytechnic Institute of Portalegre and Centre for Comparative Studies of the University of Lisbon, Lisbon, Portugal
lmcardoso@ipportalegre.pt
[3] Al- Iraqi University-Collage of Arts- English Department, Baghdad, Iraq
yasir_mohammed@aliraqia.edu.iq
[4] IAIN Curup, Bengkulu, Indonesia
eka.apriani@iaincurup.ac.id

Abstract. This quantitative methods study investigates the impact of AI-mediated language teaching on English as a Foreign Language (EFL) students' capacity to regulate their own learning as well as their feedback-response and self-regulation skills, considering the significant influence of English as a Foreign Language (EFL) teaching. Interest in AI-based educational tools and their revolutionary potential in EFL language teaching in Asia. The population in this study is junior high school class VIII, with 258 students. A total of 35 students from junior high school participated in the study and received teaching through artificial intelligence chosen by the purposive sampling technique. The self-report survey measures independent learning and second language (L2) feedback, and the pre- and post-tests evaluate proficiency in English language acquisition in the areas of vocabulary, reading comprehension, writing ability, and grammar. The study results show the results of learning English, the use of independent learning techniques, and feedback according to quantitative analysis. Qualitative analysis of five interviews with students highlights the revolutionary impact of AI platforms, which increase engagement, provide personalised learning experiences, and, ultimately, encourage drive and autonomy. These results are useful for researchers and teachers who want to use AI-powered platforms in language classes because they show that AI-mediated language teaching could improve language learning outcomes, such as better use of instructor time, lower performance-related anxiety among students, fair education for all students, and higher levels of student competency due to more constructive criticism and evaluation. The successful implementation of these AI tools depends on teachers' backgrounds, which include their biases, education, and experience. Most students think that a balanced use of these AI tools is the best way forward. Proper guidance is essential to ensuring thorough and productive use of AI tools in the classroom. Our study findings suggest that more research is needed to fully utilise these tools for the benefit of students and educators, especially about their accuracy, practicality, and efficiency in future EFL classrooms.

Keywords: Sustainable education · Artificial Intelligence · EFL · self-regulated learning and feedback

1 Introduction

In the digital era and educational transformation of the 21st century, there has been a surge of interest among language experts in maximising the use of IT for language teaching and acquisition. According to Han, et al. (2024) students can benefit from using IT in EFL classes because it allows for more personalised, interactive, engaged, interesting, and communicative learning. According to several studies (Muthmainnah et al., 2023), language instructors are starting to utilise IT to build online language classrooms that attract students' attention and simplify the learning process. Artificial intelligence (AI) is a new and exciting tool in the world of IT that promises to improve learning outcomes and effective EFL language teaching, especially in Asia-Pacific countries (Yang, 2024).

Artificial intelligence in education (AIED) refers to the integration of AI into educational contexts using specific programmes or application technologies to improve teaching, assessment, and decision-making. Computer systems could assist instructors and students by providing individualised instruction, support, and feedback. Artificial intelligence systems that mimic human intelligence to draw conclusions, offer judgments, or estimate results aid in the learning of EFL languages (Khang, et al., 2023). The use of AI has increased in EFL language classes, indicating that AI-powered tools and systems are becoming more popular among all stakeholders in the education sector. But teachers are unsure how to use technology; they are concerned about the potential benefits and harms to their students (Chan and Hu, 2023). Teachers and students should be informed about advanced AI-led tools, some of which are already being used during current learning sessions. Besides that, AI can also help minimise teacher burden (Darvishi, 2024). The field of education is one of the fields that has greatly benefited from the many applications of artificial intelligence (Li, et al. 2024), and currently, the field of artificial intelligence not only includes computer programmes that recognise faces and objects but also speech, expert systems, machine vision, and language processing experience. Therefore, educators who want to use this application with English as a Foreign Language (EFL) students need a strong foundation in artificial intelligence.

This is in line with the findings of George, (2023), who suggest that to encourage discussions about data ethics and privacy, as well as placing emphasis on student assessment and engagement, collaborative learning, and critical thinking, AI-based education tools and pedagogy curricula need to be structured better. Artificial intelligence (AI) has the potential to revolutionise education and bring about positive change (Darvishi, et al. 2024). There is great hope that advances in artificial intelligence (AI) will help teachers and students in the classroom. From the most basic functions, such as self-regulated learning and automatic feedback, to the most advanced functions, such as intelligent support, AI is capable of extraordinary things. In the past, teaching and learning in the EFL classroom required teachers to spend hours preparing all the necessary teaching materials. However, with the presence of AI, a new learning paradigm has emerged, namely a learning paradigm that makes students more involved, and teachers can facilitate a more conducive learning environment.

Even though artificial intelligence (AI) is rapidly spreading into English as a Foreign Language (EFL) classes, many educators still misunderstand the technology, even though the technology has many advantages. Even though AI is widespread in many aspects of our lives, many still associate it only with machines and robots. The use of artificial intelligence (AI) is already happening in higher education, but many educators don't know how widespread its application is or, more importantly, what its true scope is. In the modern classroom, students are expected to take an active role in their own learning and work together on collaborative projects, with the teacher taking the role of facilitator (Belda-Medina and Kokošková, 2023). It is important for educators to have the courage to confront students' misconceptions about AI and uphold ethical standards when implementing the many applications that can enhance learning (Zhou et al., 2024). Additionally, research on students' emotions (Godwin-Jones, 2024; Zhang et al., 2024a, b), collaborative interactions (Shafiee Rad, 2024), thinking and content (Alshumaimeri and Alshememry, 2023) has utilized AI applications.

Dede and Lidwell (2023) developing a next-generation model for massive digital learning. Teachers have the multitasking task of meeting their students' varying linguistic needs while also carefully recording their progress, difficulties, and absences; new developments in the field of artificial intelligence (AI) in education aim to simplify classroom management and increase teaching productivity (Mohamed Haggag, 2024). According to Dede and Lidwell, (2023), integrating AI technology into hybrid teaching platforms can improve teaching resources, foster a good learning environment for teachers and students, and ultimately result in more effective learning. Because this topic is still new, very little is known about the pros and cons, barriers, and applications of AI in education, especially as it relates to teaching English as a Foreign Language (EFL). Therefore, the aim of this research is to conduct a comprehensive study of artificial intelligence (AI) in English as a Foreign Language (ESL) classroom and to identify self-directed learning and feedback through AI-based personalization in EFL education.

2 Literature Review

2.1 The Application of AI to the Field of EFL Education (AIEd) Metaverse

The use of artificial intelligence (AI) in the metaverse classroom has the ability to completely change the way students learn and interact with teachers and machines. For the ethical integration of AI in education, it is critical to eliminate algorithmic bias, ensure data privacy, and maintain a balance between human contact and AI support (Al-Enzi et al., 2023; Husin et al., 2023; Atiyah, 2023). Students and teachers alike will benefit from the more customised, flexible, and welcoming classrooms made possible by integrating AI into education systems. There are several ways that English can be taught and learned; some examples are EFL, ELF, and ESL, the first of which is a term often used in language classrooms in Indonesia. Because EFL students in Indonesia come from a wide range of language proficiency levels, creating lessons that meet each student's needs can be a challenge. Since each skill has a unique level of difficulty, there are many obstacles to their progress in mastering it. Problems with vocabulary, grammar, tenses, and language skills such as listening, speaking, reading, and writing are difficult for students with low proficiency levels to face (Billingsley and Gardner, 2024; Arroyo-Romano, 2024; Innaci and Jona 2024). In reading, students have difficulty pronouncing words correctly and using correct grammar. Regarding their ability to speak and listen, most students have difficulty maintaining a conversation and have not mastered active listening because they do not have sufficient vocabulary (Soyoof, et al. 2024).

Teaching strategies are the main source of problems in students' speaking, but there are many other factors that play a role, such as students' limited vocabulary, difficulty pronouncing words, confusion regarding word order, and fear of making mistakes. Therefore, AI may be seen as a new approach to solving this problem and suggest the use of AI for language teaching. Modern AI systems can do more than just detect errors; they can also provide suggestions on how to improve them, which is a significant improvement over the capabilities of early systems that only detected objective errors such as spelling and grammar (Zhang, et al. 2024a, b). A product that exemplifies this "smartness" is Grammarly. While AI error detection and spelling capabilities make it suitable for English as a Foreign Language (EFL) classrooms (Al-Raimi, 2024), the utility of this technology goes beyond that. Although word processors already have built-in features for checking things like spelling, grammar, and style, new tools have emerged that help students with more sophisticated composition problems (Huete-García and Tarp (2024). The development of AI software and applications suitable for the ESL classroom provides an outline of the role of AI in EFL teaching and learning, as well as an analysis of the pros and cons of AI and examples of tools used by students and instructors in the EFL classroom. Found that when AI is fully integrated into classrooms, it will greatly improve the quality and effectiveness of EFL language teaching, (Zou, et al. 2023).

Several AI-based strategies have been developed for use in EFL classrooms. It is recommended to use an AI-based language learning platform. The use of artificial intelligence (AI) as a platform to assist EFL students in language acquisition is supported by Ravshanovna (2024). The platform enables personalized learning experiences through the use of artificial intelligence algorithms and natural language processing capabilities (Atiyah et al., 2023a, b). Students can gain many benefits from it, including the ability to plan their own unique educational course. According to Söğüt, (2024) the AI system evaluates students' level of language competency by administering an initial exam and continuously monitoring their performance. The platform analyses each student's strengths and weaknesses and then creates a personalised learning path that focuses on areas that need improvement. The use of AI applications in EFL classrooms has further potential benefits, such as helping teachers reduce their workload and increasing the efficiency of student assessment (Mohammed, 2023; Atiyah et al., 2023a, b). Junior high school students whose first language is not English can benefit from the use of AI-powered learning media and technology that provides feedback and formative assessments to improve their learning habits and make more optimal use of technology Wale and Kassahun, (2024).

Artificial intelligence systems evaluate how well students are doing and adjust the level of difficulty or content (Jokhan et al., 2022). This ensures that students are not overwhelmed and that they get help where they need it most by providing challenging but manageable assignments. Apart from being able to imitate human intelligence, AI is also able to correct and provide intelligent feedback. Students receive instant feedback on their language activities from the AI system (Kamruzzaman et al. 2023; Alnoor et al., 2023). It corrects errors in syntax, vocabulary, and pronunciation while providing context and suggestions on how to improve. In addition to assisting instructors in quickly finding errors, systems can monitor recurring errors and provide targeted feedback to students to help them overcome the problem (Mhlanga, 2023). Additionally, the platform is also equipped with artificial intelligence chatbots, also known as virtual language assistants, which not only enhance students' cultural content but also engage them in conversation simulations (Tan and Ng, 2024; Alnoor et al., 2024). Chatbots act as teachers by asking students questions, responding to their responses, and providing comments on how well they express themselves verbally. The AI system leverages natural language processing to identify and assess students' pronunciation, fluency, and speech patterns, then provide feedback on how to improve them.

2.2 Self-regulated Learning in EFL Through AI

The use of artificial intelligence (AI) is already happening in higher education, according to Southworth et al. (2023), but many educators don't know how widespread it is or, more crucially, what it comprises. In the modern classroom, students are expected to take an active role in their own learning and work together in collaborative projects, with teachers taking on the role of facilitators (Nemorin, et al. 2023). It is imperative that educators have the guts to combat student misunderstandings about AI and uphold ethical standards when implementing its many learning-enhancing applications. In addition, research on students' emotions, collaborative interactions (Koç, 2024; Ali et al., 2024), and thoughts and contents (Swargiary, 2024) has made use of AI applications and teachers already have a lot on their plates trying to meet their students' varying linguistic needs while also keeping meticulous records of their progress, struggles, and absences; new developments in the field of artificial intelligence (AI) in education aim to streamline classroom management and increase teaching productivity (Dayal, 2024; Wang et al., 2023). According to Dolunay and Temel, (2024), integrating AI technology into online teaching platforms can improve instructional resources, foster a good learning environment for both teachers and students, and ultimately lead to more effective learning. Due to the novelty of the topic, very little is known about the pros and cons, obstacles, and applications of AI in the field of education, particularly as it pertains to English as a Foreign Language (EFL) instruction, Shafiee Rad, H. (2024).

To determine the following steps in Zimmerman's self-control and self-regulation stages of self-regulated learning, students must identify various levels of assessment. A common idea in educational psychology, this research uses the Judgement of Learning as a foundation for the self-assessment of students learning EFL with the integration of AI in the classroom during the learning process. Joint Observation of Learning (JOL) is an understanding from the fields of psychology and education that describes how a person assesses their own learning Çakiroğlu and Öztürk, (2023) as will be studied in this study. It reveals how much information a person feels they have learned or retained, which influences their drives and actions when learning is mediated by media and AI technology Wang et al. (2024a, b). The level of difficulty of subject matter, students' initial knowledge and abilities, and the efficacy of learning techniques are several variables that can influence JOL, according to several studies (Pachón-Basallo, et al. 2022; Tinajero, et al. 2024). A higher JOL is associated with higher learning motivation and participation in self-regulated learning, while a lower JOL can cause a lack of interest in learning and avoidance of challenging tasks, thereby impacting students' self-regulated learning (Azevedo, et al. 2022).

JOL can also help students get feedback on their learning progress, telling them where they are lacking and how to improve them Torrington, et al. (2024) Furthermore, JOLs have the potential to influence a person's self-confidence, which in turn impacts a person's learning style (Jemstedt, 2024). According to Désiron and Schneider, (2024) the cue utilisation approach postulates that people utilise various signals, or indications, to evaluate their own learning and it is one of the most prominent theories of JOL Things like the amount of time spent studying, level of understanding, and difficulty of the content are examples of such indicators. Hoffmann and Hoffmann, (2016) argued that people are more likely to experience high JOLs when they encounter positive learning signals (such as domain-specific knowledge) and low JOLs when they encounter negative learning cues (such as feelings of unfamiliarity or difficulty). Metacognitive awareness is another important outcome of JOL, this highlights the importance of metacognitive processes, also known as higher order thinking skills in learning procedures. People direct their learning and evaluate their progress using metacognitive methods such as planning, monitoring, and evaluation, according to research. Therefore, those who score higher on JOL tend to use efficient metacognitive methods, which in turn results in better learning outcomes. A phenomenon known as "negative JOL" Azevedo et al. (2013) occurs when students become aware of their own conceptual gaps; this realisation may lead to the refinement of existing learning procedures. Imagine for a moment that students do not change their approach to this criticism. When this happens, students' metacognitive actions are categorised as "static", indicating that they are aware of gaps in their knowledge but are unwilling to make any changes. There are several JOL models proposed. Some models propose that a person's self-perception, including their abilities and self-esteem, impacts their JOLs, while other models, such as the social cognitive model, highlight the importance of social and contextual elements in learning.

The integration of AI's present educational skills with Zimmerman's theory of self-regulated learning and collaborative observation of learning holds tremendous promise for enhancing students' capacity for self-regulated learning. Current AI technology is one-sided, as we saw before about the emulation and self-control stages; that is, the user (or learner) instructs the generative AI tool to carry out its intended purpose and objective (also referred to in the following section). However, it must be stressed that the user and the AI must be able to communicate with one another in order for the classroom to function well. By enabling AI to initiate tailored learning feedback, chatbots can be integrated into many aspects of teaching and learning, leading to interactive and instructional experiences.

3 Methodology

The main aim of this research is to examine how English language learners in junior high school use and understand Chatbot-Cici-AI. To obtain measurable data about students' knowledge, use, and feelings towards these tools, this investigation used a survey methodology that was organised and flexible. A well-known platform for creating and sharing surveys, Google Forms, was used to collect data. To better understand how students interact with and feel about Cici Bot, the survey instrument has 20 Likert scale questions. To ensure everyone can understand and participate, survey questions are provided in Indonesian (Ahmed et al., 2024). This is done to help overcome language barriers and ensure the answers are accurate and reliable.

Because all participation is completely voluntary and participants are guaranteed anonymity, an atmosphere is created that encourages open and honest answers. Participants self-reported their knowledge, behaviour, and attitudes about Cic Bot-AI, which formed the basis of the data collection method. A total of 33 junior high school students (SMPN 4 Polewali Mandar) were interested in taking part in the survey. The demographic information of participants was not disclosed, but since most of the survey was distributed to second-year students, it can be assumed that they comprised the majority of respondents. Additionally, the variety of comments shows that there is a diverse spectrum of views and experiences regarding the use of AI Tutor's tool, Cici Bot (Muhsen et al., 2023).

The survey asked students to rate their level of engagement with Cicibot AI in multiple-choice categories. Questions covered topics including understanding the impact of AI Cici Bot on self-regulated learning and feedback, including frequency of use in academic settings, variables influencing the choice to accept or reject AI Cici Bot estimates of future use, and perceived benefits of AI in the classroom. If the predetermined options did not adequately capture participants' perspectives or experiences, they were also given the option to provide free-form comments in the 'other' category.

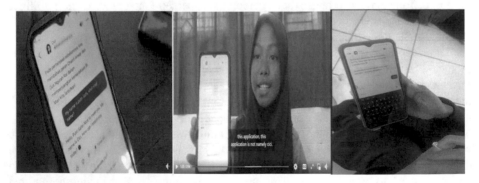

Fig. 1. Practicaly AI in EFL Class

4   Results and Discussion

4.1   Results

The results of this study show that the impact of AI-mediated language training was praised by participants, who often described it as an interesting and new way to learn the EFL language. They described how the AI platform's exercises took them into an immersive environment where they could learn at their own pace with the help of dynamic language games and instant feedback and helped students stay motivated to learn self-regulated learning which will be explained in Table 1 below.

Table 1. Descriptive Statistics of student's response on self-regulated learning and feedback in learning English

Statements	N	Minimum	Maximum	Mean	Std. Deviation
1. I set specific learning goals for myself in learning English	33	3.00	5.00	4.3636	.54876
2. I can monitor my own learning progress and performance in completing English learning assignments	33	4.00	5.00	4.6061	.49620
3. I often use strategies such as planning and organising lesson materials systematically to improve English learning	33	4.00	5.00	4.4848	.50752
4. I often look for additional resources or materials to complement English learning outside the classroom (for example, using YouTube, AI, Instagram, or TikTok)	33	3.00	5.00	4.2727	.71906
5. I actively self-reflect on my English learning progress and identify areas that need improvement	33	3.00	5.00	4.5758	.56071
6. I believe in managing time effectively to allocate sufficient study time for English learning tasks	33	3.00	5.00	4.4848	.61853

(continued)

Table 1. (*continued*)

Statements	N	Minimum	Maximum	Mean	Std. Deviation
7. I find it useful to organise learning by adjusting learning strategies based on feedback from assessments or assignments	33	3.00	5.00	4.6364	.60302
8. I apply strategies such as independent practice or repetition at home to consolidate my English learning	33	3.00	5.00	4.4848	.56575
9. I enjoy seeking clarification or help from teachers, friends, or using Cici Bot when facing challenges in learning English	33	4.00	5.00	4.6061	.49620
10. I consistently learn English, even when facing difficulties	33	4.00	5.00	4.7273	.45227
11. I believe that AI tools' personalized feedback is helpful for English learning in the classroom	33	3.00	5.00	4.3939	.60927
12. I feel that AI-based input improves understanding of material concepts and English learning theories	33	4.00	5.00	4.5758	.50189
13. I am happy with the level of specificity and detail in the feedback that AI tools provide for learning English	33	3.00	5.00	4.6667	.59512
14. I incorporate AI-generated feedback into English learning practice or revision	33	3.00	5.00	4.6667	.54006
15. AI-based input is invaluable in identifying strengths and areas for improvement in English language learning	33	3.00	5.00	4.6970	.52944

(*continued*)

Table 1. (*continued*)

Statements	N	Minimum	Maximum	Mean	Std. Deviation
16. AI-based feedback is very effective in responding to your individual learning needs and preferences in English learning	33	3.00	5.00	4.7273	.51676
17. AI-based feedback motivates me to engage more actively in English learning tasks	33	4.00	5.00	4.8182	.39167
18. The accuracy and relevance of AI-generated feedback are very reliable in guiding English learning progress	33	4.00	5.00	4.8485	.36411
19. I follow up on suggestions or recommendations given by AI tools in English learning	33	4.00	5.00	4.8788	.33143
20. Overall, I am satisfied with the feedback provided by the AI tool in supporting your English learning journey	33	4.00	5.00	4.9394	.24231
Valid N (listwise)	33				

Based on the description of statistical data in Table 1, the answers to student survey results regarding feedback and independent learning in EFL are displayed. Each statement, which covered topics such as goal planning, tracking progress, and using AI-based tools for English language learning, was rated on a scale of 3 to 5, and the sample size was 33 participants. The results showed that people generally had a positive impression of AI-based self-paced learning and feedback, with average scores ranging from 4.2727 to 4.9394. Participants will likely set concrete learning goals, track their progress, and seek out additional materials to help them improve their language skills. In addition, they demonstrated a positive attitude towards AI technology, recognising its benefits in providing personalised feedback, improving understanding of course material, and inspiring active participation in language practice. There appears to be a great deal of consensus among respondents, as indicated by the small standard deviation (0.24231 to 0.71906). In short, the research results show that students' efforts to learn English are greatly helped by self-regulated learning procedures and by feedback generated by artificial intelligence as a form of technology-based learning adaptation.

The results of statistical data on students' EFL language acquisition are shown in Table 2. The average results of the English language proficiency tests taken by students

Table 2. The mean score

	N	Minimum	Maximum	Mean	Std. Deviation
Cici Bot in EFL	33	89.00	99.00	92.4545	2.62311
Valid N (listwise)	33				

after Cici Bot, an AI-based intervention, was implemented. The dataset includes the responses of 33 participants, with English scores ranging from 89.00 to 99.00. The student standard deviation is 2.62311, and the average value is 92.4545. This research highlights the significant improvements in students' English language skills that occurred after using Cici Bot in EFL classes. Complete responses from 33 participants supported the validity of the data set and the robustness of the study.

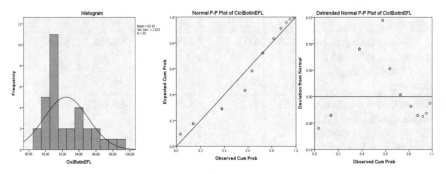

Fig. 2. The histogram and Normal P-P Plot of Cici Bot in EFL

4.2 Discussion

Metaverse classrooms are classrooms that utilize the most advanced technology, such as AI, Chatbot or ChatGPT with fast Internet connections, and student devices while learning EFL with AI. The aim is to make the classroom atmosphere and learning more efficient for both students and educators so that they are actively involved during the learning process. There are several ways in which metaverse classrooms have enhanced the teaching of English as a foreign language (EFL) and encouraged new approaches to pedagogy and technological resources. Support services and learning activities in the AI learning model enable educators to be able to design classrooms that arouse students' curiosity, self-regulated learning, prompt feedback and keep them actively involved in their own learning.

In terms of English language teaching, metaverse classrooms have the potential to offer a more accessible and versatile learning method than conventional classrooms. Understanding the relationship between self-regulated learning, classroom engagement, and the importance of feedback in learning English as a foreign language (EFL) in a metaverse classroom is the main goal of this research. Students display a wide range of academic attitudes when studying in classes that integrate AI or ChatBots, according to previous findings. There is a positive direct impact of student self-regulated learning, student academic scores on engagement in class and increased feedback as shown in Table 1 and Table 2. The next section will offer a more thorough and detailed analysis.

Several users have lauded this site for its innovative approach to language learning, which allows users to immerse themselves in the creative process of learning English as a foreign language (statements 1, 2, 3, and 4). Statement 9 states that the children found learning to be an exciting and enjoyable adventure due to the AI platform's interactive features. The ability of the AI platform to tailor lessons to each student's unique set of skills, interests, and challenges is something that students really value. They rave about how the platform's individualized approach helps users find the best learning paths for their specific needs. Thanks to the individualized instruction they receive, individuals can overcome challenges and play to their strengths. This customized learning experience supports participants' feelings of autonomy and self-assurance as they embark on their path to language acquisition. The flexibility of the platform to accommodate different user preferences and learning methods gives students a sense of agency. The AI platform can change its approach based on how fast or slow a learner learns (Fig. 1).

Participants stated that AI-mediated education was the main factor in the tremendous increase in their English language proficiency. Their increased abilities in various areas of language use, such as syntax, vocabulary, reading comprehension, and writing, are largely due to the interactive and adaptive characteristics of AI platforms (statements 10, 11, and 12). In Table 2, students demonstrate how the AI ChatBot's impact on improving their language competency facilitated their success in intensive English courses, resulting in better grades and increased self-confidence in the classroom. The use of CiciBot-AI in the classroom had a profound impact on students' enthusiasm to learn and their level of active participation in English language activities. The participants' heightened curiosity and enthusiasm continued even after class concluded. Many them actively sought out opportunities to practise and additional language materials. The AI platform ignited a need for knowledge by encouraging an authentic curiosity in the English language (Fig. 1).

In creating an atmosphere that was favourable to learning, participants believed that the AI-mediated instruction was vital. Their experience on the platform was one of constant encouragement to try new things, fail gracefully, and grow from their setbacks. A catalyst for improved learning, this welcoming environment was said to be in Table 1. However, respondents felt comfortable expressing their English performance (Fig. 1) with the help of the AI platform. Also, we observed students' anxieties were reduced in large part because of the AI platform's non-judgmental and constructive comments. According to the participants, the platform comments did more than just highlight their mistakes; it also provided context and ways to improve. They were able to grow resilient in their language learning journey and learn from their mistakes using this Cici Bot. The participants spoke about how easy and versatile the AI platform was for their English language practice. They valued the independence to study languages whenever and wherever they liked. A lot of people said that they could easily work language exercises into their regular schedules, including on the way to work or during breaks. A participant mentioned that they could practise English anytime they had a moment to spare. It transformed education from a chore to an integral part of their life.

In a metaverse classroom context, students' academic grades significantly impact their self-regulated learning and feedback, according to the statistical analysis in Fig. 2. These results indicate that in a metaverse classroom environment, students' self-regulated learning and feedback during EFL learning significantly improve EFL learning achievements. To begin with, research has shown that learning using AI is a fun learning strategy that significantly influences student engagement in English as a foreign language learning in metaverse classrooms. Students are more likely to participate actively in class if they feel good learning comfort such as enthusiasm, hope, and confidence when they learn. This is possible because, with the help of cutting-edge learning strategies and AI technology resources, metaverse classrooms can offer more opportunities for teachers in junior high schools connecting visuals, sound, and moving images into lesson plans, multimedia has the potential to improve students' mood by making previously boring subject matter more interesting, interactive and exciting (statement number 13).

Cici Bot virtual tutor provides educational opportunities that are impossible to replicate in a physical location, while generating students' interest and motivation to learn about the world around them (statement number 20). By displaying their results and providing quick feedback, interactions with Cici Bot can give students a sense of accomplishment while encouraging healthy competition. The survey results positively influenced class engagement, increased self-regulated learning and feedback significantly, according to our research results. This means that SRL AI-based feedback is very effective in responding to your individual learning needs and preferences in English learning, AI-based feedback motivates me to engage more actively in English learning tasks, The accuracy and relevance of AI-generated feedback are very reliable in guiding English learning progress (statement numbers 13,14, 15, 16, 17, 18 and 19) influences how actively they participate in class. Therefore, educators can positively integrate AI with their students by designing engaging learning environments that arouse their interest and foster their curiosity. This, in turn, increases their feedback in the metaverse class.

Student self-regulated learning and feedback increase students' active involvement in metaverse classes, increasing their confidence in their own abilities (statements 4, 5, 6, and 7). The level of student participation in class indicates the acquisition of knowledge and application skills (Fig. 1) in line with the study of Dong and Habók (2024), when students actively participate in class, they gain confidence in their learning abilities, which improves their feedback and ultimately leads to improved academic performance (Table 2) similar with Chansri, et al. (2024) findings by self-regulated learning increase the students language ability. Educators must be aware of their students' level of self-regulated learning and feedback and provide them with plenty of positive reinforcement to increase their confidence in their learning abilities and tackle problems head-on. The learning environment of the AI-based student centre learning approach significantly influences self-regulated learning. The results of this research are like Ng, et al. (2024) research by using SRLbot and Alshewiter, et al. (2024) findings also highlighted students' self-regulated learning and feedback greatly influence their level of confidence in learning more EFL, which may develop when they feel comfortable while learning (statement number 9).

Self-regulated learning also mediates the relationship between academic achievement and feedback, according to this research. Student participation in class is significantly mediated by self-regulated learning and feedback resulting from the integration of AI-Cici Bot in class, according to the findings of this study with the help of AI technology, students become more focused on what they learn in class. Students' positive emotions can indirectly increase their self-regulated learning and feedback. This means, students' cognitive abilities, evaluation of their own learning abilities, and feedback are all improved when they feel comfortable learning in class using AI Tutor Cici-Bot, which in turn increases their engagement and confidence in their own learning abilities. Building positive feedback allows students to increase their drive to learn and achieve better academic results. Consequently, to improve students' self-regulated learning and feedback in EFL learning, apart from using AI, intelligent media, and technology, teachers must foster a fun, interesting, and comfortable classroom environment to encourage active participation in metaverse classroom activities.

There are several things that need to be considered in this research, one of which is that this research may only include data from a small number of junior high schools or regional schools, and not all of them are studying to major in English. In addition, issues of subjectivity and self-report bias may remain even if the scale of the survey has been increased. Additionally, the research methodology is survey-based, which is great for analysing the impact of AI- chatbots briefly but not so great for tracking how factors change over time. A longitudinal approach could be used in future research to investigate the impact of ChatBot-Tutor AI on student academic achievement, class engagement, self-regulated learning, and feedback in the long term. Research with larger samples and multicentre studies covering several schools or regions will be more representative and can be applied to a wider audience to enrich the results of subsequent research.

5 Implications and Future Studies

a. Theoretical Implications

Metaverse classrooms that use advanced technology such as AI, Chatbot, or ChatGPT to teach English as a foreign language have many theoretical implications. To begin with, these types of classes encourage SRL, or self-regulated learning, which means that students are actively involved in their own education. Students' SRL techniques are improved by the use of AI technology, as demonstrated by ChatGPT, which offers timely and personalised feedback (statements 1, 2, 3, 4, and 10). Research by authors like Dong and Habók (2024) and Chansri et al. (2024) lends credence to the idea that students' academic performance improves when they actively participate in class and learn to control their own learning. Second, as mentioned in statements 13, 14, 15, 16, 17, 18, 19, and 20, the adaptive nature of AI platforms like Cici Bot creates a positive learning environment by lowering anxiety and boosting students' confidence in their abilities. Consistent with prior research, this supports the idea that SRL and feedback affect students' self-esteem and language acquisition (Alshewiter et al., 2024; Ng et al., 2024). Thus, theoretically speaking, students' SRL, feedback reception, and overall learning experiences are greatly impacted by metaverse classrooms' AI-driven technologies, leading to improved academic outcomes.

b. Practical Implications

A revolutionary change in pedagogy could be on the horizon, according to the practical implications of AI-enhanced metaverse classrooms for English as a foreign language education. Using ChatGPT and Cici Bot, two examples of artificial intelligence applications, teachers can create more interesting and personalised lectures (statements 9 and 10). As stated in points 9 and 10, these technologies help students become more engaged and involved in their learning by creating personalised learning materials, offering vocabulary aids, and mimicking real-life discussions. In addition, students feel safe expressing themselves since the AI-mediated instruction fosters a warm and accepting classroom climate (statements 6, 7, and 8). While AI takes care of mundane duties like grading and giving immediate feedback, teachers can concentrate on promoting active learning through conversation and activity facilitation (statements 6, 7, and 8).

c. Future Studies

To developing more efficient AI-based instructional interventions, it is important to comprehend how AI technologies impact students' feelings and perspectives on learning, as shown by their interest and excitement (statements 9 and 11). Finally, propositions 13, 14, and 15 suggest that a more thorough knowledge of the pros and cons of AI in various educational contexts might be achieved by investigating its impact on various skill levels and learning styles.

6 Conclusion

This article attempts to provide a comprehensive analysis of the use of AI in EFL classrooms that can increase self-regulated learning and feedback during the learning process in metaverse classrooms. The results outline the current status of research and also provide valuable data to academics, teachers, lecturers, and other stakeholders. The majority of research is conducted in other areas of education, so there is a need to address the research gap by investigating self-regulated learning and feedback using AI in EFL classrooms for sustainable education. Future researchers should look more deeply into the context of EFL classrooms in Indonesia. From what we saw in this review, there are many ways AI can improve the EFL classroom. Adaptive learning opportunities and personalised feedback are two ways in which AI-based intelligent tutoring systems can meet each student's unique needs. Educators, legislators, and academics can use AI to build inclusive language learning environments that prepare EFL students for success in the digital era, keeping up with 21st century development trends and educational transformation, but only if they carefully utilise the benefits of media and technology, or machine learning, effectively, wisely, and appropriately to learning needs in the education era. 6.0.

References

Ahmed, A.D., Salih, M.M., Muhsen, Y.R.: Opinion Weight criteria method (OWCM): a new method for weighting criteria with zero inconsistency. IEEE Access (2024)

Al-Enzi, S.H.Z., Abbas, S., Abbood, A.A., Muhsen, Y.R., Al-Hchaimi, A.A.J., Almosawi, Z.: Exploring Research Trends of Metaverse: A Bibliometric Analysis. In: Al-Emran, M., Ali, J.H., Valeri, M., Alnoor, A., Hussien, Z.A. (eds.) IMDC-IST 2024. LNNS, vol. 895, pp. 21–34. Springer, Cham (2023). https://doi.org/10.1007/978-3-031-51716-7_2

Ali, J., Hussain, K.N., Alnoor, A., Muhsen, Y.R., Atiyah, A.G.: Benchmarking methodology of banks based on financial sustainability using CRITIC and RAFSI techniques. Decis. Making: Appl. Manage. Eng. 7(1), 315–341 (2024)

Alnoor, A., Atiyah, A.G., Abbas, S.: Toward digitalization strategic perspective in the European food industry: non-linear nexuses analysis. Asia-Pac. J. Bus. Adm. (2023)

Alnoor, A., Atiyah, A.G., Abbas, S.: Unveiling the determinants of digital strategy from the perspective of entrepreneurial orientation theory: a two-stage SEM-ANN approach. Glob. J. Flexible Syst. Manage. 1–18 (2024)

Al-Raimi, M., Mudhsh, B.A., Al-Yafaei, Y., Al-Maashani, S.: Utilizing artificial intelligence tools for improving writing skills: exploring Omani EFL learners' perspectives. In: Forum for Linguistic Studies, vol. 6, no. 2, p. 1177 (2024)

Alshewiter, K.M., Shawaqfeh, A.T., Khasawneh, A.J., Alqudah, H., Jadallah Abed Khasawneh, Y., Khasawneh, M.A.S.: Improving the learning of language proficiency at tertiary education level through AI-driven assessment models and automated feedback systems. Migrat. Lett. 21(2), 712–726 (2024)

Alshumaimeri, Y.A., Alshememry, A.K.: The extent of AI applications in EFL learning and teaching. IEEE Trans. Learn. Technol. (2023)

Arroyo-Romano, J.E.: "My Spanish feels like my second language": addressing the challenges of the academic language of bilingual teachers. J. Latinos Educ. 23(1), 403–423 (2024)

Atiyah, A.G.: Unveiling the quality perception of productivity from the senses of real-time multisensory social interactions strategies in metaverse. In: Al-Emran, M., Ali, J.H., Valeri, M., Alnoor, A., Hussien, Z.A. (eds.) IMDC-IST 2024. LNNS, vol. 876, pp. 83–93. Springer, Cham (2023). https://doi.org/10.1007/978-3-031-51300-8_6

Atiyah, A.G., Alhasnawi, M., Almasoodi, M.F.: Understanding metaverse adoption strategy from perspective of social presence and support theories: the moderating role of privacy risks. In: Al-Emran, M., Ali, J.H., Valeri, M., Alnoor, A., Hussien, Z.A. (eds.) IMDC-IST 2024. LNNS, vol. 876, pp. 144–158. Springer, Cham (2023a). https://doi.org/10.1007/978-3-031-51300-8_10

Atiyah, A.G., Faris, N.N., Rexhepi, G., Qasim, A.J.: Integrating ideal characteristics of Chat-GPT mechanisms into the metaverse: knowledge, transparency, and ethics. In: Al-Emran, M., Ali, J.H., Valeri, M., Alnoor, A., Hussien, Z.A. (eds) IMDC-IST 2024. LNNS, vol. 895 pp. 131–141. Springer, Cham (2023b). https://doi.org/10.1007/978-3-031-51716-7_9

Azevedo, R., et al.: Lessons learned and future directions of MetaTutor: Leveraging multichannel data to scaffold self-regulated learning with an intelligent tutoring system. Front. Psychol. 13, 813632 (2022)

Azevedo, R., et al.: Using trace data to examine the complex roles of cognitive, metacognitive, and emotional self-regulatory processes during learning with multi-agent systems. Int. Handb. Metacogn. Learn. Technol. 427–449 (2013)

Belda-Medina, J., Kokošková, V.: Integrating chatbots in education: insights from the Chatbot-Human Interaction Satisfaction Model (CHISM). Int. J. Educ. Technol. High. Educ. 20(1), 62 (2023)

Billingsley, F., Gardner, S.: Utilizing generative AI with second-language users and bilingual students. In: Facilitating Global Collaboration and Knowledge Sharing in Higher Education With Generative AI, pp. 109–132. IGI Global (2024)

Çakiroğlu, Ü., Öztürk, M.: Microanalytic evaluation of students' self-regulated learning in flipped EFL instruction. J. Comput. High. Educ. 1–28 (2023)

Chan, C.K.Y., Hu, W.: Students' voices on generative AI: perceptions, benefits, and challenges in higher education. Int. J. Educ. Technol. High. Educ. **20**(1), 43 (2023)

Chansri, C., Kedcham, A., Polrak, M.: The relationship between self-regulated learning strategies and English language abilities and knowledge of undergraduate students. LEARN J.: Lang. Educ. Acquisit. Res. Netw. **17**(1), 286–307 (2024)

Darvishi, A., Khosravi, H., Sadiq, S., Gašević, D., Siemens, G.: Impact of AI assistance on student agency. Comput. Educ. **210**, 104967 (2024)

Dayal, G., Verma, P., Sehgal, S.: A comprehensive review on the integration of artificial intelligence in the field of education. In: Leveraging AI and Emotional Intelligence in Contemporary Business Organizations, pp. 331–349 (2024)

Désiron, J.C., Schneider, S.: Listen closely: prosodic signals in podcast support learning. Comput. Educ. 105051 (2024)

Dede, C., Lidwell, W.: Developing a next-generation model for massive digital learning. Educ. Sci. **13**(8), 845 (2023)

Dolunay, A., Temel, A.C.: The relationship between personal and professional goals and emotional state in academia: a study on unethical use of artificial intelligence. Front. Psychol. **15**, 1363174 (2024)

Dong, H.M., Habók, A.: Investigating the relationship between self-regulated learning and language proficiency among EFL students in Vietnam (2024)

George, A.S.: Preparing students for an AI-driven world: rethinking curriculum and pedagogy in the age of artificial intelligence. Partners Univ. Innov. Res. Publ. **1**(2), 112–136 (2023)

Godwin-Jones, R.: Distributed agency in second language learning and teaching through generative AI. arXiv preprint arXiv:2403.20216 (2024)

Han, R., Alibakhshi, G., Lu, L., Labbafi, A.: Digital communication activities and EFL learners' willingness to communicate and engagement: Exploring the intermediate language learners' perceptions. Heliyon **10**(3) (2024)

Hoffmann, M., Hoffmann, M.: Memory syndromes. In: Cognitive, Conative and Behavioral Neurology: An Evolutionary Perspective, pp. 99–130 (2016)

Huete-García, Á., Tarp, S.: Training an AI-based writing assistant for Spanish learners: the usefulness of chatbots and the indispensability of human-assisted intelligence. Lexikos **34**, 21–40 (2024)

Husin, N.A., Abdulsaeed, A.A., Muhsen, Y.R., Zaidan, A.S., Alnoor, A., Al-mawla, Z.R.: Evaluation of metaverse tools based on privacy model using fuzzy MCDM approach. In: Al-Emran, M., Ali, J.H., Valeri, M., Alnoor, A., Hussien, Z.A. (eds.) IMDC-IST 2024. LNNS, vol. 895, pp. 1–20. Springer, Cham (2023)

Innaci, D.L., Jona, P.H.: AI in second language learning: leveraging automated writing assistance tools for improving learners' writing task assessment. Jamal Acad. Res. J.: Interdisc. **5**(1) (2024)

Jemstedt, A.: Enhancing learning with a two-page study manual. Learn. Instr. **90**, 101852 (2024)

Jokhan, A., Chand, A.A., Singh, V., Mamun, K.A.: Increased digital resource consumption in higher educational institutions and the artificial intelligence role in informing decisions related to student performance. Sustainability **14**(4), 2377 (2022)

Kamruzzaman, M.M., et al.: AI-and IoT-assisted sustainable education systems during pandemics, such as COVID-19, for smart cities. Sustainability **15**(10), 8354 (2023)

Khang, A., Muthmainnah, M., Seraj, P.M.I., Al Yakin, A., Obaid, A.J.: AI-Aided teaching model in education 5.0. In: Handbook of Research on AI-Based Technologies and Applications in the Era of the Metaverse, pp. 83–104. IGI Global (2023)

Koç, F.Ş.: The development of listening and speaking skills in EFL via an artificially intelligent chatbot application: a quasi-experimental design study (2024)

Li, Z., Liang, C., Peng, J., Yin, M.: The value, benefits, and concerns of generative AI-powered assistance in writing. arXiv preprint arXiv:2403.12004 (2024)

Mhlanga, D.: Open AI in education, the responsible and ethical use of ChatGPT towards lifelong learning. In: Mhlanga, D. (ed.) FinTech and Artificial Intelligence for Sustainable Development: The Role of Smart Technologies in Achieving Development Goals, pp. 387–409. Springer, Cham (2023). https://doi.org/10.1007/978-3-031-37776-1

Mohamed Haggag, H.: Using a mobile-based learning management system (M-LMS) for developing english technical report writing skills of computers and artificial intelligence students. مجلة البحث في التربية وعلم النفس 39(1), 361–400 (2024)

Mohamed, A.M.: Exploring the potential of an AI-based Chatbot (ChatGPT) in enhancing English as a Foreign Language (EFL) teaching: perceptions of EFL Faculty Members. Educ. Inf. Technol. 1–23 (2023)

Muhsen, Y.R., Husin, N.A., Zolkepli, M.B., Manshor, N.: A systematic literature review of fuzzy-weighted zero-inconsistency and fuzzy-decision-by-opinion-score-methods: assessment of the past to inform the future. J. Intell. Fuzzy Syst. 45(3), 4617–4638 (2023)

Muthmainnah, Cardoso, L., Al Yakin, A., Tasrruddin, R., Mardhiah, M., Yusuf, M.: Visualization creativity through shadowing practices using virtual reality in designing digital stories for EFL classroom. In: Al-Emran, M., Ali, J.H., Valeri, M., Alnoor, A., Hussien, Z.A. (eds.) IMDC-IST 2024. LNNS, vol. 895, pp. 49–60. Springer, Cham (2023). https://doi.org/10.1007/978-3-031-51716-7_4

Nemorin, S., Vlachidis, A., Ayerakwa, H.M., Andriotis, P.: AI hyped? A horizon scan of discourse on artificial intelligence in education (AIED) and development. Learn. Media Technol. 48(1), 38–51 (2023)

Ng, D.T.K., Tan, C.W., Leung, J.K.L.: Empowering student self-regulated learning and science education through ChatGPT: a pioneering pilot study. Br. J. Educ. Technol. (2024)

Pachón-Basallo, M., de la Fuente, J., González-Torres, M.C., Martínez-Vicente, J.M., Peralta-Sánchez, F.J., Vera-Martínez, M.M.: Effects of factors of self-regulation vs. factors of external regulation of learning in self-regulated study. Front. Psychol. 13, 968733 (2022)

Ravshanovna, K.L.: Enhancing foreign language education through integration of digital technologies. Miasto Przyszł. 44, 131–138 (2024)

Shafiee Rad, H.: Revolutionizing L2 speaking proficiency, willingness to communicate, and perceptions through artificial intelligence: a case of Speeko application. Innov. Lang. Learn. Teach. 1–16 (2024)

Söğüt, S.: Generative artificial intelligence in EFL writing: a pedagogical stance of pre-service teachers and teacher trainers. Focus ELT J. 6(1), 58–73 (2024)

Southworth, J., et al.: Developing a model for AI Across the curriculum: transforming the higher education landscape via innovation in AI literacy. Comput. Educ.: Artif. Intell. 4, 100127 (2023)

Soyoof, A., Reynolds, B.L., Shadiev, R., Vazquez-Calvo, B.: A mixed-methods study of the incidental acquisition of foreign language vocabulary and healthcare knowledge through serious game play. Comput. Assist. Lang. Learn. 37(1–2), 27–60 (2024)

Swargiary, K.: How AI Revolutionizes Regional Language Education. Scholar Press (2024)

Tan, S.N., Ng, K.H.: Gamified mobile sensing storytelling application for enhancing remote cultural experience and engagement. Int. J. Hum.-Comput. Interact. 40(6), 1383–1396 (2024)

Tinajero, C., Mayo, M.E., Villar, E., Martínez-López, Z.: Classic and modern models of self-regulated learning: integrative and componential analysis. Front. Psychol. 15, 1307574 (2024)

Torrington, J., Bower, M., Burns, E.C.: Elementary students' self-regulation in computer-based learning environments: how do self-report measures, observations and teacher rating relate to task performance? Br. J. Edu. Technol. 55(1), 231–258 (2024)

Wale, B.D., Kassahun, Y.F.: The transformative power of AI writing technologies: enhancing EFL writing instruction through the integrative use of writerly and google docs. Hum. Behav. Emerg. Technol. **2024** (2024)

Wang, S., et al.: Large language models for education: a survey and outlook. arXiv preprint arXiv: 2403.18105 (2024a)

Wang, T., et al.: Exploring the potential impact of artificial intelligence (AI) on international students in higher education: generative AI, chatbots, analytics, and international student success. Appl. Sci. **13**(11), 6716 (2023)

Wang, Z., Zeng, J., Ardasheva, Y., Zhang, P.: Previewing test items prior to learning and receiving decorative pictures during testing: Impact on listening comprehension for English as a Foreign Language students. Appl. Cogn. Psychol. **38**(2), e4183 (2024b)

Yang, A.: Challenges and opportunities for foreign language teachers in the era of artificial intelligence. Int. J. Educ. Humanit. **4**(1), 39–50 (2024)

Zhang, C., Liu, Z., Aravind, B.R., Hariharasudan, A.: Synergizing language learning: SmallTalk AI In industry 4.0 and Education 4.0. PeerJ Comput. Sci. **10**, e1843 (2024a)

Zhang, C., Meng, Y., Ma, X.: Artificial intelligence in EFL speaking: Impact on enjoyment, anxiety, and willingness to communicate. System **121**, 103259 (2024b)

Zhou, K.Z., Kilhoffer, Z., Sanfilippo, M.R., Underwood, T., Gumusel, E., Wei, M. (2024)

Zou, B., Guan, X., Shao, Y., Chen, P.: Supporting speaking practice by social network-based interaction in artificial intelligence (AI)-assisted language learning. Sustainability **15**(4), 2872 (2023)

Enabling Sustainable Learning Through Virtual Robotics Machine Mediation of Social Interactions Between Teachers, Students, and Machines Based on Sociology Lens

Ahmad Al Yakin[1], Luis Cardoso[2], Ali Said Al Matari[3], Muthmainnah[1], and Ahmed J. Obaid[4(✉)]

[1] Universitas Al Asyariah Mandar, Polewali, Sulawesi Barat, Indonesia
[2] Polytechnic Institute of Portalegre and Centre for Comparative Studies of the University of Lisbon, Lisbon, Portugal
lmcardoso@ipportalegre.pt
[3] A'Sharqiyah University, Ibra, Oman
ali.almatari@asu.edu.om
[4] Faculty of Computer Science and Mathematics, University of Kufa, Najaf, 54001 Kufa, Iraq
ahmedj.aljanaby@uokufa.edu.iq

Abstract. This academic study analyses the use of Monica Bot in the metaverse for educational purposes, looking at how it can change the way students, teachers, and virtual robot machines interact and how it affects learning in general. The existing literature on artificial intelligence (AI) in education tends to treat AI and its interactions to investigate human behaviour in education separately. Instead, this paper delves deeply into Monica Bot's unique function in the metaverse, examining its impact on communication dynamics, interactions, student engagement, and the creation of individualised learning paths. Using a quantitative approach based on survey design, this research collects data using observation techniques and questionnaire instruments. IBM SPSS version 26 was used for data analysis. The research population was 207 students of professional teacher education (PPG) at Universitas Al Asyariah Mandar, teachers of Pancasila and citizenship education, and a sample of 32 was taken from random selection with the characteristics of respondents who had interacted with AI during learning. This study explores the many benefits of this technology and examines its limitations in terms of technology, ethics, and education by utilising real-world case studies. In the end, everything depends on the significant conclusions drawn by educational leaders and lawmakers, who must fight for equitable distribution of AI resources, strict ethical standards, and creative and innovative teaching methods. The importance of a balanced learning model in using AI capabilities, as well as proactive steps to resolve ethical issues and make them accessible in a practical and efficient manner, are at the heart of the discourse.

Keywords: Sustainable learning · Virtual Robotics · Social Interactions · Machine Learning · AI and Sociology

1 Introduction

There is increasing interest among researchers in the potential benefits of using AI-virtual robots in the classroom, as shown in research by Muthmainnah et al. (2023a, b). The current study focuses on virtual classroom robots with computational fidelity that work autonomously at multiple levels with basic educational goals. Although many robotic applications can encourage various forms of learning, such as programming, the main goal is to examine the role of these robots in the context of image interaction. According to Nanavati (2024), "the extent to which a robot can sense its environment, make plans based on that environment, and act on that environment with the intent of achieving specific task goals (either given or created by the robot) without external control" is what we mean when we talk about autonomy. Wu, et al. (2024) state that for this type of technology to provide effective pedagogical answers, it is necessary to know not only the learner's knowledge but also his emotional nature. Based on research in the field of science learning, classroom robots are programmed to detect and respond to students' emotional and affective states in learning contexts to imitate the behaviour of the teachers involved (Schiavo et al., 2024).

Some academics have proposed that stakeholders should be involved in the design process for robots to be socially acceptable, which can be seen as an overly technology-focused design process (Germain, 2024). Teachers, students, and parents are the most important groups to consider when making decisions about education. Teachers will be the main topic of this study. Technical advances in education have the potential to disrupt long-established teaching methods, making educators key constituents in the process of developing new intelligent tools such as AI for integration in the classroom. Teachers have a responsibility to their students to guide their learning, and they must also be the ones who decide whether and how to use technology in the classroom. Previous research on the topic of educational robotics found that instructors' views about robot usability predicted their propensity to use and interact with virtual robots (Muthmainnah, et al., 2023) found that teachers perceive robots as useful when they have an instrumental function rather than a relational function during the learning process in hybrid classrooms. Robots must be able to blend into the social dynamics of the classroom while still performing operational tasks to help teachers make the most of their time.

Previous research has provided some ideas for robot design, but it has not truly considered the moral and ethical consequences of technology that aims to consciously foster social and emotional engagement with students as social agents. Although information and communication technology (ICT) in the past primarily mediated or facilitated emotional bonds between humans, contemporary robotic technology increasingly makes the bond between humans and robots the main issue to be examined in this study. As a case study of ethics in research with humans, Torresen (2018) programmed their classroom robot to say things like "I like our classroom teacher" and other made-up phrases in an attempt to get students to engage with it over the long term. Additionally, a pseudo-development mechanism was implemented, allowing the robot to exhibit more behaviours and communicate more personal information with the child as the duration of their interaction with the robot increased. Designed intentionally to generate social bonds and fulfil the need for social interaction, robots don't just function; they also mediate interactions between humans technologically, human-to human or machine-to human social interaction, de Graaf, (2016).

The impact of the use of technology in the era of society 5.0 has resulted in policymakers looking for technological solutions for their application as agents in higher education. According to Adel (2024), this kind of "technological solutionism" always pervades education policy. Policies are being formulated around the world to ensure that national education systems are better equipped to address societal problems through the integration of AI technology with metaverse or hybrid learning environments. Some examples of the many demands that create challenges are ensuring affordable schools, learning facilities, internet availability, meeting economic needs, and at the same time overcoming social and educational disparities (Gladden, 2019; Macgilchrist, 2020). Educational technology sponsors, manufacturers, and suppliers often have influence and are ultimately responsible for the policy objectives. Artificial intelligence has recently become central to effective intelligent technology solutions. In this study, we need new solutions to address the growing pressures on our school systems—from excessive teacher workload to lack of social mobility—and many of these AI tools have the potential to significantly improve our school systems.

However, educational policy in many countries has been determined by this statement for a long time, although there is a strong argument that education does not have to respond to calls to accelerate the use of new technologies and that adaptation of these technologies is an obligation for the acceleration of modern education. (Al Yakin et. al., 2023a, b). Our investigation of this AI mobilisation uses an innovative mix of theory and methodology. By combining theory and methods, we can take a broader perspective on educational technology, going beyond "what works" questions that centre on individual interventions and criticising those who do not utilise technology properly. The aim of this article is to provide a sociological perspective on the current belief that artificial intelligence (AI) will be the key to improving the education of various sectors in schools. In terms of understanding the conceptualization and practice of educational technology in educational systems, by constructing knowledge graphs—a relatively unique tool in sociology—we aim to make a methodological and theoretical contribution to the expansion of this research by studying the rise of AI in education for the expansion of social interactions. We use this, along with Bourdieusean theory (Warwick, et al., 2017) to investigate the logic and incentives in the field, looking at the positions taken by various AI and educational technology stakeholders, determining the core ideas these groups support, and suggesting their reasons for doing so.

2 Literature Review

2.1 Bourdieusean Visualisation of AI in Education

According to Bourdieu (Warde, 2004), a field is a socially formed, semi-autonomous, structured domain or space with a clear but dynamic development history. According to Bourdieu (Lin, 2002), human agents identify and act according to the value they place on certain commodities, assets, talents, or resources in contested fields. When combined with farms, these assets, often called "capital", become a competitive advantage.

The term "capital" refers to any resource that is useful in a particular social arena and allows agents in that arena to benefit from their involvement and competition (Atiyah et al., 2023a, b, c). Economic capital includes things like money and physical possessions, cultural capital includes things like rare symbolic objects, skills, and titles, and social capital is things like group membership (Alnoor et al., 2024a, b). Examples of social capital include business relationships with family and friends, cultural capital is a person's position in the academic community, and economic capital is a person's ability to attract investors. It is possible to trade or raise capital. As a result, inequality and interdependence arise from the unequal distribution of capital and pooling of capital in land Tsing (2013). Fourth, we will examine symbolic capital, which is defined as "all forms of capital whether physical, economic, cultural or social (Carpiano, 2006; Ali et al., 2024). When a social agent with the ability to understand, know, and recognize becomes symbolically efficient, it acts like a magical power, a property that responds to socially constructed collective expectations. The potential and usefulness of these capitals depend on the characteristics of the field of use; they are not static assets that automatically translate into power and influence anywhere (Bourdieu, 2014; Atiyah et al., 2023a, b, c).

Therefore, farms can facilitate the conversion and accumulation of capital and serve as an incentive to do so. Industries that rely on patronage networks in addition to commercial performance to facilitate business contracts are prime examples of industries that prioritize social capital. According to Trigg, (2013) field attributes can be objective, such as commercial performance, and subjective, such as the norms and values that players on the field express, respond to, or try to shape. It is possible that these subjective and objective qualities overlap. Subjective sentiments surrounding an artist's background, for example, can increase a painting's objective value on the art market. The 'logic of practice' in this field is based on these standards and principles. The logic of these practices is most likely to be influenced by actors with the most resources in the field, including combinations of resources, and less likely to be followed or negotiated by actors with fewer resources (Atiyah et al., 2023a, b, c; Alnoor et al., 2023). The agents depicted in the graph are involved in the fields of artificial intelligence and educational technology; we argue that Bourdieu's analytical lens can explain how they adapt to the logic of practice in the sector to succeed.

This approach to knowledge graph visualization and social theory seeks to put into practice what Halford and Savage (2017) call "symphonic social science". This approach "applies visualization as a deliberate analytical strategy rather than a technocratic method of data presentation, as is the case in big data analysis" and "combines rich theoretical awareness with carefully selected data to answer social questions". After that, the knowledge graph provides an understandable picture of the field, quantitative analysis shows key aspects of the field such as which concepts are most frequently discussed by entities, and Bourdieu's theory clarifies the structure of the graph and the reasons for the frequency of mention of certain concepts.To review, the aim of this article is to offer an original theoretical and methodological framework for understanding the current optimism that artificial intelligence (AI) can improve social interaction in education.

2.2 Human-Machine Communication Through AI-Virtual Robotics

Many new, diverse, and increasingly social technologies have emerged as a direct result of technological advances in the computer sector. This social tool is also gaining popularity and usage in various fields, including education. As a result of these advancements, social technologies are mediating an increasing amount of our communications. Natale, (2021). It includes computer-mediated communication (CMC) and human-machine communication (HMC), where social technology operates as an interaction.

Opportunities and challenges arise for communication academics due to the rapid pace of technological breakthroughs, product development, and adoption of media and computer technology (Gil de Zúñiga, et al., 2024 Spence et al., 2024; Alnoor et al., 2024a, b; Cummings and Wertz, 2023; Atiyah and Zaidan, 2022). We have the advantage of being able to see popular social elements to our experience and existing methodology for analysing offline and online communication processes that occur in classrooms by integrating the AI-virtual robot machine like Monica. Social presence affordance-based methods can find the overall effect that works across features, platforms, and media during the learning process by looking into the ideas behind an AI feature or use, like the ability to record or get attention. ss. In this publication, we try to show, using theories from communication and social psychology, how contextual elements of digital HMC can complement social affordance-based approaches to studying social AI technologies. In the evolution of HMC theory, Edwards et al. (2019) outlines several ways that HMC theory can evolve in a concise format. Although Spence acknowledges that "theories that are important to HMC" "do exist and are being developed, tested, and refined", Hong, (2022) frames the discussion around one such path of the process of applying existing theories from the field of communications and related disciplines as an independent venture.

HMC further on this, in this paper, we present an angle specifically, communication is in a prime position to build interactions that occur with an HMC theoretical approach by looking at how various communication and relationship theories explain contextual elements in HMC. From this perspective, we build theory to explain socio-technological phenomena more thoroughly and to address concerns about current communication and relationship theory. An important contextual aspect recognized by researchers in Human Computer Robot Interaction (HCRI) is that computers are not humans, and most people believe that computers do not need or deserve social treatment. But in our opinion, understanding the contemporary technological situation of AI by dividing things into social and non-social categories is too simplistic and may reveal clear, non-generalizable differences between humans and computers or variations in how each group views capabilities. Socialize. One common explanation for the apparent gap between humans and media agents is the existence of heuristics, which are mental shortcuts developed in social cognition systems (Burr et al., 2018). It is generally believed that heuristics such as these explain why humans and digital interlocutors appear different; in fact, they take into account a set of characteristics and capabilities that may be unique to the object under investigation.

Díaz Boladeras, M. (2017) bonding with robotic pets to investigate children's cognitions, emotions and behaviors towards pet-robots. Applications in a robot assisted quality of life intervention in a pediatric hospital. Our aim in this study is to contribute to how to analyse AI technology as a social agent by increasing the accuracy of existing human and machine engagement and also to build social interaction in the digital era. Using Bandura's social learning theory as a framework (Bandura, 1969; Dodgers, 2023), we begin by discussing socialization in its broadest sense. Albert Bandura expanded Robert Sears' research to form social learning theory Jiang et al. (2024) and Al-Najjar, et al. (2024) both argued that people grow and change as a result of their interactions with others, but both also highlighted the importance of observational learning and how it can influence the adoption of norms and practices in behaviour. According to research by Bandura (Atra, 2024) and Martin, (2024), children act aggressively after watching violent films or TV shows. Furthermore, its showed that vicarious reinforcement, that is, aggressive actions in which a person is praised or punished, influences the tendency to act aggressively. Children are less likely to act aggressively when they see aggressive behaviour being punished, as expected, and more likely to act aggressively when they see aggressive behaviour being praised. Bandura suggests that individuals may learn an action and its results by watching others and then applying the principles of vicarious reinforcement.

Research conducted by Al Yakin, et al. (2023a, b) and colleagues shows that virtual environments can also be an effective place for observational learning and technology-based reinforcement of learning behavior using social cognitive approach. A study by van der Sloot (2024) and Bozkurt, et al., (2023) showed that students interacted with AI-friends more when they could interact in real-time, in particular, if the depiction of the robot avatar was more human-like, this effect would be more pronounced. The varied and expanded forms that individuals can assume in virtual environments, whether intentional or not, are qualitatively different from our natural contexts, and virtual environments offer new opportunities for exploring students' identities and learning performance.

We also think about the impact of social developments that occur in digital HCRI outside the virtual environment, in addition to social learning that occurs in the virtual environment, namely in the classroom during the learning process. Using computers, we can build models to navigate social environments by inferring social features in virtual environments (Bönsch, et al. 2024; Ali et al., 2024; Atiyah et al., 2023a, b, c), but we cannot directly see or engage with actions and their outcomes. Social events are easier to witness in the real world or during learning behavior, and students access technology and carry out interactions that can be assessed through the classic observational learning process, which is related to social learning. Developmental psychology researchers study the limitations caused by devices. Engagement with personal computers, such as cell phones, can also influence students' learning and interaction behavior and make them more active in seeking information. We conclude that interacting with computers teaches us many wonderful things that we hope for. The presence of humans in cyberspace reduces the impact of these limitations, but when interacting with machines, these limitations remain, although to a lesser extent due to shared influences use of these machines for example virtual assistants like Alexa (Wenzel, et al. 2024; Sharevski, et al. 2021). Although some exchanges may occur in public, most will occur privately outside the classroom, and it is possible that these interactions lack the social importance necessary for successful vicarious reinforcement. There are social development consequences of digital HCRI's reduced reliance on observed learning. And strengthening representation. A shift towards a more experience-based mental calculus (i.e. the ratio of experience to observational learning) may occur. It occurred to us that the scripts, models, or schemas formed through these digital HCRI interactions and relationships may be very personal in nature, considering that the quantity of digital HCRI that occurs is based on experience and has different results.

To evaluate the accuracy of the above statements and any conclusions drawn from the data, substantial empirical investigation is required. Different behavioural norms may emerge from a developmental calculus that relies less on conventional socialisation processes, for example, especially when students manifest themselves verbally in interactions with virtual creatures or virtual machines such as the Monica AI investigated in this study. According to research conducted by Rapp, et al. (2021), when people talk online with chatbots, their initial messages are very open, pleasant, extroverted, and thorough compared to when they chat with human friends. At first glance, these results are troubling and may pose dangers to our interpersonal interactions when viewed within the broader, social learning theory-based context of this behavior.

The problem with virtual robotics-based learning is that it can cause more people to use language that is not appropriate for social situations (Akalin and Loutfi, 2021; Atiyah, 2023; Scaradozzi, et al., 2019). This may make such activities more important, although we would argue that the chances of them occurring in a learning situation are smaller. Because of this imbalance, an individual's exposure to aggressive or frustrated behaviour, as well as the use of dehumanising language when interacting with nonhuman beings, can have a profound impact on others around them. Second, these communication patterns may be the result of more individualised scripting and direct learning. Third, meetings involving digital HCRI may give less consideration to interpersonal norms due to the absence of human presence, even when the person feels socially present. When used alone, such technology (such as Alexa) may not be as ingrained in a person's habits and routines as when used in a group (such as when friends and family are in the same room). Subversion has its benefits, but in the highly saturated digital HCRI environment, it can be more difficult to find or even see specific interpersonal activities that are beneficial. Scripts or schemas formed through digital HCRI may not be a problem if this behaviour is the result of a learning calculus based on learning experiences to enrich knowledge or updated learning material content in a virtual environment through AI in the learning process, as shown in Fig. 1. The aim of this study is to investigate the role of AI virtual robot machines in social interactions in the classroom between professional teacher education students, lecturers, and Monica AI in university.

Fig. 1. Human Computer Robot Social Interaction

3 Methodology

3.1 Research Design

The purpose of this quantitative research is to find out the perspective of educators who take part in professional, pre-service teacher education programmes regarding the use of virtual robot machines in the classroom to determine the impact on social interactions between educators, students, and Monica AI during the learning process. A comprehensive theoretical framework regarding the impact of robots on social contexts. The perspective of pre-service teachers should also be considered, as they are almost professionally active in the field of education, in contrast to our previous research, which mainly focused on teachers' practices (Al Yakin et al., 2023a, 2023b; Ahmed et al., 2024). This research focus on future technologies means that pre-service teachers will one day have to face the reality of robots in the classroom (Husin et al., 2023; Muhsen et al., 2023). Additionally, educational systems and experiences within them vary between cultures. For a more comprehensive and useful understanding of the phenomenon under investigation, we will focus on the 32 pre-service teachers who interacted with Monica AI.

Focus groups help us answer research questions by allowing us to create honest and normal conversations that discuss, in depth, a chosen topic. This group allows participants to have more control over the conversation and be more involved with each other during treatment. Teachers must take the lead in shaping the debate based on their shared practice and theoretical background; therefore, this is considered important. Rather than trying to draw broad conclusions about the phenomenon under study, our method is structured around four research questions that will help us create a thorough theoretical explanation of the phenomenon, namely, considering the design and presence of robots in the classroom and how pre-service teacher educators view robots as being able to enhance and complement the social interaction process during learning.

H1c: There is a significant influence of vividness in metaverse over digital marketing.

3.2 Participants

The researchers looked for experienced educators and those with previous teaching experience who were registered in professional teacher education program to participate in this research. The total research population, namely educators in professional teacher education program, was 207 people. This research involved 32 pre-service teachers who were selected using purposive random sampling. Figure 1 displays participant demographics. At the time the research was conducted (which occurred in mid-2024) most participants had teaching experience and had teaching degrees. Along with demographic information and consent forms, we collected information regarding participants' experiences of interacting with AI technology through validated questionnaires (Al-Enzi et al., 2023). The questionnaire consists of 20 questions with a Likert scale regarding the participant's current use and perception of social interactions that occur through the mediation of the Monica AI virtual machine robot using smartphones or mobile phones and the Internet. Most participants liked technology and appreciated access to AI technology and its impact on social interactions in this hybrid classroom.

3.3 Procedures

Beforehand, in accordance with national and institutional ethical norms, participants were briefed on the aims of the study. The next step is for them to be involved in learning based on Monica's AI intelligent technology for eight meetings. Researchers observed the behaviour and interaction patterns that occurred in the classroom between pre-service teachers, lecturers, and AI virtual robots. Then, at the end of the meeting, they were asked to fill out a questionnaire survey including demographic data and complete a survey on the use of AI technology for social engagement to get their positive opinion about Monica AI. The steps taken by the participants were to listen to material from the lecturer, who discussed the latest advances in social robots. In the video, we see how they interact in the classroom and compare several robots, including telepresence and autonomous humanoids, also featured in the video from YouTube, all of which are now used in schools in different countries.

To raise ethical issues and encourage participants to consider beyond their current experiences with technology. After that, researchers will ask participants to download Monica AI via the Google Play Store application and observe their interactions with the robot in class. The goal of this sketch is to illustrate a specific use case for educational robots by showing how robots can function as social agents and modify learning activities appropriately. To demonstrate potential behaviour related to social interaction. Pre-service teachers are given the opportunity to freely express their opinions regarding the use of robots in the classroom in each focus group discussion and guide the discussion. We deliberately kept the questions broad and open-ended so that no one could lead the discussion.

3.4 Statistical Approach

We surveyed participants to determine the influence and magnitude of each social interaction component on their learning satisfaction with Monica AI. Metaverse learning scenarios are available to participants. Participants are given one week to acclimatise to this learning environment before formal experiences begin. To eliminate any device-related consequences, the event requires all attendees to use cellular. This scenario incorporates elements of social learning theory (SCT) and gives participants real-time feedback while monitoring their academic progress using Monica AI to construct the teaching material. A participant commitment of at least one hour to the formal experience is required to ensure social change learning. After learning, participants were asked to fill out a survey. Participants evaluated the efficacy of each AI component after a metaverse learning session. All combined reliability values were greater than 0.7, and the average value of the extracted variance was greater than the minimum criterion of 0.5 Walter and Andersen, (2013)., indicating strong construct reliability and convergent validity using a 5-point Likert scale analysed using the SPSS version 26 application.

4 Results and Discussion

4.1 Data from Survey

The results of this research describe the impact of the Monica-AI virtual robot machine as a subject of social interaction in digital HCRI, which in turn determines the appropriate main objectives, as shown in Table 1. That is, in digital HCRI, that prioritises the same set of goals as face-to-face conversations between humans and machines because of the inherent or perceived differences between human and computer capabilities. General assumptions regarding the capabilities of computers and media agents for mutual interaction between lecturers, students, and virtual robots form the basis of our discussion, which will be discussed as follows:

Focusing on prospective teachers' perspectives regarding the integration of virtual robotics machines in educational environments, Table 1 provides descriptive statistics for respondents. The study used a Likert scale to assess 32 participants' opinions on a variety of topics, including the following: the impact of virtual robotics machines on collaboration and the provision of personalised feedback; the extent to which participants enjoy learning and engaging in social interactions enabled by these machines; and the extent to which participants believed that the machine improved communication between teachers and students. With average scores ranging from 4.0938 to 4.4375, the data shows that prospective teachers consistently have good opinions and experiences. Virtual robotics machines, according to participants, encourage diversity and inclusion in learning communities, strengthen bonds of friendship and a sense of belonging, and inspire cooperation and collaboration. Additionally, they see the impact of the virtual robot Monica AI on increasing students' passion for learning, shaping societal standards, and levelling the playing field in terms of access to resources and opportunities. With a standard deviation between 0.61484 and 0.93109, it can be said that the majority of respondents agree. The overall data shows that virtual robotics machines have a major impact on many areas of learning and will play an important role in shaping education in the digital era.

Table 1. Descriptive Statistic of respondents answering the questionnaire

Pre-service teachers' statement	N	Minimum	Maximum	Mean	Std. Deviation
1. I enjoy learning and engaging in social interactions with teachers, students, and virtual robotics machines in the learning environment	32	3.00	5.00	4.0938	.68906
2. I view virtual robotics machines as facilitating communication between teachers and students in EFL educational environments	32	2.00	5.00	4.3125	.73780
3. I frequently collaborate with Monica AI's virtual robotics engine to complete learning assignments or projects	32	3.00	5.00	4.2188	.70639
4. I believe that the feedback that virtual robotics machines provide on academic performance and progress is invaluable	32	2.00	5.00	4.1875	.93109
5. I like to seek help or clarification from virtual robotics machines when facing difficulties in my studies and confirm with the lecturer in class	32	3.00	5.00	3.9375	.71561
6. I believe virtual robotics machines contribute to creating a sense of togetherness and belonging in the classroom learning environment	32	3.00	5.00	4.3125	.69270
7. In my opinion, the role of virtual robotics machines in encouraging teamwork and collaboration among students in the pre-service teacher education profession	32	3.00	5.00	4.2188	.65915

(continued)

Table 1. (*continued*)

Pre-service teachers' statement	N	Minimum	Maximum	Mean	Std. Deviation
8. I feel more comfortable expressing my thoughts and opinions to a virtual robotics machine than interacting with a teacher or peers	32	3.00	5.00	4.4063	.61484
9. I am often involved in group discussions or activities facilitated by virtual robotics machines	32	3.00	5.00	4.4062	.66524
10. Learning with virtual robotics (AI) engines impacts the learning experience by providing personalised feedback and guidance	32	3.00	5.00	4.2187	.70639
11. I believe that the dynamics of social interaction between teachers, students, and virtual robotics machines in a learning environment increase learning motivation	32	3.00	5.00	4.4375	.61892
12. I feel the impact of virtual robotics machines on the distribution of power and authority in the classroom	32	3.00	5.00	4.3750	.65991
13. The presence of virtual robotics machines influences social norms and interactions between students and lecturers during the learning process	32	3.00	5.00	4.2500	.67202
14. I think virtual robotics machines contribute to promoting inclusivity and diversity in learning communities	32	3.00	5.00	4.4063	.61484

(*continued*)

Table 1. (*continued*)

Pre-service teachers' statement	N	Minimum	Maximum	Mean	Std. Deviation
15. I feel the contribution of virtual robotics machines in building trust and good relationships between teachers and students during learning	32	2.00	5.00	4.1563	.76662
16. Virtual robotics machines influence the formation of social bonds and friendships among students in the teacher education profession	32	3.00	5.00	4.2812	.72887
17. Virtual robotics machines help academic achievement and learning success in the digital era of the 21st century	32	3.00	5.00	4.1563	.67725
18. Virtual robotics machines influence the distribution of resources and opportunities among students greatly	32	3.00	5.00	4.1875	.73780
19. How virtual robotics machines influence the sense of autonomy and independence in learning experiences and social interactions	32	3.00	5.00	4.3125	.69270
20. Virtual robotics machines influence overall satisfaction, engagement in the learning process and I comfortable interacting with virtual robotics machines to access learning materials and other knowledge	32	3.00	5.00	4.2500	.71842
Valid N (listwise)	32				

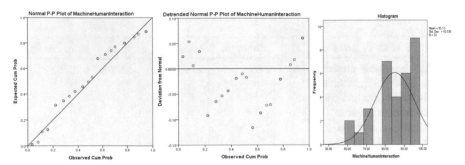

Fig. 2. The histogram and P-P Plot of Virtual Robotics Machine mediation of social interactions between teachers, students, and machines.

Table 2. The Pre-Service teachers survey on HCRI in the class

	N	Minimum	Maximum	Mean	Std. Deviation
MachineHumanInteraction	32	61.00	98.00	85.1250	10.53642
Valid N (listwise)	32				

Based on the data in Table 2 and Fig. 2, it displays data collected from the pre-service teacher survey on human-computer-Robot interaction (HCRI) in the classroom. The results show that of the 32 people who took the survey, the scores ranged from 61.00 to 98.00, with a mean of 85.1250 and a standard deviation of 10.53642. It is clear from these numbers that there is great variation in how respondents encounter and think about HCRI in the classroom. The results are reliable because the analysis is robust and the sample size is 32, which is a valid number. Further research and understanding of this phenomenon is needed in modern pedagogical methods, as data shows that pre-service teachers have different views of how humans, computers, and robots interact in the classroom.

Table 3. Reliability Statistics

Cronbach's Alpha	N of Items
.959	20

The results of the statistical reliability data for this research instrument are presented in Table 3. The very high level of consistency between the items that make up the instrument is shown by the Cronbach's alpha coefficient of 0.959, which is a measure of internal consistency dependence. This coefficient shows that this research questionnaire instrument is reliable for assessing the construct under study, considering that the survey or questionnaire consists of a total of 20 items. A high Cronbach's alpha value indicates that the components of the instrument are well connected to each other so that it can be trusted and consistent as a measuring tool. With a high level of internal consistency, we are confident that the data we use to draw conclusions is credible and trustworthy.

4.2 Discussion

Based on the research results, it is known that overall respondents stated that social interaction using Monica Bot significantly increased, and interaction with lecturers and classmates also occurred. All interactions that occur are situational, and this effort is made to provide more thorough and practical details about how to interact with Monica Bot, starting with identification as a human through logging in, verifying goals, and introducing oneself before accessing information according to the content of the material. We found that the academic environment tends to generate less than optimal interaction if the learning model implemented is not interesting.

The data results in Table 1 show descriptive statistics about how prospective educators view the use of virtual robotics in the classroom. The data displays twenty statements with Likert scale answers ranging from three to five, indicating varying levels of agreement with each statement. With average ratings ranging from 3.9375 to 4.4375, educators are generally optimistic about the potential of virtual robotics machines to improve interaction in educational settings. In particular, pre-service instructors praise the social interactions these tools enable, praising their capacity to improve student-teacher communication, group work, and assessment. Respondents also noted that virtual robotics machines impact classroom social dynamics, such as the allocation of power and authority, encouraging diversity and inclusion, and the development of student-teacher relationships. This research confirms what other studies have shown: that classrooms that utilise technology can improve student engagement, teamwork, and classroom performance (Tang, 2024; Balalle, 2024). A small standard deviation and a high average value support these findings, showing that respondents support and agree with the integration of AI during the learning process. Virtual robotics machines can improve educational outcomes and encourage inclusive learning environments, and this research adds to existing knowledge by offering empirical evidence regarding their positive impact on reciprocal interactions between humans and machines, between fellow students and lecturers, as well as between students and Monica Bot teaching and learning processes.

On the other hand, this research covers most of the academic community, namely Table 2 learning achievement, which has also been significantly improved due to the implementation of this AI. The results of this study provide data that challenges the stigma that sociologists do not always understand the logic of practice in the field, especially when it comes to AI-powered personalization. For example, the emphasis on customisation to encourage "social mobility" is indicative of this. Policy advocates see personalization in AI and education as a way to invest in human capital and empower people to become more economically and civically engaged (Boyte and Ström, 2020). However, according to sociologists (Kuhn, et al. 2023), there are problems with an overly simplified and instrumentalist view of education that views students only as potential human resources who can overcome systemic inequalities using technology.

The results of this research indicate that educational institutions are not one of several fields where technological interventions based on this logic have previously been successful. The push for technological adaptation after COVID-19 accelerates human social styles grows with developing technology and changes in increasingly flexible social interactions in the era of Society 5.0. Education, whether technologically advanced or not, does not magically produce a social class that is more attractive for interaction and communication and that has great personal and democratic benefits. There are additional, external, and structural forces in educational institutions that can cause AI-led interventions to backfire, and the lack of sociological thinking in the field is critical. There is a risk that existing teachers could become targets for exploitation if, for example, AI-powered personalised learning is implemented in an education sector that already severely lacks facilities to support AI and lacks qualified specialists. AI supporters state that this technology will have a multiplicative impact on the education sector if it eases the burden on teachers and modernises the curriculum in line with the transformation of education in the AI era. Likewise, schools that have large funds and can afford to purchase, install, maintain, upgrade, train, and use AI-augmented personalisation technology to equip highly qualified teachers will be the ones who will benefit from this technology, while highly underfunded or poorly managed schools will face the greatest need for intervention and the greatest challenges in facing the transformation and digitalization of education.

Based on a sociological point of view, the educational gap created by digitalization has an overall impact on our social community. Changes in learning and teaching patterns that are dominated by technology or LMS-based, or e-learning higher education management have an impact on parents who have a lot of social, cultural, and economic capital, often applying techniques such as hiring private teachers to help their children develop in school, in line with scholars investigation (Reay, 2020). Therefore, it is hoped that artificial intelligence (AI) can level the playing field, especially if artificial intelligence (AI) especially helps students who do not have privileges equally to participate in group work and can improve student achievement (statement numbers 6, 7, 8, 9, and 10).

According to Williamson (2023) research, in the sociology of education the rise of artificial intelligence and educational technology is part of a broader political economy that questions the purpose of education. There has recently been a shift in the public sector that has made educational institutions more receptive to measurement, evaluation, comparison, auditing, ranking, and accountability with the help of AI. The results of this research show that the adaptation and harmonisation of AI-based technology do not separate us from each other and our shared experiences; in fact, the presence of AI increases the interaction and awareness built by respondents in the classroom to share information, share material, and practice together for the purpose of education. Respondents did not show feelings of alienation between each other caused by AI interaction during the learning process in class (statements 15, 16, and 17).

The reason this threat exists is because AI is designed to provide clear and measurable results that align with the logic of current human practices. Therefore, educational interventions driven by AI reduce the gap in educational models in the 21st century, unless these interventions are based more on sociology. By taking a sociological stance towards education policy, we can ensure that AI is part of a broader plan to create a more just society by clarifying its boundaries and addressing the ways in which certain societies use technology to deal with structural pressures originating outside their control starting with studies that are interdisciplinary, nondeterministic, locally based, and designed to examine the recursive relationships between human actions and broader organisational and system contexts, which should include educational and structural conditions and community impact, is the first step in strengthening the evidence base for AI mobilisation that can reduce the gap between the relationship between educational technology and educational inequality in achieving more dynamic and efficient learning outcomes and future knowledge.

5 Implications and Future Studies

5.1 Theoretical Implications

The results of this study illuminate the theoretical implications of using artificial intelligence, and Monica Bot in particular, in the classroom. It stresses the significance of AI-mediated social connections by focusing on how pre-service teachers and their professors or classmates are able to communicate and participate more effectively through the use of Monica Bot (statements 15, 16, and 17). This adds to what is already known about the effects of AI on classroom interactions by giving proof that these technologies can improve two-way communication and collaboration between instructors and their students. The study also casts doubt on the idea that learning models driven by technology are infallible, arguing instead that an interesting learning environment is necessary for the best possible interaction (statements 6, 7, and 8). This is in line with the views of sociologists who warn against simplifying AI's function in the classroom and instead highlight the importance of looking at the bigger picture to determine how beneficial it.

5.2 Practical Implications

The study's real-life consequences suggest that artificial intelligence (AI), and more especially Monica Bot, could revolutionized current methods of teaching and learning. According to the study, AI has the potential to greatly improve learning outcomes. Respondents said that their academic performance improved after implementing AI (Table 2). According to points 6, 7, 8, 9, and 10, AI has the potential to create personalized learning experiences, which can help close educational disparities and increase social mobility. The study does caution against ignoring the social and cultural implications of AI integration, though, and stresses the need to think about these things when putting AI into practice (Reay, 2020). This means that schools, particularly those with fewer resources, need to plan ahead when implementing AI in order to make sure that all students have equal access and that it is used effectively (statements 13, 14, and 15). It is important to include AI-based interventions in larger educational strategies that address systemic inequities (Williamson, 2023) if we want to get the most out of AI in creating inclusive learning environments.

5.3 Future Studies

This study lays the groundwork for future research by pointing researchers in several directions. Statements 4, 5, and 6 suggest that in order to measure the long-term effects of AI integration in classrooms, researchers should follow students over time to see how their engagement, learning, and social connections evolve. Statements 13, 14, and 15 suggest that greater sample numbers from multicenter studies would improve the results' generalizability and shed light on a variety of educational settings. The ways in which AI influences students' learning experiences, such as how it promotes autonomy, motivation, and self-efficacy (statements 9, 10, and 11), can also benefit from additional investigation. Understanding how AI technology affects students' emotions and attitudes toward learning can help create more effective AI-based educational interventions (statements 9 and 11). To make sure that AI interventions are in line with the larger social aims of making a more equal society, the study also demands research that approaches education policy from a sociological viewpoint (statements 15, 16, and 17). This includes looking at how AI may help with resource distribution and teacher assistance, as well as how it might lessen educational inequality and increase social mobility (statements 13, 14, and 15).

Finally, the study sheds light on how to best incorporate AI, and Monica Bot in particular, into classroom settings. It highlights the value of AI-enabled social interactions and the technology's ability to reduce achievement gaps in schooling. Taking into account the pros and cons of AI from a social standpoint is essential for a comprehensive understanding of its function in education, as shown by the theoretical implications. The study emphasizes the revolutionary power of AI in making classrooms more welcoming to all students, but it cautions against ignoring systemic issues. We can learn more about how AI affects students' social connections and learning experiences if we do longitudinal and multicenter studies in the future. To maximize AI benefits while addressing societal disparities, a sociological approach to its integration into education is necessary.

6   Conclusion

This initial report investigates the current level of adoption, use, and perceptions regarding the current use of virtual machine-Monica AI robots among pre-service teachers that contribute to their knowledge, usage patterns, and social interactions. Data results that explore various possibilities indicate the need for additional investigation and study to understand the function of AI-based instruments in educational contexts. Identifying the social interactions that influence pre-service teachers' uptake and use of the Monica AI ChatBot as the field continues to develop becomes increasingly important. Improving the perceived usefulness of these tools while addressing concerns regarding their accuracy, reliability, and ethical implications can revolutionized the academic landscape and encourage a more seamless incorporation of AI-powered technologies in educational environments.

The findings of this research can guide future studies, policies, and practices. This study paves the way for a more complete understanding of the function of GenAI in the classroom and its sociological impact on academia, social society 5.0, and addressing educational disparities. However, it is important to emphasize the need for continuous data collection and observation in future research, as AI technology is developing rapidly. The use of AI Chatbots tools other than Monica in the classroom will increase, and with it, it is likely that students' viewpoints, habits, social interactions, communication patterns, and actions will change. To monitor these changes and understand the long-term impact of integrating AI into education, it is critical to conduct research repeatedly, preferably regularly and widely. To keep up with the ever-changing development of GenAI in the classroom and to anticipate the difficulties students may face when adapting to new technology, it is important to conduct learning continuously. To keep up with these changes and ensure their policies, procedures, and tactics adapt to student needs and new technologies, researchers and educators must continue to collect data. Additionally, long-term research will provide a more comprehensive picture of the long-term impact of AI on student academic performance. Growth in analytical and problem-solving abilities, 21st century skills, literacy, participation, enthusiasm, and academic achievement can fall into this category. To maximize the use of AI in educational settings and ensure that these tools are useful aids in effective, engaging, motivating, and student-centered learning approaches rather than just novelty toys used for entertainment purposes, it is important to understand these consequences.

References

Adel, A.: The convergence of intelligent tutoring, robotics, and IoT in smart education for the transition from industry 4.0 to 5.0. Smart Cities 7(1), 325–369 (2024)

Ahmed, A.D., Salih, M.M., Muhsen, Y.R.: Opinion weight criteria method (OWCM): a new method for weighting criteria with zero inconsistency. IEEE Access (2024)

Akalin, N., Loutfi, A.: Reinforcement learning approaches in social robotics. Sensors 21(4), 1292 (2021)

Al Yakin, A., Khang, A., Mukit, A.: Personalized social-collaborative iot-symbiotic platforms in smart education ecosystem. In: Smart Cities, pp. 204–230. CRC Press (2023a)

Al Yakin, A., Obaid, A.J., Muthmainnah, Shnawa, A.H., Haroon, N.H.: Unlocking the potential of mobile computing for infusing computational thinking using social cognitive approach in higher education institutes. In In: Swaroop, A., Polkowski, Z., Correia, S.D., Virdee, B. (eds.) ICDAM 2023. LNNS, vol. 786, pp. 105–114. Springer, Singapore (2023b). https://doi.org/10.1007/978-981-99-6547-2_9

Al-Enzi, S.H.Z., Abbas, S., Abbood, A.A., Muhsen, Y.R., Al-Hchaimi, A.A.J., Almosawi, Z.: Exploring research trends of metaverse: a bibliometric analysis. In: Al-Emran, M., Ali, J.H., Valeri, M., Alnoor, A., Hussien, Z.A. (eds.) IMDC-IST 2024. LNNS, vol. 895, pp. 21–34. Springer, Cham (2023). https://doi.org/10.1007/978-3-031-51716-7_2

Ali, J., Hussain, K.N., Alnoor, A., Muhsen, Y.R., Atiyah, A.G.: Benchmarking methodology of banks based on financial sustainability using CRITIC and RAFSI techniques. Decis. Making: Appl. Manage. Eng. 7(1), 315–341 (2024)

Al-Najjar, N., Al Bulushi, M.Y., Al Seyabi, F.A., Al-Balushi, S.M., Al-Harthi, A.S., Emam, M.M.: Perceived impact of initial Student teaching practice on teachers' teaching performance and professional skills: a retrospective study. Pedag.: Int. J. 1–20 (2024)

Alnoor, A., Atiyah, A.G., Abbas, S.: Toward digitalization strategic perspective in the European food industry: non-linear nexuses analysis. Asia-Pac. J. Bus. Adm. (2023)

Alnoor, A., Atiyah, A.G., Abbas, S.: Unveiling the determinants of digital strategy from the perspective of entrepreneurial orientation theory: a two-stage SEM-ANN approach. Glob. J. Flexible Syst. Manage. 1–18 (2024a)

Alnoor, A., et al.: How positive and negative electronic word of mouth (eWOM) affects customers' intention to use social commerce? A dual-stage multi group-SEM and ANN analysis. Int. J. Hum.-Comput. Interact. 40(3), 808–837 (2024b)

Atiyah, A.G.: Unveiling the quality perception of productivity from the senses of real-time multisensory social interactions strategies in metaverse. In: Al-Emran, M., Ali, J.H., Valeri, M., Alnoor, A., Hussien, Z.A. (eds.) IMDC-IST 2024. LNNS, vol. 876, pp. 83–93. Springer, Cham (2023). https://doi.org/10.1007/978-3-031-51300-8_6

Atiyah, A.G., Zaidan, R.A.: Barriers to using social commerce. In: Alnoor, A., Wah, K.K., Hassan, A. (eds.) Artificial Neural Networks and Structural Equation Modeling, pp. 115–130. Springer, Singapore (2022). https://doi.org/10.1007/978-981-19-6509-8_7

Atiyah, A.G., Alhasnawi, M., Almasoodi, M.F.: Understanding metaverse adoption strategy from perspective of social presence and support theories: the moderating role of privacy risks. In: Al-Emran, M., Ali, J.H., Valeri, M., Alnoor, A., Hussien, Z.A. (eds.) IMDC-IST 2024. LNNS, vol. 876, pp. 144–158. Springer, Cham (2023a). https://doi.org/10.1007/978-3-031-51300-8_10

Atiyah, A.G., All, N.D.A., Zaidan, A.S., Bayram, G.E.: Understating the social sustainability of metaverse by integrating adoption properties with users' satisfaction. In: Al-Emran, M., Ali, J.H., Valeri, M., Alnoor, A., Hussien, Z.A. (eds.) IMDC-IST 2024. LNNS, vol. 895, pp. 95–107. Springer, Cham (2023b). https://doi.org/10.1007/978-3-031-51716-7_7

Atiyah, A.G., Faris, N.N., Rexhepi, G., Qasim, A.J.: Integrating ideal characteristics of chat-GPT mechanisms into the metaverse: knowledge, transparency, and ethics. In: Al-Emran, M., Ali, J.H., Valeri, M., Alnoor, A., Hussien, Z.A. (eds.) IMDC-IST 2024. LNNS, vol. 895, pp. 131–141. Springer, Cham (2023c). https://doi.org/10.1007/978-3-031-51716-7_9

Atra, M.: Teachers' self-efficacy when instructing students with emotional and behavioral disorders in inclusive classrooms (2024)

Balalle, H.: Exploring student engagement in technology-based education in relation to gamification, online/distance learning, and other factors: a systematic literature review. Soc. Sci. Humanit. Open **9**, 100870 (2024)

Bandura, A.: Social-learning theory of identificatory processes. Handb. Soc. Theory Res. **213**, 262 (1969)

Bönsch, A., Ehret, J., Rupp, D., Kuhlen, T.W.: Wayfinding in immersive virtual environments as social activity supported by virtual agents. Front. Virtual Reality **4**, 1334795 (2024)

Bourdieu, P.: The Habitus and the space of life-styles: (1984). In: The People, Place, and Space Reader, pp. 139–144. Routledge (2014)

Boyte, H.C., Ström, M.L.: Agency in an AI avalanche: education for citizen empowerment. Eidos. J. Philos. Cult. **4**(2), 142–161 (2020)

Bozkurt, A., et al.: Speculative futures on ChatGPT and generative artificial intelligence (AI): a collective reflection from the educational landscape. Asian J. Dist. Educ. **18**(1), 53–130 (2023)

Burr, C., Cristianini, N., Ladyman, J.: An analysis of the interaction between intelligent software agents and human users. Mind. Mach. **28**(4), 735–774 (2018)

Carpiano, R.M.: Toward a neighborhood resource-based theory of social capital for health: can Bourdieu and sociology help? Soc Sci Med **62**(1), 165–175 (2006)

Cummings, J.J., Wertz, E.E.: Capturing social presence: concept explication through an empirical analysis of social presence measures. J. Comput.-Mediat. Commun. **28**(1), zmac027 (2023)

de Graaf, M.M.: An ethical evaluation of human–robot relationships. Int. J. Soc. Robot. **8**, 589–598 (2016)

Dodgers, S., Cordoba, S., Coe, J.: Examining the role of childhood experiences in gender identity and expression: an interpretative phenomenological analysis using social learning theory. Gend. Issues **40**(2), 255–274 (2023)

Edwards, C., et al.: Human-machine communication: what does/could communication science contribute to HRI?. In: 2019 14th ACM/IEEE International Conference on Human-Robot Interaction (HRI), pp. 673–674. IEEE (2019)

Germain, E.: Teachers' descriptions of robot educators. Doctoral dissertation, Grand Canyon University (2024)

Gil de Zúñiga, H., Goyanes, M., Durotoye, T.: A scholarly definition of artificial intelligence (AI): advancing AI as a conceptual framework in communication research. Polit. Commun. **41**(2), 317–334 (2024)

Gladden, M.E.: Who will be the members of Society 5.0? Towards an anthropology of technologically posthumanized future societies. Soc. Sci. **8**(5), 148 (2019)

Halford, S., Savage, M.: Speaking sociologically with big data: symphonic social science and the future for big data research. Sociology **51**(6), 1132–1148 (2017)

Hong, J.W.: Living with the most humanlike nonhuman: understanding human-ai interactions in different social contexts. Doctoral dissertation, University of Southern California (2022)

Husin, N.A., Abdulsaeed, A.A., Muhsen, Y.R., Zaidan, A.S., Alnoor, A., Al-mawla, Z.R.: Evaluation of metaverse tools based on privacy model using fuzzy MCDM approach. In: Al-Emran, M., Ali, J.H., Valeri, M., Alnoor, A., Hussien, Z.A. (eds.) IMDC-IST 2024. LNNS, vol. 895, pp. 1–20. Springer, Cham (2023). https://doi.org/10.1007/978-3-031-51716-7_1

Jiang, S., Chen, Y., Wang, L.: Effectiveness of community-based programs on aggressive behavior among children and adolescents: a systematic review and meta-analysis. Trauma Violence Abuse 15248380241227986 (2024)

Kuhn, C., et al.: Understanding digital inequality: a theoretical kaleidoscope. In: Jandrić, P., MacKenzie, A., Knox, J. (eds.) Constructing Postdigital Research. Postdigital Science and Education, pp. 333–373. Springer, Cham (2023). https://doi.org/10.1007/978-3-031-35411-3_17

Lin, N.: Social Capital: A Theory of Social Structure and Action, vol. 19. Cambridge University Press, Cambridge (2002)

Macgilchrist, F., Allert, H., Bruch, A.: Students and society in the 2020s. Three future 'histories' of education and technology. Learn. Media Technol. **45**(1), 76–89 (2020)

Martin, R.: The Sociology of Power. Taylor & Francis (2024)

Muhsen, Y.R., Husin, N.A., Zolkepli, M.B., Manshor, N.: A systematic literature review of fuzzy-weighted zero-inconsistency and fuzzy-decision-by-opinion-score-methods: assessment of the past to inform the future. J. Intell. Fuzzy Syst. **45**(3), 4617–4638 (2023)

Muthmainnah, M., Khang, A., Al Yakin, A., Oteir, I., Alotaibi, A.N.: An innovative teaching model: the potential of metaverse for English learning. In: Handbook of Research on AI-Based Technologies and Applications in the Era of the Metaverse, pp. 105–126. IGI Global (2023a)

Muthmainnah, Obaid, A.J., Al Yakin, A., Brayyich, M.: Enhancing computational thinking based on virtual robot of artificial intelligence modeling in the English language classroom. In: Swaroop, A., Polkowski, Z., Correia, S.D., Virdee, B. (eds.) ICDAM 2023. LNNS, vol 787, pp. 1–11. Springer, Singapore (2023b)

Nanavati, A.: Achieving deployable autonomy through customizability and human-in-the-loop: a case study in robot-assisted feeding. In: Companion of the 2024 ACM/IEEE International Conference on Human-Robot Interaction, pp. 136–138 (2024)

Natale, S.: Communicating through or communicating with: approaching artificial intelligence from a communication and media studies perspective. Commun. Theory **31**(4), 905–910 (2021)

Rapp, A., Curti, L., Boldi, A.: The human side of human-chatbot interaction: a systematic literature review of ten years of research on text-based chatbots. Int. J. Hum. Comput. Stud. **151**, 102630 (2021)

Reay, D.: Education and cultural capital: the implications of changing trends in education policies. In: Bourdieu and Education, pp. 92–105. Routledge (2020)

Scaradozzi, D., Screpanti, L., Cesaretti, L.: Towards a definition of educational robotics: a classification of tools, experiences and assessments. Smart Learn. Educ. Robot.: Using Rob. Scaffold Learn. Outcomes 63–92 (2019)

Schiavo, F., Campitiello, L., Todino, M.D., Di Tore, P.A.: Educational robots, emotion recognition and ASD: new horizon in special education. Educ. Sci. **14**(3), 258 (2024)

Sharevski, F., Jachim, P., Treebridge, P., Li, A., Babin, A., Adadevoh, C.: Meet Malexa, Alexa's malicious twin: malware-induced misperception through intelligent voice assistants. Int. J. Hum. Comput. Stud. **149**, 102604 (2021)

Spence, P.R., Kaufmann, R., Lachlan, K.A., Lin, X., Spates, S.A.: Examining perceptions and outcomes of AI versus human course assistant discussions in the online classroom. Commun. Educ. 1–22 (2024)

Tang, F.: Understanding the role of digital immersive technology in educating the students of English language: does it promote critical thinking and self-directed learning for achieving sustainability in education with the help of teamwork? BMC Psychol. **12**(1), 144 (2024)

Torresen, J.: A review of future and ethical perspectives of robotics and AI. Front. Robot. AI **4**, 75 (2018)

Trigg, A.B.: Deriving the Engel curve: Pierre Bourdieu and the social critique of Maslow's hierarchy of needs. In: Consuming Symbolic Goods, pp. 119–132. Routledge (2013)

Tsing, A.: Sorting out commodities: how capitalist value is made through gifts. HAU: J. Ethnogr. Theory **3**(1), 21–43 (2013)

van der Sloot, B.: Regulating the Synthetic Society: Generative AI, Legal Questions, and Societal Challenges (2024)

Walter, M., Andersen, C.: Indigenous Statistics: A Quantitative Research Methodology, p. 159. Taylor & Francis (2013)

Warde, A.: Practice and field: revising Bourdieusian concepts, vol. 65. Centre for Research on Innovation & Competition, The University of Manchester, Manchester (2004)

Warwick, R., McCray, J., Board, D.: Bourdieu's habitus and field: implications on the practice and theory of critical action learning. Action Learn.: Res. Pract. **14**(2), 104–119 (2017)

Wenzel, K., Kaufman, G.: Designing for harm reduction: communication repair for multicultural users' voice interactions. arXiv preprint arXiv:2403.00265 (2024)

Williamson, B., Eynon, R., Knox, J., Davies, H.: Critical perspectives on AI in education: political economy, discrimination, commercialization, governance and ethics. In: Handbook of Artificial Intelligence in Education, pp. 553–570. Edward Elgar Publishing (2023)

Wu, T.T., Lee, H.Y., Li, P.H., Huang, C.N., Huang, Y.M.: Promoting self-regulation progress and knowledge construction in blended learning via ChatGPT-based learning aid. J. Educ. Comput. Res. **61**(8), 3–31 (2024)

The Role of Artificial Intelligence in Improving Customer Service and Retaining Human Resources: Digital Sustainability as a Mediating Variable

Abbas Ali Mohammed[1], Hadi AL-Abrrow[2], Krar Muhsin Thajil[3], Alhamzah Alnoor[4], and Sammar Abbas[5(✉)]

[1] Department of Business Administration, Directorate of Education of Basrah, Basrah, Iraq
[2] Department of Business Administration, College of Administration and Economics, University of Basrah, Basrah, Iraq
[3] Department of Business Administration, Mazaya University College, Nasiriyah, Thi-Qar, Iraq
[4] Southern Technical University, Basrah, Iraq
Alhamzah.malik@stu.edu.iq
[5] Institution of Business Studies, Kohat University of Science and Technology, Kohat, Pakistan
sabbas@kust.edu.pk

Abstract. Artificial intelligence (AI) has recently been heavily relied upon to provide better services, and there has been a long discussion about the opportunities that AI can offer to secure competitive advantage. The present paper, in turn, tries to present a new model for using artificial intelligence to improve customer service (CS) and to facilitate the retention of organizational human resources. This study also evaluates digital sustainability as a mediator in the relationship between artificial intelligence and customer service improvement. Data was collected from 560 individuals attending Iraqi hospitals in southern Iraq. This research can scientifically contribute to further understanding the role of AI in the context of Iraqi hospitals and also can help to locate potential future research avenues.

Keywords: Artificial intelligence · Customer service · Health sector · Digital sustainability

1 Introduction

Artificial intelligence (AI) is referred to as the automated intelligence resulting from the use of robots, human or non-human, that behave like humans (Prentice et al., 2020). AI applications are aimed at increasing organizational outputs (Prentice & Nguyen, 2020; Atiyah et al., 2023a, b, c), maximizing their profits, and improving employees as well as firms' efficiency and effectiveness (Wirtz, 2020; Atiyah et al., 2023a, b, c). AI is widely recognized and used technology in enhancing employees' desire to provide the best levels of performance (Abdeldayem & Aldulaimi, 2020; Shahzad et al., 2023) as it contributes to the automation of routine tasks, which enhances employees' immersion

© The Author(s), under exclusive license to Springer Nature Switzerland AG 2024
A. Alnoor et al. (Eds.): AIRDS 2024, LNNS 1033, pp. 77–92, 2024.
https://doi.org/10.1007/978-3-031-63717-9_5

in their work for a longer period and provides services in an ideal manner (Wassan, 2021; Nguyen & Mali, 2022). Artificial intelligence has contributed to the healthcare field in many countries by adopting many and varied innovations (Berhil et al., 2020). Thus, the healthcare sector needs to automate a range of activities that aim at helping employees acquire talent and creative skills (Garikapati et al., 2022). These activities include obtaining accurate information about job applicants (Mikhailova & Sharova, 2023). Artificial intelligence is now considered the most important aspect that affects employees' ability to provide services in a better way (Arora, 2020). It is also used in providing training, support and development programs for employees, which enhances digital sustainability (Pan & Nishant, 2023).

Artificial Intelligence (AI) and digital sustainability can significantly contribute to the improvement of customer service in Iraqi healthcare institutions (Sapienza et al., 2022). This is because it creates positive competition that contributes to the improvement of customer service (Damoah et al., 2021) as employees become proactive in providing services. Improving customer service is vital to organization's overall achievements (Siala & Wang, 2022), organizations may gain a competitive advantage by applying artificial intelligence and digital sustainability (Pigola et al., 2021), and are more likely to achieve job satisfaction that artificial intelligence technology provides (Akinola & Telukdarie, 2023). Job satisfaction makes employees compete to provide customer service in a more distinguished manner (Sajwani, 2024). Artificial intelligence has a prominent role in retaining human resources (Li et al., 2023). This becomes clear through the application of digital technology (Rao & Punjabi, 2020) because the way of work is conducted according to artificial intelligence technology will clearly push employees to stay in the organization and stay away from negative thinking that includes trying to leave the organizations in which they work and seeking to move to other organizations (Shahzad et al., 2023). It may provide them with more services and enhance their job satisfaction (Tewari & Pant, 2020). Hence, organizations must be open to innovation and knowledge exchange (Vrontis et al., 2022), and the need to reduce the stress at work that employees may be exposed to if they practice activities in a traditional way (Sakka et al., 2022), and as a result, it is necessary to rely on artificial intelligence technology if health organizations want to preserve their human resources and enhance their capabilities and skills, which are difficult to dispense with if the organizations want to succeed and grow (Tambe et al., 2019; Pillai & Sivathanu, 2020; Nahar et al., 2024).

2 Literature Review

2.1 Artificial Intelligence

Artificial intelligence (AI) deals with transferring anthropomorphic intelligence and reasoning to computers with the objective of information-based well-informed decisions (Bock et al., 2020). It was John McCarthy (1956) who coined the term AI. AI has been used across different disciplines including engineering, mathematics, physics, and technology (Bender & Cortés-Ciriano, 2021; Atiyah et al., 2023a, b, c). It gradually appeared and gained strength (Hassani et al., 2020). AI has completely transformed the ways of doing thnings (Yang, 2022; Alnoor et al., 2024a, b, c). It means that that a machine can acquire intelligence (Langer et al., 2021). They include areas in which machines have become smart enough to acquire intelligence, adjust to the given circumstance, and rectify mistakes (Mouloodi et al., 2021). In other words, machines can think on their own without being encoded with instructions (Michaeli et al., 2022).

In recent times AI has enabled machines to perform a human-like variety of functions in both familiar and new scenarios (Du-Harpur et al., 2020; Alnoor et al., 2023). The term "artificial intelligence" is currently used interchangeably as "machine learning (ML)" or "deep learning (DL)", with deep learning as another dimension of machine learning (Ahmed et al., 2022) that we discuss in the following (Table 1). The term "machine learning" denotes the use of data and algorithms to recognize data patterns and make inferences (Chowdhary, 2020).

Table 1. AI & ML related important terms.

Artificial intelligence (AI)	The capacity of machines to mimic human intelligence, such as computers
Machine learning (ML)	Refers to the use of algorithms and statistical models to identify data patterns. ML also facilitates machines to perform certain functions without human aid
Supervised learning	It denotes machine learning in locating the appropriate functions that match a set of inputs machine inputs with outputs. This type of training or learning is based on pre-batched pairs. In contrast, unsupervised learning identifies unexpected patterns in data, such as groupings or "clusters", without the aid of labeling or previous knowledge
Overfitting	It is a common ML-related issue where a model is accurate when run using training data but the operation of the model cannot be generalized to other data sources
Neural network	A supervised machine learning technique based on biological principles that involves data passing through a network of interconnected neurons that are each given a weight in order to generate predictions. In training, the network iteratively processes the data and modifies the weightings to maximize the network capacity to synchronize labels to data
Deep learning	It describes a neural network that has several layers of "neurons" with movable weights, or mathematical functions
Convolutional neural network	It is a kind of neural network in which the layers give regions of image filters tailored to particular features

2.2 Customer Service

The conventional definition of excellent customer service calls for cordial and productive in-person interactions between salespeople and clients (Lee & Lee, 2020). The development of digital technology is causing this opinion to shift (Van der Goot et al., 2020;

Alnoor et al., 2024a, b, c). Many businesses heavily rely on AI, particularly the services sector (Schanke et al., 2021). Due to the wider implementation of AI in the service sector, there has been a significant change in users' behavior (e.g., marketing of online services and offers, logistics and payment preferences, financing, etc.) (Danaher et al., 2024). The way the provider provides value and experiences to customers must also alter in response to the customer's omnichannel service purchase behavior (Aburayya et al., 2020). Healthcare organizations are under a lot of pressure to continue providing high-quality care while also lowering costs and raising safety (Javaid et al., 2023). The aging population, an increase in chronic illnesses, increased patient expectations, and shortage of trained medical personnel have all made it more difficult for healthcare organizations to carry out their duties (Lee & Lee, 2021). AI has captured an interest of academics, the general public, and organizations (Zygiaris et al., 2022; Ali et al., 2024). Artificial intelligence technologies have made a major breakthrough in society due to recent advancements in this field, as well as shifts in public perception, hardware advancements, the automation of tasks, and the integration of AI into customer service delivery (Chatterjee et al., 2021). By implementing these technologies, organizations hope to reap benefits and protect themselves from the risk of providing subpar customer service (Endeshaw, 2020). Artificial intelligence technologies are widely used in many different industries (Babroudi et al., 2021), but the health industry is one area where they are particularly useful (Haleem et al., 2022).

2.3 Retaining Human Resources

One of an organization's top responsibilities is to retain its human resources, particularly its middle and upper-level managers (Bilan et al., 2020). Identifying gifted workers, nurturing their abilities, and leveraging them to satisfy the expanding needs of the company are all long-term goals of human resource retention (Thomas et al., 2018). Talented workers are seen as exceptionally talented or possess competencies unique to the firm and contribute significantly to its intellectual capital (Chiţu & Russo, 2020). Organizations make HR-related strategic decisions that encompass the prediction of organizational future HR needs and then the development of human capital accordingly (Oster & Jonze, 2013; Atiyah, 2023). High-keeping performance work systems are put in place by organizations to retain skilled middle and upper-level managers and to boost organizational commitment and job happiness (Fitz-Lewis, 2018). A collection of HR procedures known as "high-performance work systems" are intended to function methodically and cooperatively with the systems of a company in order to boost worker productivity via artificial intelligence (Gyurak Babeľova et al., 2020).

Innovation and development in information technology have led to abrupt changes in the business environment and require organizations to be flexible enough to accommodate these technologies for market survival (Asad et al., 2022; Atiyah and Zaidan, 2022). This has also necessitated the need for trained and technology-equipped human resources (Alves et al., 2019). Human beings who possess high skills in performing tasks are a precious asset for any organization (Arvand et al., 2022).

3 Hypothesis Development

3.1 Theoretical Development and Hypotheses

Health services are greatly affected by the fast advancements in digital technologies such as AI (Esmaeilzadeh et al., 2021). One major field of interest in applying artificial intelligence to provide customers online is the field of chatbots based on artificial intelligence and digital sustainability (Uzir et al., 2021). These chatbots have been introduced as digital assistants in customer service delivery environments in the health sector via the internet to enrich customers' experience and achieve expectations through real-time interactions (Li et al., 2023). Artificial intelligence and digital sustainability technologies actually play a major role in facilitating the process of providing services in the health sector (Musbahi et al., 2021), for example, saving time, reducing costs, saving data, and making it easy to retrieve from knowledge reservoirs prepared for this purpose (Xu et al., 2020). On the other hand, artificial intelligence technologies enhance employees' immersion and immersion in work due to the attractiveness of using these technologies (Pelau et al., 2021). This, of course, is likely to enhance the organization's ability to retain human resources, develop their capabilities (Amjad et al., 2023), and stimulate positive feelings that push them towards providing services in a distinctive manner (Nicolescu & Tudorache, 2022).

Numerous studies have illuminated the role of AI in healthcare (Kavitha et al., 2023), particularly in the field of medical informatics, where AI can improve patient care and diagnose and interpret medical data (Cannavale et al., 2022). AI applications used in consulting have reduced human detection errors and improved customer service, especially with digital sustainability in place (Strohm et al., 2020). However, some of the interconnected factors of moral and social trust as well as reliance on AI have yet to be developed (Rane, 2023). Previous research has shown that clinical applications of clinical AI sometimes perform better than human responders and specialists in detecting diseases such as Alzheimer's disease using natural language processing techniques (Hmoud, 2021), and skin cancer and arrhythmia using deep neural networks with AI technology (Sakka et al., 2022). Its performance is comparable to, and sometimes better than, that of human responders and experts (Hmoud & Várallyai, 2020). The sensitive nature of health information and consumers' high exposure to potential medical errors are the main reasons why people working in AI technology are naturally immersed in advanced tasks and their services will improve as digital sustainability exists (Budhwar et al., 2023).

The above presented literature leads us to the following hypotheses.

H1. AI significantly and positively influences CS.

H2. AI has a significant and positive effect on the retention of human resources (RHR).

H3. AI has a significant and positive effect on digital sustainability (DS).

The current business environment is much more volatile and highly unpredictable (Shaheen et al., 2018), As a result, competition in the workplace intensifies (Sweeney et al., 2023). The availability of modern technology has enhanced employees' well-being and increased motivation to improve services (Hammedi et al., 2017). Organizations need to retain employees who are emotionally attached and committed (Salifu et al., 2022), as this is expected to improve performance and contribute significant value to the organization (Qureshi & Syed, 2014). Occasionally, it becomes challenging for the managers to ensure employees' motivation and commitment given the tough situation (Jeffreys, 2015). Latest technologies such as AI can facilitate to intact employees' motivation and commitment (Minkiewicz et al., 2014). But unfortunately, there is little research on employees' engagement at turbulent and volatile workplaces (Beauvais et al., 2017). In light of this, positive psychologists are of the view that higher levels of employees' commitment and motivation can be achieved with the better application of modern computer technologies like AI (Tom Dieck & Han, 2022; Wang et al., 2023).

Employee turnover, the inability to retain human resources, is often identified as a cause of poor organizational performance (De Oliveira, & Da Silva, 2015), In fact, employees are more likely to move to other organizations in search of better financial rewards and better opportunities for career development (Finocchiaro, 2022). The old-fashioned ways of retaining employees through financial rewards and incentives do not work anymore (Victor & Hoole, 2017). One of the most important reasons that motivate employees to stay in the organization is the use of the latest business technologies, such as digital sustainability (Parry & Battista, 2023) which contributes to providing a more immersive employee experience, which in turn increases the organization's ability to retain talent (Greer, 2021). Therefore, it is necessary to provide advanced human resources management infrastructure and use artificial intelligence to increase digital sustainability in the organization (Yalabik et al., 2017). An appropriate application of these technologies ensures employees; commitment (Eickhoff et al., 2012). This effect can be achieved by enhancing in-role (i.e. organizational commitment) and extra-role (i.e. organizational citizenship) behaviors, which can boost morale and encourage the retention of human resources as employees become more engaged in their work, as shown in Fig. 1 (Veth et al., 2019).

We narrate the following hypotheses on the basis of the above-presented literature.

H4. Digital sustainability directly and positively influences customer service.

H5. Digital sustainability directly and positively influences the retention of human resources.

H6. Digital sustainability as a mediator positively affects the relationship between artificial intelligence and customer service.

H7. Digital sustainability as a mediator positively affects the relationship between artificial intelligence and retention of human resources.

Following is the depiction of conceptual model of the study.

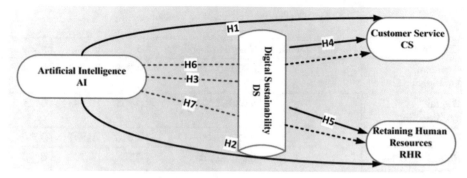

Fig. 1. Research Framework.

4 Methodology

The sample of this study comprised of 560 patients who visited health facilities in southern Iraq and these facilities provided them with services. This number was obtained from 600 questionnaires directed to patients (clients who benefit from service provision). The response rate was 93%. Off-the-shelf scales were relied upon because they already existed and there was insufficient time to attempt to create a researcher-specific scale for the current study.

AMOS version 25 and SPSS version 25 were used to test the proposed hypotheses and analyze the data. The reliability and stability of the measurements were verified, and descriptive statistics and correlation coefficients were presented.

5 Results

Prior to hypothesis testing, a confirmatory factor analysis (CFA) was conducted to assess the factor structure of self-report items measuring the four central constructs of this study: artificial intelligence (AI), digital sustainability (DS), customer service (CS), and retaining human resources (RHR) (Hair et al., 2017; Abbas et al., 2023). In addition to measuring the internal consistency of these constructs Cronbach's alpha reliability was examined. Notably, all constructs showed strong reliability with values above 0.7 (Hair et al., 2020; Muhsen et al., 2023) (see Table 2).

To establish convergent validity, average variance extracted (AVE) and composite site reliability (CR) were evaluated (Fornell and Larcker, 1981) both AVE and CR values were above the threshold of 0.5, highlighting robust convergent validity (Table 3).

Next, discriminant validity was assessed by calculating the square root of AVE for each construct (see Table 3) and comparing these values to the inter-constituent correlation (Fornell & Larcker, 1981; Farrell & Rudd, 2009; Ahmed et al., 2024; Alnoor et al., 2024a, b, c). This analysis clearly confirmed the discriminant validity of the constructs.

Finally, a common method bias (CMB) assessment was performed using the Harmon single-factor method. The results showed that this index was 0.33, which is significantly

Table 2. Reliability and Validity Measurement.

Construct	Number of items	Range of Loadings	α	C.R	AVE
Artificial intelligence	7	0.765–0.879	0.775	0.902	0.654
Digital sustainability	6	0.689–0.897	0.805	0.895	0.609
Customer service	6	0.632–0.889	0.846	0.867	0.582
Retaining human resources	7	0.587–0.976	0.765	0.799	0.567

lower than the threshold of 0.5 (Husin et al., 2023; Al-Enzi et al., 2023). This result confirms that there is no substantial common method bias in the results of this study. Table 3 presents the correlation matrix (including demographic variables). It can be observed that the independent, mediating, and dependent variables are significantly correlated with each other.

Table 3. Means, Standard Deviations, and correlations among the Variables.

	AI	DS	CS	RHR
AI	1			
DS	.198*	1		
CS	.289**	.564**	1	
RHR	.456**	.632**	.455**	1

Notes: *$p < 0.05$; **$p < 0.01$; AI = Artificial intelligence; DS = Digital sustainability; CS = Customer service; RHR = Retaining human resources.

Through a closer look at the results testing the hypotheses shown in Table 4, it becomes clear that all direct and indirect hypotheses are accepted, showing the existence of a positive influence relationship between all variables of the study directly, as well as through the mediating role of employee immersion.

6 Contributions and Future Research

6.1 Theoretical Implications

The current study explains the meaning of artificial intelligence and how to best use and employ it in Iraqi health organizations. So that it is easy for researchers to recognize the significance of using AI in the service sector in general and the health sector in particular, as well as the mutual and complementary relationship between AI techniques and achieving competitive advantage in providing services. The current study has added a

Table 4. Hypothesis testing.

Hypothesis	Path	β	S. E	T-value	P-value	Result
H1	AI → CS	.331	.013	10.189	0.000	Accepted
H2	AI → RHR	.432	.019	18.403	0.000	Accepted
H3	AI → EW	.336	.032	8.167	0.000	Accepted
H4	DS → CS	.543	.033	12.125	0.000	Accepted
H5	DS → RHR	.550	.026	17.654	0.000	Accepted
H6	AI → DS → CS	.185	.027	3.443	0.000	Accepted
H7	AI → DS → RHR	.184	.013	10.821	0.000	Accepted

contribution that may be important to the future of studies in service sectors by pointing at the implementation of AI in service sectors in developing countries. The results indicated that the usage AI would significantly increase an ability of health organizations to providing better services. In addition, artificial intelligence techniques and technology will enhance the organization's ability to retain human resources.

6.2 Practical Implications

The world of renewal and continuous development and change requires organizations to adapt to the many and rapid developments, and it is necessary to use the best and latest technologies if health organizations are after the ability to continue, compete and succeed in providing services. So, the current study highlighted the role of modern technologies represented by artificial intelligence, because it will have clear importance in keeping pace with developments and changes taking place in the world at the level of service provision in a way that is compatible with the environment in which these organizations operate. This study provides a framework that can benefit researchers, managers, and practitioners in general. The study showed the variables that can serve as elements of success for using artificial intelligence techniques and exploiting them to obtain the best results.

6.3 Future Studies

Traditional methods of providing services have begun to disappear because of the increasing trend towards the use of digital technology in general and artificial intelligence technology in particular, at various organizational levels, and in all service sectors, especially the health sector, as it is directly linked to human lives. In light of the fierce and intense competition for organizations to provide better services, they must use the latest technologies. This requires more research effort to find mechanisms and strategies through which artificial intelligence technologies can be exploited to achieve maximum benefit and obtain competitive advantage. Future studies can take into account the application of the study in other service sectors, for example the banking sector. Future studies can also use additional variables such as the impact of artificial intelligence on job performance and/or job commitment, and a longitudinal study can also be presented if time is available for that.

7 Conclusion

Methods and strategies for working with emotional intelligence will undoubtedly surpass traditional methods of job execution. Artificial intelligence integrates a suite of technologies developed to perform tasks effectively, such as the use of robots in telecommunications companies for automated responses, or the use of modern technology to examine diseases and use advanced X-rays to diagnose damage and sometimes cure incurable diseases. Artificial intelligence programs provide digital interfaces and networks that are easy for employees to work with and interact with. Artificial intelligence is contributing to a qualitative leap in the ability of organizations to deliver services in a high-performance manner. Artificial intelligence deserves more attention from researchers, practitioners, and managers alike. There is an urgent need to integrate artificial intelligence technologies with strategies and activities to maintain the best services to society beneficiaries and to develop long-term plans and strategies to ensure that all standards and values are upheld. For a sustainable environment technology is developing rapidly. In addition, the use of AI technology in various service activities will result in an immersive service experience for customers.

The results of the current study have proven the necessity of working using artificial intelligence techniques, as they contribute to raising the level of performance, improving customer service, and achieving a state of job immersion that will push employees towards job commitment and even reach organizational citizenship behavior and remain in the organizations in which they work, which reduces the state of work turnover, which is what organizations seek. This is because organizations aim to retain human resources and enhance available skills and energies as a knowledge reserve that contributes to enhancing the organization's competitive advantage.

References

Abbas, S., et al.: Antecedents of trustworthiness of social commerce platforms: a case of rural communities using multi group SEM & MCDM methods. Electron. Commer. Res. Appl. **62**, 101322 (2023)

Abdeldayem, M.M., Aldulaimi, S.H.: Trends and opportunities of artificial intelligence in human resource management: aspirations for public sector in Bahrain. Int. J. Sci. Technol. Res. **9**(1), 3867–3871 (2020)

Aburayya, A., Marzouqi, A., Alawadhi, D., Abdouli, F., Taryam, M.: An empirical investigation of the effect of employees' customer orientation on customer loyalty through the mediating role of customer satisfaction and service quality. Manage. Sci. Lett. **10**(10), 2147–2158 (2020)

Ahmed, A.D., Salih, M.M., Muhsen, Y.R.: Opinion weight criteria method (OWCM): a new method for weighting criteria with zero inconsistency. IEEE Access (2024)

Akinola, S., Telukdarie, A.: Sustainable digital transformation in healthcare: advancing a digital vascular health innovation solution. Sustainability **15**(13), 10417 (2023)

Al-Enzi, S.H.Z., Abbas, S., Abbood, A.A., Muhsen, Y.R., Al-Hchaimi, A.A.J., Almosawi, Z.: Exploring research trends of metaverse: a bibliometric analysis. In: Al-Emran, M., Ali, J.H., Valeri, M., Alnoor, A., Hussien, Z.A. (eds.) IMDC-IST 2024. LNNS, vol. 895, pp. 21–34. Springer, Cham (2023). https://doi.org/10.1007/978-3-031-51716-7_2

Ali, J., Hussain, K.N., Alnoor, A., Muhsen, Y.R., Atiyah, A.G.: Benchmarking methodology of banks based on financial sustainability using CRITIC and RAFSI techniques. Decis. Making: Appl. Manage. Eng. 7(1), 315–341 (2024)

Alnoor, A., Atiyah, A.G., Abbas, S.: Toward digitalization strategic perspective in the European food industry: non-linear nexuses analysis. Asia-Pac. J. Bus. Adm. (2023)

Alnoor, A., Atiyah, A.G., Abbas, S.: Unveiling the determinants of digital strategy from the perspective of entrepreneurial orientation theory: a two-stage SEM-ANN approach. Glob. J. Flexible Syst. Manage. 1–18 (2024)

Alnoor, A., Chew, X., Khaw, K.W., Muhsen, Y.R., Sadaa, A.M.: Benchmarking of circular economy behaviors for Iraqi energy companies based on engagement modes with green technology and environmental, social, and governance rating. Environ. Sci. Pollut. Res. 31(4), 5762–5783 (2024)

Alnoor, A., et al.: How positive and negative electronic word of mouth (eWOM) affects customers' intention to use social commerce? A dual-stage multi group-SEM and ANN analysis. Int. J. Hum.-Comput. Interact. 40(3), 808–837 (2024)

Alves, D., Dieguez, T., Conceição, O.: Retaining talents: impact on innovation. In: ECMLG 2019 15th European Conference on Management, Leadership and Governance, p. 36. Academic Conferences and publishing limited. (2019)

Amjad, A., Kordel, P., Fernandes, G.: A review on innovation in healthcare sector (telehealth) through artificial intelligence. Sustainability 15(8), 6655 (2023)

Arora, A.: Conceptualising artificial intelligence as a digital healthcare innovation: an introductory review. Med. Devices: Evidence Res. 223–230 (2020)

Arvand, H., Angazi Ghods, A.A.: Designing a model for recruiting and retaining human resources capable of case study (Nezaja Ranger Training Center). Defense-Hum. Capit. Manag. 1(4), 143–166 (2022)

Asad, A., Hidayati, S., Fridiyanto, F.: Education and human resources: retaining future human resources' behaviours to nature through environmental education. J. High. Educ. Theory Pract. 22(2) (2022)

Atiyah, A.G.: Unveiling the quality perception of productivity from the senses of real-time multisensory social interactions strategies in metaverse. In: Al-Emran, M., Ali, J.H., Valeri, M., Alnoor, A., Hussien, Z.A. (eds.) IMDC-IST 2024. LNNS, vol. 876, pp. 83–93. Springer, Cham (2023). https://doi.org/10.1007/978-3-031-51300-8_6

Atiyah, A.G., Zaidan, R.A.: Barriers to using social commerce. In: Alnoor, A., Wah, K.K., Hassan, A. (eds.) Artificial Neural Networks and Structural Equation Modeling, pp. 115–130. Springer, Singapore (2022). https://doi.org/10.1007/978-981-19-6509-8_7

Atiyah, A.G., Alhasnawi, M., Almasoodi, M.F.: Understanding metaverse adoption strategy from perspective of social presence and support theories: the moderating role of privacy risks. In: Al-Emran, M., Ali, J.H., Valeri, M., Alnoor, A., Hussien, Z.A. (eds.) IMDC-IST 2024. LNNS, vol. 876, pp. 144–158. Springer, Cham (2023a). https://doi.org/10.1007/978-3-031-51300-8_10

Atiyah, A.G., All, N.D.A., Zaidan, A.S., Bayram, G.E.: Understating the social sustainability of metaverse by integrating adoption properties with users' satisfaction. In: Al-Emran, M., Ali, J.H., Valeri, M., Alnoor, A., Hussien, Z.A. (eds.) IMDC-IST 2024. LNNS, vol. 895, pp. 95–107. Springer, Cham (2023b). https://doi.org/10.1007/978-3-031-51716-7_7

Atiyah, A.G., Faris, N.N., Rexhepi, G., Qasim, A.J.: Integrating ideal characteristics of chatGPT mechanisms into the metaverse: knowledge, transparency, and ethics. In: Al-Emran, M., Ali, J.H., Valeri, M., Alnoor, A., Hussien, Z.A. (eds.) IMDC-IST 2024. LNNS, vol. 895, pp. 131–141. Springer, Cham (2023c). https://doi.org/10.1007/978-3-031-51716-7_9

Babroudi, N.E.P., Sabri-Laghaie, K., Ghoushchi, N.G.: Re-evaluation of the healthcare service quality criteria for the Covid-19 pandemic: Z-number fuzzy cognitive map. Appl. Soft Comput. **112**, 107775 (2021)

Beauvais, B., Richter, J., Brezinski, P.: Fix these first: how the world's leading companies point the way toward high reliability in the Military Health System. J. Healthc. Manag. **62**(3), 197–208 (2017)

Bender, A., Cortés-Ciriano, I.: Artificial intelligence in drug discovery: what is realistic, what are illusions? Part 1: ways to make an impact, and why we are not there yet. Drug Discov. Today **26**(2), 511–524 (2021)

Berhil, S., Benlahmar, H., Labani, N.: A review paper on artificial intelligence at the service of human resources management. Indon. J. Electr. Eng. Comput. Sci. **18**(1), 32–40 (2020)

Bilan, Y., Mishchuk, H., Roshchyk, I., Joshi, O.: Hiring and retaining skilled employees in SMEs: problems in human resource practices and links with organizational success. Bus.: Theory Pract. **21**(2), 780–791 (2020)

Bock, D.E., Wolter, J.S., Ferrell, O.C.: Artificial intelligence: disrupting what we know about services. J. Serv. Mark. **34**(3), 317–334 (2020)

Budhwar, P., et al.: Human resource management in the age of generative artificial intelligence: perspectives and research directions on ChatGPT. Hum. Resour. Manag. J. **33**(3), 606–659 (2023)

Cannavale, C., Esempio Tammaro, A., Leone, D., Schiavone, F.: Innovation adoption in inter-organizational healthcare networks–the role of artificial intelligence. Eur. J. Innov. Manag. **25**(6), 758–774 (2022)

Chatterjee, S., Goyal, D., Prakash, A., Sharma, J.: Exploring healthcare/health-product ecommerce satisfaction: a text mining and machine learning application. J. Bus. Res. **131**, 815–825 (2021)

Chiṭu, E., Russo, M.: The impact of employer branding in recruiting and retaining human resources. Int. J. Commun. Res. **10**(2), 202–212 (2020)

Chowdhary, K.R.: Fundamentals of Artificial Intelligence, pp. 603–649. Springer, New Delhi (2020). https://doi.org/10.1007/978-81-322-3972-7

Damoah, I.S., Ayakwah, A., Tingbani, I.: Artificial intelligence (AI)-enhanced medical drones in the healthcare supply chain (HSC) for sustainability development: a case study. J. Clean. Prod. **328**, 129598 (2021)

Danaher, T.S., Danaher, P.J., Sweeney, J.C., McColl-Kennedy, J.R.: Dynamic customer value cocreation in healthcare. J. Serv. Res. **27**(2), 177–193 (2024)

De Oliveira, L.B., Da Silva, F.F.R.A.: The effects of high-performance work systems and leader-member exchange quality on employee engagement: evidence from a Brazilian non-profit organization. Procedia Comput. Sci. **55**, 1023–1030 (2015)

Du-Harpur, X., Watt, F.M., Luscombe, N.M., Lynch, M.D.: What is AI? Applications of artificial intelligence to dermatology. Br. J. Dermatol. **183**(3), 423–430 (2020)

Eickhoff, C., Harris, C.G., de Vries, A.P., Srinivasan, P.: Quality through flow and immersion: gamifying crowdsourced relevance assessments. In: Proceedings of the 35th International ACM SIGIR Conference on Research and Development in Information Retrieval, pp. 871–880 (2012)

Endeshaw, B.: Healthcare service quality-measurement models: a review. J. Health Res. **35**(2), 106–117 (2020)

Esmaeilzadeh, P., Mirzaei, T., Dharanikota, S.: Patients' perceptions toward human–artificial intelligence interaction in health care: experimental study. J. Med. Internet Res. **23**(11), e25856 (2021)

Farrell, A.M., Rudd, J.M.: Factor analysis and discriminant validity: a brief review of some practical issues. In: Australia and New Zealand Marketing Academy Conference 2009. Anzmac (2009)

Finocchiaro, P.A.: What is the role of place attachment and quality of life outcomes in employee retention? Worldwide Hospit. Tour. Themes **14**(3), 261–273 (2022)

Fitz-Lewis, T.: Human resources strategies for retaining employees in St. Lucian banks. Walden University (2018)

Fornell, C., Larcker, D.F.: Evaluating structural equation models with unobservable variables and measurement error. J. Mark. Res. **18**(1), 39–50 (1981)

Garikapati, K., Shaw, K., Shaw, A., Yarlagadda, A.: Digital society artificial intelligence in health care: issues of legal ethical and economical sustainability. In: 2nd International Conference on Sustainability and Equity (ICSE-2021), pp. 131–137. Atlantis Press (2022)

Greer, C.R.: Strategic human resource management. Pearson Custom Publishing (2021)

Gyurak Babeľova, Z., Starecek, A., Koltnerova, K., Cagáňová, D.: Perceived organizational performance in recruiting and retaining employees with respect to different generational groups of employees and sustainable human resource management. Sustainability **12**(2), 574 (2020)

Hair, J.F., Jr., Howard, M.C., Nitzl, C.: Assessing measurement model quality in PLS-SEM using confirmatory composite analysis. J. Bus. Res. **109**, 101–110 (2020)

Hair, J.F., Jr., Matthews, L.M., Matthews, R.L., Sarstedt, M.: PLS-SEM or CB-SEM: updated guidelines on which method to use. Int. J. Multivariate Data Anal. **1**(2), 107–123 (2017)

Haleem, A., Javaid, M., Singh, R.P., Suman, R.: Medical 4.0 technologies for healthcare: features, capabilities, and applications. Internet Things Cyber-Phys. Syst. **2**, 12–30 (2022)

Hammedi, W., Leclerq, T., Van Riel, A.C.: The use of gamification mechanics to increase employee and user engagement in participative healthcare services: a study of two cases. J. Serv. Manag. **28**(4), 640–661 (2017)

Hassani, H., Silva, E.S., Unger, S., TajMazinani, M., Mac Feely, S.: Artificial intelligence (AI) or intelligence augmentation (IA): what is the future? Ai **1**(2), 8 (2020)

Hmoud, B.: The adoption of artificial intelligence in human resource management and the role of human resources. In: Forum Scientiae Oeconomia, vol. 9, no. 1, pp. 105–118. Wydawnictwo Naukowe Akademii WSB (2021)

Hmoud, B.I., Várallyai, L.: Artificial intelligence in human resources information systems: investigating its trust and adoption determinants. Int. J. Eng. Manag. Sci. **5**(1), 749–765 (2020)

Husin, N.A., Abdulsaeed, A.A., Muhsen, Y.R., Zaidan, A.S., Alnoor, A., Al-mawla, Z.R.: Evaluation of metaverse tools based on privacy model using fuzzy MCDM approach. In: Al-Emran, M., Ali, J.H., Valeri, M., Alnoor, A., Hussien, Z.A. (eds.) IMDC-IST 2024. LNNS, vol. 895, pp. 1–20. Springer, Cham (2023). https://doi.org/10.1007/978-3-031-51716-7_1

Javaid, M., Haleem, A., Singh, R.P.: ChatGPT for healthcare services: an emerging stage for an innovative perspective. BenchCouncil Trans. Benchmarks Stand. Eval. **3**(1), 100105 (2023)

Jeffreys, M.R.: Teaching Cultural Competence in Nursing and Health Care: Inquiry, Action, and Innovation. Springer, Heidelberg (2015)

Kavitha, M., Roobini, S., Prasanth, A., Sujaritha, M.: Systematic view and impact of artificial intelligence in smart healthcare systems, principles, challenges and applications. Mach. Learn. Artif. Intell. Healthc. Syst. 25–56 (2023)

Langer, M., et al.: What do we want from Explainable Artificial Intelligence (XAI)?–a stakeholder perspective on XAI and a conceptual model guiding interdisciplinary XAI research. Artif. Intell. **296**, 103473 (2021)

Lee, S.M., Lee, D.: "Untact": a new customer service strategy in the digital age. Serv. Bus. **14**(1), 1–22 (2020)

Lee, S.M., Lee, D.: Opportunities and challenges for contactless healthcare services in the post-COVID-19 Era. Technol. Forecast. Soc. Chang. **167**, 120712 (2021)

Li, P., Bastone, A., Mohamad, T.A., Schiavone, F.: How does artificial intelligence impact human resources performance? Evidence from a healthcare institution in the United Arab Emirates. J. Innov. Knowl. **8**(2), 100340 (2023)

Michaeli, T., Romeike, R., Seegerer, S.: What students can learn about artificial intelligence–recommendations for K-12 computing education. In: Keane, T., Lewin, C., Brinda, T., Bottino,

R. (eds.) WCCE 2022. IFIP Advances in Information and Communication Technology, vol. 685, pp. 196–208. Springer, Cham (2022). https://doi.org/10.1007/978-3-031-43393-1_19

Mikhailova, A.A., Sharova, D.E.: Artificial intelligence ethics code in healthcare. Sustainability of artificial intelligence systems: why do we talk about their impact on the environment? Digit. Diagn. **4**(1S), 93–95 (2023)

Minkiewicz, J., Evans, J., Bridson, K.: How do consumers co-create their experiences? An exploration in the heritage sector. J. Mark. Manag. **30**(1–2), 30–59 (2014)

Mouloodi, S., Rahmanpanah, H., Gohari, S., Burvill, C., Tse, K.M., Davies, H.M.: What can artificial intelligence and machine learning tell us? A review of applications to equine biomechanical research. J. Mech. Behav. Biomed. Mater. **123**, 104728 (2021)

Muhsen, Y.R., Husin, N.A., Zolkepli, M.B., Manshor, N.: A systematic literature review of fuzzy-weighted zero-inconsistency and fuzzy-decision-by-opinion-score-methods: assessment of the past to inform the future. J. Intell. Fuzzy Syst. **45**(3), 4617–4638 (2023)

Musbahi, O., Syed, L., Le Feuvre, P., Cobb, J., Jones, G.: Public patient views of artificial intelligence in healthcare: a nominal group technique study. Digit. Health **7**, 20552076211063680 (2021)

Nahar, K., Akhter, S., Shaturaev, J.: Impact of artificial intelligence on the human resources management (2024)

Nguyen, T.M., Malik, A.: Impact of knowledge sharing on employees' service quality: the moderating role of artificial intelligence. Int. Mark. Rev. **39**(3), 482–508 (2022)

Nicolescu, L., Tudorache, M.T.: Human-computer interaction in customer service: the experience with AI chatbots—a systematic literature review. Electronics **11**(10), 1579 (2022)

Oster, H., Jonze, J.: Employer branding in human resource management: the importance of recruiting and retaining employees (2013)

Pan, S.L., Nishant, R.: Artificial intelligence for digital sustainability: an insight into domain-specific research and future directions. Int. J. Inf. Manage. **72**, 102668 (2023)

Parry, E., Battista, V.: The impact of emerging technologies on work: a review of the evidence and implications for the human resource function. Emerald Open Res. 1(4) (2023)

Pelau, C., Dabija, D.C., Ene, I.: What makes an AI device human-like? The role of interaction quality, empathy and perceived psychological anthropomorphic characteristics in the acceptance of artificial intelligence in the service industry. Comput. Hum. Behav. **122**, 106855 (2021)

Pigola, A., da Costa, P.R., Carvalho, L.C., Silva, L.F.D., Kniess, C.T., Maccari, E.A.: Artificial intelligence-driven digital technologies to the implementation of the sustainable development goals: a perspective from Brazil and Portugal. Sustainability **13**(24), 13669 (2021)

Pillai, R., Sivathanu, B.: Adoption of artificial intelligence (AI) for talent acquisition in IT/ITeS organizations. Benchmarking: Int J. **27**(9), 2599–2629 (2020)

Prentice, C., Nguyen, M.: Engaging and retaining customers with AI and employee service. J. Retail. Consum. Serv. **56**, 102186 (2020)

Prentice, C., Dominique Lopes, S., Wang, X.: Emotional intelligence or artificial intelligence–an employee perspective. J. Hosp. Market. Manag. **29**(4), 377–403 (2020)

Qureshi, M.O., Syed, R.S.: The impact of robotics on employment and motivation of employees in the service sector, with special reference to health care. Saf. Health Work **5**(4), 198–202 (2014)

Rane, N.: Enhancing customer loyalty through artificial intelligence (AI), internet of things (IoT), and big data technologies: improving customer satisfaction, engagement, relationship, and experience. Internet of Things (IoT), and Big Data Technologies: Improving Customer Satisfaction, Engagement, Relationship, and Experience (2023)

Rao, S., Chitranshi, J., Punjabi, N.: Role of artificial intelligence in employee engagement and retention. J. Appl. Manag.-Jidnyasa 42–60 (2020)

Sajwani, R.A.: Artificial intelligence for sustainability development in healthcare. In: Al Marri, K., Mir, F.A., David, S.A., Al-Emran, M. (eds.) BUiD Doctoral Research Conference 2023. Lecture Notes in Civil Engineering, vol. 473, pp. 264–272. Springer, Cham (2024). https://doi.org/10.1007/978-3-031-56121-4_26

Sakka, F., El Maknouzi, M.E.H., Sadok, H.: Human resource management in the era of artificial intelligence: future HR work practices, anticipated skill set, financial and legal implications. Acad. Strateg. Manag. J. **21**, 1–14 (2022)

Salifu, A., Yemofio, D., Kulondwa, S.: Medifem hospital–efficient customer journey/experience. Doctoral dissertation, Department of Business Administration, Ashesi University (2022)

Sapienza, M., Nurchis, M.C., Riccardi, M.T., Bouland, C., Jevtić, M., Damiani, G.: The adoption of digital technologies and artificial intelligence in urban health: a scoping review. Sustainability **14**(12), 7480 (2022)

Schanke, S., Burtch, G., Ray, G.: Estimating the impact of "humanizing" customer service chatbots. Inf. Syst. Res. **32**(3), 736–751 (2021)

Shaheen, M., Zeba, F., Mohanty, P.K.: Can engaged and positive employees delight customers? Adv. Dev. Hum. Resour. **20**(1), 103–122 (2018)

Shahzad, M.F., Xu, S., Naveed, W., Nusrat, S., Zahid, I.: Investigating the impact of artificial intelligence on human resource functions in the health sector of China: a mediated moderation model. Heliyon **9**(11) (2023)

Siala, H., Wang, Y.: SHIFTing artificial intelligence to be responsible in healthcare: a systematic review. Soc Sci Med **296**, 114782 (2022)

Strohm, L., Hehakaya, C., Ranschaert, E.R., Boon, W.P., Moors, E.H.: Implementation of artificial intelligence (AI) applications in radiology: hindering and facilitating factors. Eur. Radiol. **30**, 5525–5532 (2020)

Sweeney, J.C., Frow, P., Payne, A., McColl-Kennedy, J.R.: How does a hospital servicescape impact the well-being and satisfaction of both health care customers and professionals? J. Serv. Mark. **37**(9), 1120–1131 (2023)

Tambe, P., Cappelli, P., Yakubovich, V.: Artificial intelligence in human resources management: challenges and a path forward. Calif. Manage. Rev. **61**(4), 15–42 (2019)

Tewari, I., Pant, M.: Artificial intelligence reshaping human resource management: a review. In: 2020 IEEE international conference on advent trends in multidisciplinary research and innovation (ICATMRI), pp. 1–4. IEEE (2020)

Thomas, A., Uitzinger, D., Chrysler-Fox, P.: Perceptions of human resource professionals of challenges to and strategies for retaining managers. Acta Commercii **18**(1), 1–10 (2018)

Tom Dieck, M.C., Han, D.I.D.: The role of immersive technology in customer experience management. J. Mark. Theory Pract. **30**(1), 108–119 (2022)

Uzir, M.U.H., et al.: Applied artificial intelligence and user satisfaction: smartwatch usage for healthcare in Bangladesh during COVID-19. Technol. Soc. **67**, 101780 (2021)

van der Goot, M.J., Hafkamp, L., Dankfort, Z.: Customer service chatbots: a qualitative interview study into the communication journey of customers. In: Følstad, A., et al. (eds.) CONVERSATIONS 2020. LNCS, vol. 12604, pp. 190–204. Springer, Cham (2020). https://doi.org/10.1007/978-3-030-68288-0_13

Veth, K.N., Korzilius, H.P., Van der Heijden, B.I., Emans, B.J., De Lange, A.H.: Which HRM practices enhance employee outcomes at work across the life-span? Int. J. Hum. Resour. Manag. **30**(19), 2777–2808 (2019)

Victor, J., Hoole, C.: The influence of organisational rewards on workplace trust and work engagement. SA J. Hum. Resour. Manag. **15**(1), 1–14 (2017)

Vrontis, D., Christofi, M., Pereira, V., Tarba, S., Makrides, A., Trichina, E.: Artificial intelligence, robotics, advanced technologies and human resource management: a systematic review. Int. J. Hum. Resour. Manag. **33**(6), 1237–1266 (2022)

Wang, R., et al.: Transparency in persuasive technology, immersive technology, and online marketing: facilitating users' informed decision making and practical implications. Comput. Hum. Behav. **139**, 107545 (2023)

Wassan, S.: How artificial intelligence transforms the experience of employees. Turk. J. Comput. Math. Educ. (TURCOMAT) **12**(10), 7116–7135 (2021)

Wirtz, J.: Organizational ambidexterity: cost-effective service excellence, service robots, and artificial intelligence. Organ. Dyn. **49**(3), 1–9 (2020)

Xu, Y., Shieh, C.H., van Esch, P., Ling, I.L.: AI customer service: task complexity, problem-solving ability, and usage intention. Australas. Mark. J. **28**(4), 189–199 (2020)

Yalabik, Z.Y., Swart, J., Kinnie, N., Van Rossenberg, Y.: Multiple foci of commitment and intention to quit in knowledge-intensive organizations (KIOs): what makes professionals leave? Int. J. Hum. Resour. Manag. **28**(2), 417–447 (2017)

Yang, W.: Artificial Intelligence education for young children: why, what, and how in curriculum design and implementation. Comput. Educ.: Artif. Intell. **3**, 100061 (2022)

Zygiaris, S., Hameed, Z., Ayidh Alsubaie, M., Ur Rehman, S.: Service quality and customer satisfaction in the post pandemic world: a study of Saudi auto care industry. Front. Psychol. **13**, 842141 (2022)

Modelling Intelligent Agriculture Decision Support Tools to Boost Sustainable Digitalization: Evidence from MCDM Methods

Yousif Raad Muhsen[1](✉) and Ahmed Abbas Jasim Al-hchaimi[2] (iD)

[1] Civil Department, College of Engineering, Wasit University, Kut, Wasit, Iraq
yousif@uowasit.edu.iq
[2] Thiqar Technical College, Southern Technical University, Basrah, Iraq
ahmed.alhchaimi@stu.edu.iq

Abstract. Intelligent agriculture decision support tools (IADSTs) are essential tools that help farmers manage their farms under sustainable conditions. However, the assessment of IADSTs is rather complicated; it is difficult for farmers to select suitable and efficient farm-level DSTs because of the many assessment criteria, the fluctuation of data, and the large number of available IADSTs. Thus, this study evaluates and discusses different IADSTs to identify the best one that satisfies the given criteria and assessment at the farm level. MCDM is an approach used in expert systems and artificial intelligence, and many past and current researchers have recommended it for use in selecting the best IADSTs from a group of IADSTs. This study is considered distinct because it succeeded in integrating the Opinion Weight Criteria Method (OWCM) and TODIM to extract the weights of the criteria and rank IADSTs, respectively. The methodology is outlined in three phases: 1 - building the IADST decision matrix, 2 - weighting criteria, and 3 - evaluating the IADSTs. The analysis results showed that the best tool was "FSA," which had the highest IADST score of (1), followed by "RISE 3.0," with a score of 0.84936. This study also provides important recommendations for farmers and policymakers to improve productivity under sustainable conditions.

Keywords: MCDM · OWCM · TODIM · sustainable · decision-making

1 Introduction

Developed citizens see sustainable development as the key concept for passing the global climate change crisis and socioeconomic disparities (Arulnathan et al., 2020). Sustainable development is an ambivalent, normative, and subjective process that cannot be univocally formulated or implemented (Coteur et al., 2020). At its core, it is an ethical issue that involves three important moral imperatives—meeting human needs, social justice, and environmental limits for the well-being of people among and between generations (Trabelsi et al., 2019).

The goal of sustainable decision-making should be based on strong prescriptive, experimental, and quantitative foundations (Alnoor et al., 2024; Smith et al., 2013). In

© The Author(s), under exclusive license to Springer Nature Switzerland AG 2024
A. Alnoor et al. (Eds.): AIRDS 2024, LNNS 1033, pp. 93–105, 2024.
https://doi.org/10.1007/978-3-031-63717-9_6

this way, sustainability indicators are useful to decision-makers in identifying, planning, and evaluating performance (Sadeghfam & Abadi, 2021). More than 500 sustainability indicators have been documented by (Mele et al., 2021). The concept of human well-being is based on food security. Thus, food production and consumption form the core of the majority of sustainable development discourses (Zhang et al., 2021). The scale and intensity of production needed to meet the demands of more than 7 billion people today often result in significant environmental costs to agricultural and food systems (Kareem et al., 2023; Todaro et al., 2021).

Like some sectors, such as Metaverse and financial services (Husin et al., 2023), agriculture is also moving in the direction of agricultural Industry 4.0, influenced by technological advances such as big data, artificial intelligence, cloud computing, and the Internet of Things. These technologies also improve agricultural productivity (Abbasi et al., 2022). It is difficult for stakeholders and farmers to convert a large amount of data into useful information so that they can make sound judgments about agricultural management (Sari & Sari, 2021). They need a set of IADSTs that will enable them to make the right evidence-based decisions. IADSTs are considered a specific class of intelligent information systems that support management decision-making (Duan & Wibowo, 2021). Due to the large number of IADSTs, other issues arise, such as how to choose the optimal one under the umbrella of sustainability. Therefore, it is vital to assess the diversity of IADSTs.

BellagioSTAMP is an acronym for the Bellagio Sustainability Assessment and Measurement Principles, which are new guidelines used to measure and evaluate progress toward sustainability. These principles have been developed because many people have come to realize the need for greater integration of society with the natural environment, as well as the need to maintain the well-being of current and future generations (Arulnathan et al., 2020). BellagioSTAMP offers a complete system for evaluating sustainability on different fronts, including environmental, social, and economic aspects. The principles emphasize the interconnectedness of these dimensions and recommend a comprehensive approach to sustainability assessment. All IADSTs were evaluated in terms of their compatibility with the Bellagio STAMPs. All IADSTs were labelled "Consistent, "Not Consistent," or "Partially Consistent" for all principles of the Bellagio. There are many issues of assessment, including (i) criterion importance and (ii) data variation (Pariwat & Seresangtakul, 2021), that should be resolved in the assessment and benchmarking of IADSTs; therefore, to bridge the identified research gap, many previous and ongoing studies have suggested the use of MCDM (Ali et al., 2024; Bozanic et al., 2023; Chew et al., 2023).

MCDM is one of the most commonly used approaches in expert systems and decision science domains for solving multicriteria decision problems and is particularly concerned with making suitable decisions based on available data (Abbas et al., 2023; Al-Hchaimi et al., 2023; Muhsen, Husin, Zolkepli, Manshor, & Al-Hchaimi, 2023). Compared with traditional methods, the ability of MCDM to improve the quality of choices through a more rational, more transparent, and more efficient process is highly valuable, so the use of MCDM is becoming fast and popular (Al-Enzi et al., 2023; Al-Hchaimi et al., 2022; Khairan et al., 2022; Salih et al., 2022). Over the last few decades, many discrete problem-solving techniques have been developed; each has a specific objective in solving discrete problems and can be applied in multiple domains, such as engineering, economics, management, sociology (Al-Enzi et al., 2023; Husin et al., 2023; Kareem et al., n.d.; Muhsen, Husin, Zolkepli, & Manshor, 2023; Muhsen, Husin, Zolkepli, Manshor, Al-Hchaimi, et al., 2023). This paper prescribes the evaluation of different IADSTs to realize sufficient and appropriate sets and types of pluses and minuses that satisfy the criteria in the model. The OWCM MCDM method is one of the most modern and effective technologies for achieving a set of weights (Ahmed et al., 2024), and it was able to remove comparisons, define fair and easily understandable comparisons, and exclude inconsistencies.

The initial purpose of OWCM was to measure nine assessment criteria of machine learning algorithms (Ahmed et al., 2024). The disadvantage of OWCM is that an external method needs to be used to assign a rank of alternatives. The TODIM method is a famous method of decision-making and was presented by Gomez and Lima. This approach is associated with the prospect theory (PT) developed by Leoneti (Leoneti & Gomes, 2021), where its value function is presented in terms of gains and losses from a psychologically neutral reference point (Singh et al., 2022).

Over the years, the TODIM method has attracted attention from the scientific community (Alali & Tolga, 2019; He et al., 2021; Wu et al., 2022) and has been used in several multicriteria problems. For instance, the TODIM method was applied in selecting the locations for rented residential properties in Volta Redonda, Brazil (Gomes, 2009), in situating natural gas reserves discovered in the Alberta basin, Brazil, and in ranking chemical industries and thermal power station. This article makes the following major contributions.

1. This study addresses the gap in assessing IADSTs and determines the optimal IADST.
2. This paper is the first to use an MCDM methodology in the context of sustainability to evaluate IADST performance.
3. This research presented a new formulation of weighted criteria called the OWCM.
4. The present study is the first to combine OWCM and FDOSM.

2   Methodology

The methodology of this paper includes 3 phases (see Fig. 1). The first phase involves discovering decision-making, the second phase involves weighing the criteria, and the last phase involves ranking the alternatives.

2.1   Decision Matrix

Within this study, our decision matrix is a structured framework that incorporates eight separate criteria developed from the BellagioSTAMP methodology, focusing on comprehensive analysis and evaluation. These criteria cover different dimensions important for our investigation, from technology feasibility to socioeconomic aspects (Arulnathan et al., 2020). In addition to these criteria, we critically consider nine other smart agriculture decision support tools, which themselves provide different methods and functions for dealing with the issues at hand. Through the juxtaposition of these alternatives with

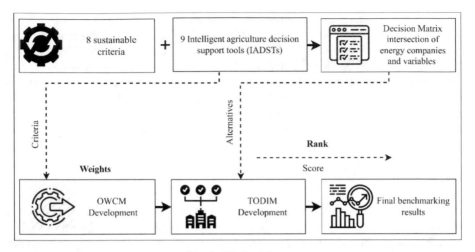

Fig. 1. Methodology.

the given criteria, we strive to identify the best solution that implies efficient reconciliation with the complexity of agricultural decision-making. With our systematic methods, we seek to provide informed decisions that fuel innovation and sustainability in the agriculture field. Table 1 shows the description of the main criteria.

Table 1. Criteria description (Arulnathan et al., 2020)

No:	Criteria	Brief
1	Guiding vision	Achieving well-being in light of capabilities within the capacity of the biosphere helps in assessing sustainability
2	Essential considerations	Sustainability assessment must relate to the decision-making process, drivers of change, risks, multiple dimensions of sustainability, and uncertainties
3	Adequate scope	The time horizon of decisions is a vital factor in assessing sustainability
4	Frameworks and indicators	Effective sustainability assessment is based on pre-defined indicators with clear standards and objectives
5	Transparency	A trustworthy sustainability assessment depends on transparency and information that can be accessed by stakeholders
6	Effective communication	Relying on objective information and using visual tools helps interpret the results and increases the reliability of the sustainability assessment

(*continued*)

Table 1. (*continued*)

No:	Criteria	Brief
7	Broad participation	Sustainability assessment based on appropriate opinions and methods that reflect public participation is an essential factor
8	Continuity and capacity	Sustainability assessment must be flexible and capable of being reconsidered and changed based on improvement processes and different environmental changes

Different reviews on the issue of IADSTs were applied to identify papers on integrated intelligent agriculture. To narrow the field, the following criteria were utilized:

- Only instruments were checked with Excel and the Web.
- Only IADSTs developed in the last ten years were included.
- Received IADSTs with an emphasis on either the environmental aspect of sustainability.
- Only IADSTs are available for free.
- The digital services directorates worked only in the agri-food sector.

Finally, only 9 IADSTs were used in our paper, as shown in Table 2.

Table 2. Names of IADSTs

Scope of application	NO: -	Tools name
Agri-food sector	1	COMET-Farm
	2	FSA
	3	IFSM
	4	SAFA Tool
	5	RISE
	6	SMART Farm Tool
	7	SENSE Tool
	8	CCAFS-MOT
	9	Cool Farm Tool

The decision matrix is organized from the intersection of criteria and alternatives, as shown in Table 3.

2.2 Weighting Criteria

This phase provides the construction of the suggested weighting method using new MCDM approaches. The OWCM served as the weighting criterion. The next subsections outline the major steps (Ahmed et al., 2024).

Table 3. Decision matrix

Agri-food secto	Principle 1	Principle 2	Principle 3	Principle 4	Principle 5	Principle 6	Principle 7	Principle 8
COMET-Farm 2.2	0	0.5	1	0.5	1	1	0	0.5
Farm Sustainability Assessment (FSA)	1	0.5	1	1	1	1	0.5	1
Integrated Farm Simulation Model (IFSM) 4.4	0	1	1	0.5	1	1	0	1
SAFA Tool 2.2. 40	1	1	0.5	1	0.5	1	0.5	1
RISE 3.0	1	1	1	1	1	1	0.5	1
SMART Farm Tool	1	1	1	1	1	1	0	1
SENSE	0	0.5	1	1	1	0.5	0	1
CCAFS-MOT	0	0.5	1	0.5	1	1	0.5	1
Cool Farm Tool	0	0	1	0.5	1	1	0.5	1

* (1 = Full consistency, 0.5 = Partial consistency, 0 = No consistency).

Step 1: Utilizing the five-point Likert scale, translate the linguistic terms to crisp values (see Table 4).

Table 4. The linguistic terms

Linguistic terms	Crisp value
NO_D	1
S_D	2
DI	3
B_D	4
H_D	5

Step 2: Normalize the crisp decision matrix to standardize it using Eq. (1):

$$R_{ij} = \frac{x_{ij}}{x_j^{max}} \tag{1}$$

Step 3: Calculate the average score of the standardized decision matrix via Eq. (2).

$$N = \frac{1}{N} \sum_{i=1}^{m} R_{ij} \tag{2}$$

Step 4: Equation (3) establishes the extent of preference variation and its associated numerical value.

$$\varnothing j = \sum_{i=1}^{m} \left[R_{ij} - N \right]^2 \tag{3}$$

Step 5: The next Eq. (4) is used to determine the deviation of the preference values.

$$\Omega_j = 1 - \varnothing j \tag{4}$$

Step 6: The criteria weight is identified using Eq. (5).

$$w_j = \frac{\Omega_j}{\sum_{j=1}^{n} \Omega_j} \tag{5}$$

The total number of weights for the criteria should be 1. The result of the weighting method will be used with the ranking method (TODIM) to obtain the final ranking.

2.3 Ranking Alternatives

The primary principle of the TODIM methodology is to determine the dominance of an individual alternative or degree of dominance over others with the help of the expected value function. That is, this approach measures the partial and total degree of dominance of each alternative over the other and ranks the alternatives. The calculations of the field steps in the TODIM method are given below (Sun et al., 2023).

Step 1: The decision matrix S_{ij} was normalized by useful parameters, which were suggested by high values calculated by Eq. (6). The proposed unhelpful criteria, the low values determined by Eq. (7), are as follows:

$$P_{ij} = \frac{s_{ic}}{\sum_{i=1}^{n} s_{ic}} \tag{6}$$

$$P_{ij} = \frac{1/s_{ic}}{\sum_{i=1}^{n} 1/s_{ic}} \tag{7}$$

where $P_{ij} =$ is the normalized value of S_{ij}.
Step 2: For the reference criteria, the relative weights of the criteria are calculated via Eq. (8).

$$w_{cr} = \frac{w_c}{w_r} \tag{8}$$

where $w_j =$ are the priority weights, $w_{cr} =$ are the relative weights, $w_r =$ are the weights of the reference criteria,
Step 3: The dominance degree of A_i over alternative A_j is determined using Eq. (9).

$$\left(A_i, A_j \right) = \sum_{c=1}^{m} \Phi_c \left(A_i, A_j \right) \tag{9}$$

$$\Phi_c\left(A_i, A_j\right) = \begin{cases} \sqrt{\frac{w_{c(r}\left(P_{ic}-P_{jc}\right)}{\sum_{c-1}^{m} w_{cr}}} \, if\left(P_{ic} - P_{jc}\right) > 0 \\ 0 \, if\left(P_{ic} - P_{jc}\right) = 0 \\ \frac{-1}{\theta}\sqrt{\frac{\sum_{c-1}^{m} w_{c\sigma}\left)\left(P_{ic}-P_{jc}\right)}{w_{cr}}} \, if\left(P_{ic} - P_{jc}\right) < 0 \end{cases} \tag{10}$$

In the previous conditions, $\left(P_{ic} - P_{jc}\right) > 0$ and $\left(P_{ic} - P_{jc}\right) < 0$, respectively denote the profit and loss of the ith alternative over the jth alternative, and θ represents the attenuation factor of the loss. In the negative quadrant, different values of θ provide different values of the prospect value function.

Step 4: The complete dominance degree of alternative $A_i(\xi_i)$ is obtained by Eq. (11).

$$\xi_i = \frac{\sum_{j=1}^{n}\left(A_i, A_j\right) - \min \sum_{j=1}^{n}\left(A_i, A_j\right)}{\max \sum_{j=1}^{n}\left(A_i A_j\right) - \min \sum_{j=1}^{n}\left(A_i, A_j\right)} \tag{11}$$

Step 5: In the final step, all the alternatives are ranked based on the dominance values, and the alternative with the higher value is the best.

3 Discussion Results

This section provides the evaluation and discrimination results of the IADSTs. The part is subdivided into two subparts. The "Results of criteria weighting" section illustrates the outcomes of the OWCM approach and the criteria weights created; in particular, expert judgment is transformed into overall weights within this section using mathematical conversions. The "Results of alternative ranking" section then shows the ultimate results of IADSTs based on the TODIM.

3.1 Criteria Weighting Results

The OWCM approach used for weight determination is a strong method that considers the relationships between criteria. The OWCM approach provides a detailed understanding of the interrelations and the importance of different criteria in decision-making by employing analysis and expert evaluation. Using Eqs. (1–5), one obtains the total number of criteria weights. The results are shown in Table 5.

Table 5. Final Weight

Criteria	Guiding vision	Essential considerations	Adequate scope	Frameworks and indicators	Transparency	Effective communication	Broad participation	Continuity and capacity
Weighting	0.0216	0.2227	0.1515	0.1113	0.1531	0.1515	0.0348	0.1531

The importance of the criteria in Table 5 reflects the relative importance of each criterion in assessing or evaluating IADSTs. A criterion with a higher weight is considered more important in the overall assessment. Guidance vision has the least weight (0.0216), which means that it is not as important as the other criteria. Fundamental considerations

(0.2227) and the appropriate range (0.1515) have higher values, indicating that they are more critical factors. Other important considerations, but to a somewhat lesser extent, are frameworks and indicators (0.1113), transparency (0.1531), and effective communication (0.1515). Although less prominent but still notable features are included, broad participation (0.0348) and continuity and capacity (0.1531) are among these neglected features.

It is recommended that developers focus on parameters with larger weights. In such cases, more attention should be given to the key considerations of satisfactory coverage, openness, and effective communication, as they carry greater weight in the overall evaluation process. Each criterion is important to some extent, but allocating resources and effort according to the evaluation of weights facilitates a more efficient and effective development process. In the next part, the final score is used together with the TODIM to determine the final ranking.

3.2 IADST Ranking Results

The TODIM technique is an effective tool for decision-making, particularly in situations where several criteria are involved. It offers an organized and methodical method of analysing alternatives and ranking them in terms of their performance across different criteria. Use the above equations in the TODIM section. Table 6 shows the final ranking of this process, with the results of the evaluation of the alternatives according to performance against the specified criteria. Table 6 represents a rapid view of each alternative, their rank and score, showing how all options compare to each other.

Table 6. Final rank.

Agri-food sector		Scour	Rank
1	COMET-Farm 2.2	0.561487	7
2	Farm Sustainability Assessment (FSA)	1	1
3	Integrated Farm Simulation Model (IFSM) 4.4	0.551416	8
4	SAFA Tool 2.2. 40	0.577377	6
5	RISE 3.0	0.849362	2
6	SMART Farm Tool	0.736701	5
7	SENSE	0	9
8	CCAFS-MOT	0.814714	3
9	Cool Farm Tool	0.807339	4

The final results give clear expression to IADSTs that take into account sustainability factors. The results of the alternatives ranking confirm that Farm Sustainability Assessment (FSA) and RISE 3.0 received the highest rating with score (1), while SENSE was at the end of the ranking in terms of sustainable performance with score (0). The rankings provide important insights into the advantages and disadvantages of each alternative in the agri-food sector. This data can be utilized by decision-makers to prioritize the options that best fit their needs, preferences, and objectives. Based on results, the author recommend that farmers use tools with high rank. In addition, designers should develop the rest of the tools based on sustainability factors. The final outcome is to pass validation.

4   Validation

The study frequently uses validation methods, including sensitivity analysis, to assess the validity and applicability of the results (Ahmed et al., 2024; Ali et al., 2024; Muhsen, Husin, Zolkepli, Manshor, Al-Hchaimi, et al., 2023). The researchers of the current study performed an analysis to evaluate the resilience of the desirable outcome by altering the proportions of the assessment criteria. The primary goal of validation is to evaluate the impact of the criteria weight modifications on the final ranking of IADST.

The first step is to define the highest important criteria. According to the data provided in the Table 5, the weighing of the criteria was analyzed, and it was determined that "REssential considerations" was the most vital criteria that should be taken into account. Following previous studies, the researchers increased their weight by 0.5, which is much the same as that of former academic interrogations. According to an analysis of 8 criteria, Fig. 2 shows ten possible results. The remaining criteria were weighted by using the equation below.

$$w_n : (1 - w_{z1}) = w_n^* : \left(1 - w_{z1}^*\right) \tag{12}$$

Fig. 2. Sensitivity analysis

The results of the sensitivity analysis that was carried out to determine the effect of changing the weights assigned to the criteria on the rankings of different alternatives are presented in Fig. 2. In the "initial", the rankings are shown according to the initial set of criteria weights. The rankings are arranged in columns (S1 to S10) after changing the weights in different scenarios. The final ranking observation can be performed by comparing the rankings using a number of schedules to assess the effect of changes in weights assigned to the criteria. This is the case in the second scenario, in which the weights are altered in a way different from the original weights, which results in changes in the rankings of some alternatives. Figure 2 allows for a visual comparison of the rankings under different scenarios so that researchers or even decision-makers can understand how sensitive the rankings are to changes in the weights of the criteria. The final result shows a slight change in the final order, confirming the effectiveness of the methodology applied by weighting and ranking.

5 Conclusion

This study aimed to evaluate nine IADST systems to identify the most appropriate and efficient tools for helping farmers manage their farms in the context of global climate change and sustainability challenges. This research focused on the assessment and quantification of IADST problems using the modern MCDM methods OWCM and TODIM. Eight significant criteria were used to assess nine IADSTs. The methodology process is outlined in three phases: 1- building the IADST decision matrix, 2- weighting criteria, and 3- evaluating the IADSTs. The study revealed that "FSA" had the highest IADST score (1), followed by "RISE 3.0", with a score of 0.84936. The assessment results of the IADST were conducted using sensitivity analysis and were found to be valid. The evaluation of IADSTs has many implications for scholars and practitioners. The assessment of IADSTs is useful for identifying the most valuable agricultural tool. Therefore, more research needs to examine the determinants of IADSTs. This study seeks to identify the most powerful IADSTs with sustainability. This study provided farmers with guidelines on how to choose the best tools to reduce greenhouse gases. Moreover, this research opens an avenue for further studies in the agriculture industry aimed at the exploration and promotion of the sustainability commitment strategy and its potential effect on Industry 4.0. Our research provides a guide for policymakers and governments on how to minimize food wastage and costs, thereby helping them fight hunger crises in different countries.

References

Abbas, S., et al.: Antecedents of trustworthiness of social commerce platforms: a case of rural communities using multi group SEM & MCDM methods. Electron. Commer. Res. Appl. 101322 (2023)

Abbasi, R., Martinez, P., Ahmad, R.: The digitization of agricultural industry – a systematic literature review on agriculture 4.0. Smart Agric. Technol. **2**, 100042 (2022). https://doi.org/10.1016/j.atech.2022.100042

Ahmed, A.D., Salih, M.M., Muhsen, Y.R.: Opinion weight criteria method (OWCM): a new method for weighting criteria with zero inconsistency. IEEE Access (2024)

Al-Enzi, S.H.Z., Abbas, S., Abbood, A.A., Muhsen, Y.R., Al-Hchaimi, A.A.J., Almosawi, Z.: Exploring Research Trends of Metaverse: A Bibliometric Analysis. In: Al-Emran, M., Ali, J.H., Valeri, M., Alnoor, A., Hussien, Z.A. (eds.) IMDC-IST 2024. LNNS, vol. 895, pp. 21–34. Springer, Cham (2023). https://doi.org/10.1007/978-3-031-51716-7_2

Al-Hchaimi, A.A.J., et al.: A comprehensive evaluation approach for efficient countermeasure techniques against timing side-channel attack on MPSoC-based IoT using multi-criteria decision-making methods. Egypt. Inform. J. **24**(2), 351–364 (2023)

Al-Hchaimi, A.A.J., Sulaiman, N.B., Mustafa, M.A.B., Mohtar, M.N.B., Mohd, S.L.B., Muhsen, Y.R.: Evaluation approach for efficient countermeasure techniques against denial-of-service attack on MPSoC-based IoT using multi-criteria decision-making. IEEE Access (2022)

Alali, F., Tolga, A.C.: Portfolio allocation with the TODIM method. Expert Syst. Appl. **124**, 341–348 (2019)

Ali, J., Hussain, K.N., Alnoor, A., Muhsen, Y.R., Atiyah, A.G.: Benchmarking methodology of banks based on financial sustainability using CRITIC and RAFSI techniques. Decis. Making: Appl. Manage. Eng. **7**(1), 315–341 (2024)

Alnoor, A., Chew, X., Khaw, K.W., Muhsen, Y.R., Sadaa, A.M.: Benchmarking of circular economy behaviors for Iraqi energy companies based on engagement modes with green technology and environmental, social, and governance rating. Environ. Sci. Pollut. Res. **31**(4), 5762–5783 (2024)

Arulnathan, V., Heidari, M.D., Doyon, M., Li, E., Pelletier, N.: Farm-level decision support tools: a review of methodological choices and their consistency with principles of sustainability assessment. J. Clean. Prod. **256** (2020). https://doi.org/10.1016/j.jclepro.2020.120410

Bozanic, D., Tešić, D., Puška, A., Štilić, A., Muhsen, Y.R.: Ranking challenges, risks and threats using fuzzy inference system. Decis. Making: Appl. Manage. Eng. **6**(2), 933–947 (2023)

Chew, X., Khaw, K.W., Alnoor, A., Ferasso, M., Al Halbusi, H., Muhsen, Y.R.: Circular economy of medical waste: novel intelligent medical waste management framework based on extension linear Diophantine fuzzy FDOSM and neural network approach. Environ. Sci. Pollut. Res. 1–27 (2023)

Coteur, I., Wustenberghs, H., Debruyne, L., Lauwers, L., Marchand, F.: How do current sustainability assessment tools support farmers' strategic decision making? Ecol. Indicat. **114**, 106298 (2023). https://doi.org/10.1016/j.ecolind.2020.106298

Duan, S.X., Wibowo, S.: A multicriteria analysis approach for evaluating the performance of agriculture decision support systems for sustainable agribusiness (2021)

Gomes, L.F.A.M.: An application of the TODIM method to the multicriteria rental evaluation of residential properties. Eur. J. Oper. Res. **193**(1), 204–211 (2009)

He, S., Pan, X., Wang, Y.: A shadowed set-based TODIM method and its application to large-scale group decision making. Inf. Sci. **544**, 135–154 (2021)

Husin, N.A., Abdulsaeed, A.A., Muhsen, Y.R., Zaidan, A.S., Alnoor, A., Al-mawla, Z.R.: Evaluation of metaverse tools based on privacy model using fuzzy MCDM approach. In: Al-Emran, M., Ali, J.H., Valeri, M., Alnoor, A., Hussien, Z.A. (eds.) IMDC-IST 2024. LNNS, vol. 895, pp. 1–20. Springer, Cham (2023). https://doi.org/10.1007/978-3-031-51716-7_1

Kareem, B.A., Zubaidi, S.L., Al-Ansari, N., Muhsen, Y.R.: Review of recent trends in the hybridisation of preprocessing-based and parameter optimisation-based hybrid models to forecast univariate streamflow (n.d.)

Khairan, H.E., Zubaidi, S.L., Muhsen, Y.R., Al-Ansari, N.: Parameter optimisation-based hybrid reference evapotranspiration prediction models: a systematic review of current implementations and future research directions. Atmosphere **14**(1), 77 (2022)

Leoneti, A.B., Gomes, L.F.A.M.: A novel version of the TODIM method based on the exponential model of prospect theory: the ExpTODIM method. Eur. J. Oper. Res. **295**(3), 1042–1055 (2021)

Mele, A., Paglialunga, E., Sforna, G.: Climate cooperation from Kyoto to Paris: what can be learnt from the CDM experience? Soc.-Econ. Plann. Sci. **75**, 100942 (2021). https://doi.org/10.1016/j.seps.2020.100942

Muhsen, Y.R., Husin, N.A., Zolkepli, M.B., Manshor, N.: A systematic literature review of fuzzy-weighted zero-inconsistency and fuzzy-decision-by-opinion-score-methods: assessment of the past to inform the future. J. Intell. Fuzzy Syst. 1–22 (2023a)

Muhsen, Y.R., Husin, N.A., Zolkepli, M.B., Manshor, N., Al-Hchaimi, A.A.J.: Evaluation of the routing algorithms for NoC-based MPSoC: a fuzzy multi-criteria decision-making approach. IEEE Access (2023b)

Muhsen, Y.R., Husin, N.A., Zolkepli, M.B., Manshor, N., Al-Hchaimi, A.A.J., Ridha, H.M.: Enhancing NoC-based MPSoC performance: a predictive approach with ANN and guaranteed convergence arithmetic optimization algorithm. IEEE Access (2023)

Pariwat, T., Seresangtakul, P.: Multi-stroke Thai finger-spelling sign language recognition system with deep learning. Symmetry **13**(2), 262 (2021)

Sadeghfam, S., Abadi, B.: Decision-making process of partnership in establishing and managing of rural wastewater treatment plants: using intentional and geographical-spatial location data. Water Res. **197**, 117096 (2021). https://doi.org/10.1016/j.watres.2021.117096

Salih, M.M., Al-Qaysi, Z.T., Shuwandy, M.L., Ahmed, M.A., Hasan, K.F., Muhsen, Y.R.: A new extension of fuzzy decision by opinion score method based on Fermatean fuzzy: a benchmarking COVID-19 machine learning methods. J. Intell. Fuzzy Syst. 1–11 (2022)

Sari, F., Sari, F.K.: Multi criteria decision analysis to determine the suitability of agricultural crops for land consolidation areas. Int. J. Eng. Geosci. **6**(2), 64–73 (2021). https://doi.org/10.26833/ijeg.683754

Singh, R.R., Maity, S.R., Zindani, D.: Using the TODIM method as a Multi-Criteria Decision-Making support methodology for an automobile parts manufacturing company. Mater. Today: Proc. **62**, 1294–1298 (2022)

Smith, T.F., Thomsen, D.C., Gould, S., Schmitt, K., Schlegel, B.: Cumulative pressures on sustainable livelihoods: coastal adaptation in the Mekong delta. Sustainability **5**(1), 228–241 (2013). https://doi.org/10.3390/su5010228

Sun, H., Yang, Z., Cai, Q., Wei, G., Mo, Z.: An extended Exp-TODIM method for multiple attribute decision making based on the Z-Wasserstein distance. Expert Syst. Appl. **214**, 119114 (2023)

Todaro, F., Barjoveanu, G., De Gisi, S., Teodosiu, C., Notarnicola, M.: Sustainability assessment of reactive capping alternatives for the remediation of contaminated marine sediments. J. Clean. Prod. **286**, 124946 (2021). https://doi.org/10.1016/j.jclepro.2020.124946

Trabelsi, M., Mandart, E., Le Grusse, P., Bord, J.P.: ESSIMAGE: a tool for the assessment of the agroecological performance of agricultural production systems. Environ. Sci. Pollut. Res. **26**(9), 9257–9280 (2019). https://doi.org/10.1007/s11356-019-04387-9

Wu, Q., Liu, X., Qin, J., Zhou, L., Mardani, A., Deveci, M.: An integrated generalized TODIM model for portfolio selection based on financial performance of firms. Knowl.-Based Syst. **249**, 108794 (2022)

Zhang, Y., et al.: Sustainable ex-situ remediation of contaminated sediment: a review. Environ. Pollut. **287**, 117333 (2021). https://doi.org/10.1016/j.envpol.2021.117333

"The Role of AI Applications in the Advertising and Design in Social Media Marketing: A Stride in the Direction of Marketing Sustainability"

Hadi AL-Abrrow[1], Nadia Atiyah Atshan[2], and Ali Said Jaboob[3(✉)]

[1] Department of Business Administration, College of Administration and Economics, University of Basrah, Basrah, Iraq
[2] Technical College of Management, Middle Technical University, Baghdad, Iraq
nadia.atshan@mtu.edu.iq
[3] Dhofar University, Salalah, Oman
ajaboob@du.edu.om

Abstract. The opportunities that AI applications for design and social media advertising present to managers, clients, and marketing and advertising companies have become a topic of great discussion as a result of their widespread adoption. This paper proposes a new model that explores the relationship between AI applications and social media advertising and addresses issues related to AI and social media marketing. 420 workers from Iraqi beauty centers who were chosen at random made up the research sample. Additionally, twenty managers of beauty centers participated in unstructured interviews. A checklist for examining the possible advantages of the connection between AI applications and social media marketing is also provided by the study to researchers.

Keywords: AI applications · Design · Advertising · Social Media Marketing · sustainability

1 Introduction

In the long run, artificial intelligence (AI) will become a crucial component of all businesses worldwide. Significant changes in the AI landscape are reflected in the new trends in AI-driven automation. This is apparent in the reorganized concepts, pursuits, and financial commitments made by the company in the area of AI adoption (Basri, 2020). Everyone is aware that the world is transitioning from manual to digital transmission at this time. Things are changing quickly in both their functionality and movement (Capatina et al., 2020; Al-Enzi et al., 2023). In the world of computers, the term "artificial intelligence" has gained rapid popularity. "Artificial intelligence" is a term that was first used in Computer Science in 1956 by John McCarthy, an American computer expert (Liu et al., 2023).

AI adoption by marketing managers and consumers is increasing at an exponential rate (Nair & Gupta, 2021). Where the use of artificial intelligence technology is

A. Alnoor et al. (Eds.): AIRDS 2024, LNNS 1033, pp. 106–122, 2024.
https://doi.org/10.1007/978-3-031-63717-9_7

beneficial in social media monitoring because it provides a comprehensive picture of the interactions that take place on social media (Albinali & Hamdan, 2021). This technology is also useful in marketing content based on the habits of social media users. Many studies show that data-driven marketers have emerged thanks to AI, eventually exploring new marketplaces or marketing dimensions based on the insights gained from AI-driven social media marketing (Dwivedi et al., 2021). It has also been documented why the majority of social media marketers chose to use AI despite its seemingly basic principles (Arasu et al., 2020). Several studies agree that understanding human psychology has been the driving force behind the adoption and application of AI-driven social media marketing, with startups and SMEs being no exception (Zhang & Song, 2022). Online AI applications should be used to promote the hedonic and utilitarian value of products and services across various online forums, according to the social media literature. These apps are suggested to play a significant role in the promotion of goods and services (Benabdelouahed & Dakouan, 2020).

Utilizing AI applications in design and social media advertising aims to provide exclusive access to products that combine artificial intelligence, mixed reality, virtual reality, and augmented reality with a focus on innovation for customers (Gkikas & Theodoridis, 2019). The use of AI applications in design and social media advertising aims to provide exclusive access to products that combine artificial intelligence, mixed reality, virtual reality, and augmented reality with a focus on innovation for customers (Haleem et al., 2022). The purpose of this research is to shed light on how artificial intelligence applications can be applied to social media advertising and design, as well as to improve sustainability in the field of marketing these services in a group of Iraqi beauty centers. It also hopes to provide these centers with a foundation for artificial intelligence applications when they begin their marketing research.

2 Literature Review

2.1 Underpinning Theory

This study contextualizes AI-driven social theory by Mökander and Schroeder (2022), which focuses exclusively on the widespread use of AI-based models to solve business problems. Furthermore, the theory asserted that AI-based models can synthesize knowledge from incredible sources. Then, that enhanced knowledge will be meaningful in solving business problems systematically. It is due to the enhanced social understanding of the business due to AI tools that come from the real-time use of massive data and algorithms to synthesize social behaviors on social media platforms (Bjola, 2022) Furthermore, AI technologies potentially enhance the social understanding of organizations by leveraging data and algorithms to analyze social behaviors and interactions on social media platforms further (Cheng & Jiang, 2020).

Moreover, AI-driven social theory emphasizes the usage of social media marketing tools, such as advertising, by transforming the approach in which businesses interact with their audience. In addition, the AI approaches that include data analysis and advanced algorithms will help to get deeper explorations and insights to explore user behavior, experiences, and preferences (Mökander & Schroeder 2022). Henceforth, these advancements allow organizations to devise novel and improved marketing strategies. For example, exceptional accuracy in content and personalized delivery that truly represents the intended audience demographics (Meyer & Lunnay, 2013) In addition, AI tools maximize the reach to the audience by providing real-time content, where the business may get quicker feedback about ongoing trends(Donaldson & Crano, 2011). Ultimately, tools not only improve the overall marketing landscape but significantly improve the relationship between organizations and consumers, fostering brand engagement and loyalty in the long term (Albinali & Hamdan, 2021).

2.2 Artificial Intelligence Leads to Sustainability in Various Contexts

AI has gained significant and dramatic attention worldwide from almost every sector (Simay et al., 2023) and is believed to be a game-changer in driving innovation and efficiency of organizations at strategic and operational levels as well. In this view, Nazir et al. (2023) explained in their recent study that AI pertains to machines that simulate human intellect to carry out everyday routine tasks such as learning, problem-solving, and planning. Similarly, AI-based systems induced emerging insights in a wide range of sectors, including education, hospitality, healthcare, and business performance, and especially uplifted the social media role in various online platforms, that's led to improving sustainability (Esmaeilzadeh, 2020; Imran et al. 2020; Mishra et al. 2022; Fathahillah et al. 2023; Nazir et al. 2023; Terranova et al. 2024).

The study of Hopcan et al. (2023) highlights that AI has a significant role in enhancing teaching styles with multiple learning experiences in various contexts. Interestingly, personalized learning is the considerable contribution of AI, where AI algorithms assess the data of individual candidates to provide instructions according to their strengths, weaknesses, and learning styles. This personalized approach assures that every learner advances at their ideal pace; thereby, learning takes place at their optimum level. In the same way, the hospitality sector has a different use of AI-based software that could be utilized differently, such as empowering their clients to make the best decision based on sufficient data through AI applications (Nazir et al. 2023). Fostering sustainability through interaction with stakeholders, and improving the customer experience. Furthermore, Esmaeilzadeh (2020) explored how AI-based applications work in health care and provide an immediate course of action to cure patients. Similarly, patients can obtain customized medical advice and information remotely through virtual health assistants that utilize natural language processing, encouraging patient participation and self-management of health (Terranova Cestonaro et al. 2024).

Firm performance is one of the eminent phenomena of businesses that need to be enhanced at their optimum level with their different ways forward (Mishra, Ewing et al. 2022; Alnoor et al., 2023). AI is considered a prominent and pivotal distinct technological system that makes businesses more competitive in the digital world (Wamba-Taguimdje, Wamba et al. 2020). Studies have shown that AI can significantly influence firm performance by improving efficiency, decision-making processes, and customer service. By leveraging AI-based transformation projects, firms can enhance their business value, leading to improved performance metrics (Wamba-Taguimdje et al., 2020; Chen et al., 2022; Atiyah et al., 2023). In the same context, Imran, Ofli et al. (2020)

found that the field of social media marketing is predicted to witness a widespread use of novel AI technologies, making the integration of AI a big opportunity for marketers. AI-driven software for social media marketing uses human-like cognitive powers to evaluate messages exchanged between businesses and prospective clients, boosting presentation quality and increasing client interaction (Garg & Pahuja 2020; Alnoor et al., 2024; Ali et al., 2024).

2.3 Artificial Intelligence Application Positively and Using Design in Social Media Marketing.

AI applications involve the use of AI technologies to perform tasks that typically necessitate human intelligence. According to Duan, Edwards et al. (2019), AI-based products have widespread applications and use in different fields and industries that show immense transformational usefulness across various products. In the context of social media marketing, it is important to devise a plan that contains attractive and vibrant designs that not only capture the social media fraternity but also build a positive image of the business (Muhsen et al., 2023). On the same note, the proper use of graphics and visually tempting design captivates the audience's attention and is also considered an effective way to transfer messages to viewers. (Micu, Capatina et al. 2018, Capatina, Kachour et al. 2020; Alnoor et al., 2024). Moreover, a recent study by Lyu, Wang et al. (2022) revealed that vibrant colors, designs, and visuals play an important role in attaining and engaging user attention in transmitting product information. With specific to increase brand recognition, Mitrović and Holland (2020) suggested that creative design is critical to creating consistent brand identities across the channels. From the consumer perspective, the visualizations in the form of videos, images, and other visual content increased positive brand perceptions, thus improving consistent engagement(Mitrović & Holland 2020; Atiyah, 2023; Abbas et al., 2023).

From these studies, it is, therefore, deduced that AI-powered social media marketing design and strategies help to improve more customized user experiences and provide novel ways to integrate with consumers' preferences (Alnoor et al., 2024; Husin et al., 2023). In this way, Shaik Vadla et al. (2024) stated that marketplaces like Amazon use AI technologies to attain considerable amounts of consumer data to develop an understanding of their customer's preferences, behaviors, trends, and mindset and make strategies to perform targeted advertisements that directly pitch social media users. Based on these research-oriented arguments, the present study anticipated that the integration of AI-driven design strategies into Amazon's social media marketing would boost conversion rates, improve user engagement metrics, and improve brand perception, all of which will contribute to increased profitability and competitiveness in the market. Based on the literature as mentioned earlier, the present study hypothesized that:

H1: Artificial Intelligence Application positively leads to using design in social media marketing and improves marketing sustainability.

2.4 Artificial Intelligence Applications Positively Lead to the Use of Advertising in Social Media Marketing.

Considering its potential to improve advertising efficacy, the introduction of AI applications into social media marketing has gained significant attention from social media users (Capatina et al. 2020; Atiyah et al., 2023). Notably et al. (2024) found that AI technologies can be utilized at their optimum level to provide enough benefits, including audience segmentation, data analysis, and targeted content distribution that improve social media platform advertising techniques and sustainability. Additionally, AI technologies are emerging with updated software that produces revolutionary advertising practices by enabling more efficient and effective audience targeting on social media platforms (Yu, 2022; Atiyah and Zaidan, 2022). Businesses may optimize work automation, improve competitor analysis, and raise the overall efficacy of their marketing efforts by utilizing AI-driven technologies for social media monitoring (Jha, 2024). Organizations can only attain and sustain a competitive advantage in the market through the mode of interactions with their intended audience. Also, the reliance on a data-driven approach to advertising has increased significantly. In this context, Li (2019) stressed that AI tools facilitate businesses and consumers with real-time advertising content where both parties can be responsive and, thus, optimize the user input and improve sustainability in marketing (Atiyah et al., 2023).

Consequently, these customized tools and techniques significantly improve the accuracy of user searches and may reach conscious clicks. Thus, it positively influences brand exposure, new client recruitment, and achieve overall financial gain. Consequently, the literature witnessed that the integration of AI tools with the social media marketing landscape dramatically benefits consumers through accurate results, and related content, and also results favorable in the account of businesses.

Henceforth, the present study hypothesized as follows:

H2: Artificial Intelligence Application positively leads to using advertising in social media marketing and improve marketing sustainability.

3   Research Methodology

3.1   Research Design

The primary aim of the research is to ascertain the impact of artificial intelligence applications on design and advertising through social media marketing in beauty centers in Iraq. It is noteworthy that the easy and affordable access to information provided by the integration of secondary data into primary data is advantageous (Abbott & McKinney, 2013). When secondary data are combined with primary data to create a qualification (or other) basis, it also guarantees that the conclusions drawn from the combined data are meaningful (Marczyk et al., 2010; Ahmed et al., 2024). Therefore, it is evident that the purpose of secondary data is to inform future research methods, which are accomplished by gathering and analyzing primary data, before determining the accuracy and dependability of the conclusions drawn from the primary research (Groenewald, 2004). It is anticipated that the utilization of primary and secondary data in this study will shed light on the subject in other nations and commercial enterprises with characteristics akin to AI applications in Iraq and that the related findings in these contexts may be comparable to those in the selected research setting.

Inferentially, it has been shown that both primary and secondary data are useful and insightful in that they provide a variety of perspectives on the opinions of the research participants (Berman et al., 2000). Several gaps in the literature were found during the review, which was appropriate given the equivocal nature of the subject of artificial intelligence applications in design and social media advertising in beauty centers in Iraq. Further, the research approach has been linked to the supply of more comprehensive data, which enhances the representativeness of a research process and makes the data valuable for extrapolation to comparable sociocultural, economic, and political contexts as those the current study focuses on.

3.2 Data Collection, Population, and Sample

The study's sample comprised 450 workers from a group of cosmetic centers in Iraq who used artificial intelligence applications for social media advertising and design. 420 of the 450 workers who were contacted overall had their responses taken into account for the study, yielding a response rate of.93.3%. Participants are chosen by purposive sampling from a sizable pool of workers in Iraqi cosmetic centers. The reason for this is that these centers have a high customer demand, which leaves them open to fierce competition. As a result, they must develop their services and the technology they employ in the present. For this reason, artificial intelligence applications are a crucial topic because they provide a means of continuous and quick development across all domains. The variables' measures were created with the current study in mind.

3.3 Data Analysis

After gathering and categorizing primary data according to themes like their demographic characteristics and possible answers to the study's objectives, this research will use both descriptive and inferential statistical techniques. Certainly, these methodologies have been utilized as they facilitate the interpretation of the unprocessed data collected from subjects. Statistical tables, graphs, and charts will therefore be used to present the primary data that is collected. The process of analysis will come before the sections on the discussion, conclusion, and recommendations. Based on questionnaires with a variety of questions, all the variables are measured. The current study used measures that had already been developed. In terms of analysis methods, the respondents' profiles are used to extract results. Most of the time, the five-point Likert scale questions are used to determine the results. Additionally, partial least square-structure equation modeling (PLS-SEM) was used to analyze the data.

4 Result and Discussion

4.1 Primary Data Presentation

Presenting the secondary data had two main goals: it set the stage for gathering first-hand information and figuring out whether there was any concurrence. It is imperative to note that the study was conducted using a survey methodology, with employees of beauty centers in Iraq receiving direct questionnaires. And unstructured interviewing officials in these centers.

4.2   Demographic Information

A total of 450 participants in the study were given questionnaires. 30 of these participants did not send in their completed questionnaires for review and interpretation. It is implied that 420 participants completed the completed questionnaires. Since the study's confidence interval was set at 90%, it can be assumed that the data it collected from enough participants led to statistically significant and pertinent results. The gender of the chosen employees was one of the demographic factors that was looked at.

This option's primary objective was to determine the benefit that artificial intelligence applications in design and advertising offer to employees by forecasting whether these applications will influence social media usage, particularly in light of the role these applications play in beauty centers in Iraq. to ease the working processes. Of the participants, 55.32% of the participants were male, and 44.68% of the participants were female.

4.3   In Relation to the Study's Aim and Objectives

Artificial Intelligence (AI) has emerged as a trend in several fields recently, including education, business, science, medicine, and the automobile, Marketing has also embraced AI (Basri, 2020). Social media marketing can greatly benefit from the use of artificial intelligence (AI) apps. It supports the development of complex and sophisticated algorithms, the proliferation of information and data sources, and the enhancement of software's data management features. Brands and users are interacting differently as a result of artificial intelligence (Liu et al., 2023). With a greater customer-focused approach, marketers can now promptly address customers' needs. Because AI generates and collects data through algorithms, they can quickly decide which channel to use at any given time and what content to target customers with. When AI is used to personalize user experiences, users feel more at ease and are more likely to purchase what is offered (Nair et al., 2021).

The principal aim of the research was to investigate the effects of artificial intelligence applications in advertising and design through social media in Iraqi beauty centers. It featured a sizable cluster of beauty centers that use social media to advertise their services and deal with clients directly. Respondents answered the questionnaire's questions on a Likert scale.

For example, a high percentage of respondents said their beauty center had fully implemented AI social media applications, compared to a somewhat lower percentage who said their beauty center had only partially implemented AI applications. Figure 1 shows the conclusion drawn from the data, which shows the proportion of research participants and their status in beauty centers. Beauty centers in Iraq are using artificial intelligence applications to help with social media marketing. AI applications are being adopted and implemented across social media platforms by most beauty centers, indicating a growing trend in this regard.

According to the results of the current study, there has been a noticeable increase in the use of artificial intelligence applications in social media for advertising and design,

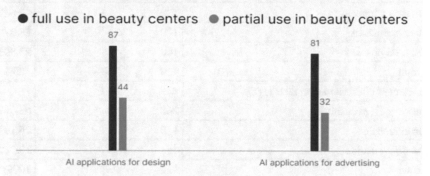

Fig. 1. AI applications and participants and status in beauty centers in Iraq.

particularly in the service industry (Nair & Gupta, 2021). Which plays a crucial role in drawing attention to and advertising the services it offers on social media, which enhances the sustainability of its success (Albinali & Hamdan, 2021). It's also evident that Iraq is using artificial intelligence applications at a rate that is significantly higher than what the current study indicates. Because artificial intelligence applications offer so many benefits, businesses are compelled to address these marketing strategies. Expanding prospects for expansion and financial gain (Zhang & Song, 2022). Social media advertising is greatly impacted by using artificial intelligence applications because these tools offer a multitude of choices for manipulating and displaying images and videos in three dimensions or simulating reality (Benabdelouahed & Dakouan, 2020). Since these programs allow for the addition of numerous features and the use of ready-made templates for design, their use has also had a significant impact on social networking site design (Theodoridis & Gkikas, 2019). And Positive effects accruing from the use of AI applications in beauty centers in Iraq are shown in Table 1.

Table 1. Participant's responses on the Likert scale

Benefits of using AI applications through social media marketing in Iraq	Number of the Participants	Percentage
facilitating work processes and achieving quicker outcomes		

(continued)

Table 1. (*continued*)

Benefits of using AI applications through social media marketing in Iraq	Number of the Participants	Percentage
In strong agreement	200	47.61%
In agreement	120	28.57%
Neutral position	100	23.80%
Total	420	100%
Get more clients and earn their loyalty		
In strong agreement	265	63.09%
In agreement	110	26.19%
Neutral position	45	10.71%
Total	420	100%
increased organizational capacity for organizing		
In strong agreement	241	57.38%
In agreement	132	31.42%
Neutral position	47	11.19%
Total	420	100%
Increased opportunities for competition and marketing on a large scale		
In strong agreement	218	51.90%
In agreement	183	43.57%
Neutral position	19	4.52
Total	420	100%

4.4 Evidence Through Structural Equation Modeling (SEM)

Prior to hypothesis testing, the factor structure of the self-report items measuring the three main constructs in this study—AI applications, using design in social media marketing, and using ads in social media marketing—was evaluated using a confirmatory factor analysis (CFA) (Hair et al., 2010). A reliability analysis using Cronbach's alpha was also conducted to assess the internal consistency of these constructs. Significantly, every construct showed values higher than 0.7, which suggests strong reliability (Hair et al., 2014) (refer to Table 2). The average variance extracted (AVE) and composite reliability (CR) were assessed to determine convergent validity (Fornell and Larcker, 1981). Strong convergent validity was highlighted by both AVE and CR values exceeding the 0.5 threshold, as shown in Table 2.

The matrix of correlation is shown in Table 3. The observation indicates a significant correlation between the independent and dependent variables. Furthermore, there is a strong correlation between social media marketing and AI applications and demographic factors like age, gender, and level of education.

Table 2. Reliability and Validity of Measurement

Construct	Number of items	Range of Loadings	α	C.R	AVE
AI applications	10	0.534–0.810	0.890	0.821	0.506
Using design in social media marketing and improving marketing sustainability	10	0.711–0.903	0.800	0.830	0.559
Using advertising in social media marketing and improving marketing sustainability	10	0.621–0.844	0.811	0.799	0.543

Table 3. The observation indicates a significant correlation

	Mean	S. D	Gender	Education	Age	AI	DSMM	ASMM
Gender	3.45	329	1					
Education	2.22	.520	.054	1				
Age	3.11	.319	.032	.023	1			
AI	3.25	.810	.002	.010	.012	1	1	
DSMM	3.78	.543	.054	.011	.025	.345**	.337**	1
ASMM	3.99	.592	.041	.035	.019	.440**		

Notes: * $p < 0.05$; ** $p < 0.01$; AI = Artificial intelligence applications; DSMM = Design in social marketing; ASMM = Advertising in social media marketing

Table 4 presents an overview of the diverse correlations among the variables being examined.

Table 4. Hypothesis testing

Hypothesis	Path	β	S.E	T-value	P-value	Result
H1	AI→ DSMM	.399	.020	10.323	0.000	Accepted
H2	AI→ ASMM	.384	.017	13.344	0.000	Accepted

All study hypotheses were confirmed by the data displayed in the above table. The first hypothesis, which shows that artificial intelligence applications have a positive effect on the use of design on social networking sites, is supported by the results. The second hypothesis, which shows a positive correlation between artificial intelligence applications and the use of social networking site advertising, was also accepted.

4.5 Construction of Interview Schedule

Interview questions concerning artificial intelligence applications and social media marketing. These items determined the fundamental components that were created for the interview schedule. Twenty managers from Iraqi cosmetic centers were interviewed (refer to Table 5).

Table 5. Construction of Interview Schedule for AI applications and social media marketing.

Categories of AI applications and social media marketing	Item no	Scheduled question	Reference
AI applications	1	Could you tell me about the advantages your company has seen from using AI in social media?	Peters et al., 2020
Social media marketing	2	What steps has your company taken to employ AI applications and engage social media users?	Sanches et al., 2019
design	3	Do you believe that users in advertising and design have benefited from the AI applications used in your organization?	D'Alfonso, 2020
advertising	4	How can you make sure AI isn't harmful to your organization?	D'Alfonso, 2020
	5	Could you describe a set of artificial intelligence tools that your business uses for social media design and advertising?	Peters et al., 2020

These interviews discuss AI applications, which begin with the development of ideas for social media marketing in advertising and design. This paper also discussed the role of AI applications in design and advertising for social media marketing in Iraqi beauty centers. We also examined the validity (i.e., whether the study's findings covered the intended topics), reliability (i.e., consistency of the results), and generalizability (i.e., ability to extrapolate study findings to larger populations) of the interview results.

4.5.1   Analyzing and Discussing Interviews

To protect the respondents' privacy, their names and positions were withheld so that they could freely express their opinions. They were given letters that represented them rather than their names, such as (A, B, C, D). It was discovered that there is wide agreement among the managers of the cosmetic centers in the study sample on the topic of using artificial intelligence applications in the field of design and advertising via social media, as manager (A) expressed on this topic, saying:

"Our center's adoption of artificial intelligence applications was like a magic fix for the majority of the issues we had with sharing our work on social media in the past because it saved time, money, and effort"

Regarding their utilization of a collection of applications related to design and advertising on social media, managers (D and E) responded as follows:

" In fact, we employ a variety of AI tools in advertising and design, including (midjourney, khroma, fronty, adobe sensei, leonardo.ai, hotpot.ai, jasper art)".

In terms of the level of customer confidence in the applications used by the centers in the research sample, Director (B) responded by saying:

"To be honest, we saw a lot of interest in using these apps and sharing our work on social media. We are now completing our work without delay or fear of making additional efforts. Previously, our concerns were about how to create something appropriate and market it. The problem has been solved thanks to artificial intelligence".

These findings lead us to the conclusion that artificial intelligence applications are required at work and that organizations should use these applications to develop their work. The topic extends beyond social media marketing and affects all aspects of work.

5   Implications of the Study

5.1   Theoretical Implications

This paper conducted a methodical discussion and offered a set of recommendations for using artificial intelligence applications in social media design and advertising development. The study then made recommendations about how well the artificial intelligence applications would work in Iraqi beauty centers to deploy a variety of AI applications on social media. A group of officials in these centers are also interviewed to find out how beneficial artificial intelligence applications are in this field.

Regarding this, I thought that the most crucial components of social networking sites were their design and advertising. In addition, this paper made a significant contribution to the applications of AI and how it is relevant from an organizational standpoint to social media marketing. It implies that an organization must use artificial intelligence advertising and design applications, particularly in beauty centers and their services, if it hopes to have a long-lasting relationship with customers, compete on a larger scale, and draw in many clients.

The most significant elements and motivators of customers' propensity to buy goods or services are speed and accuracy (Theodoridis & Gkikas, 2019). This paper examines how an individual's desire to purchase and search for services through social media is influenced by the ease of use of applications. Without secure applications that guarantee users' trust and understanding, it is impossible to obtain high-quality information on social media, particularly when it comes to the exchange of accurate and dependable information (Morandé & Amini, 2023). Applications of artificial intelligence, which show that people are more likely to engage in an organization's social media activities when they perceive potential benefits, are crucial to the use of social media to draw in and keep customers as well as to develop advertising and design.

5.2   Practical Implications

The competition to publish advertisements or design services offered by organizations via social networking sites has become extremely fierce in this era of rapid technological development. According to the current study, social media can effectively use artificial intelligence applications (Stone et al., 2020). The impact of social media in conjunction with AI applications should not be disregarded, and people's reactions, feelings, and thoughts on social media should be closely monitored. As per the views expressed by the interviewees, artificial intelligence technology and its applications allow businesses to prioritize quality and speed when distributing beauty center services to clients, protect copyrights, and use applications' technological capabilities to execute advertising effectively. The endeavors of beauty centers ought to align with the utilization of artificial intelligence and its function in drawing clientele and engaging in wider competition. Our research revealed some new insights into what artificial intelligence applications should offer in terms of elucidating customer preferences and marketing strategies through social networking sites utilized by Iraqi beauty centers. AI applications can foster trust and openness toward website design and management by utilizing AI and related knowledge to meet customized customer desires. Beauty centers can develop various options for AI applications for advertising and design to engage customers from diverse backgrounds in fulfilling their requirements; The goals and design of AI should put the needs of the user first.

5.3   Limitations and Future Directions

Every study has barriers and difficulties that limit the scope of the investigation and the study itself. The hardest part of the current study was interviewing officials at the Iraqi beauty center because they are very busy and reluctant to provide detailed information about their work system, which took a lot of time the work involved in conducting these interviews. The other challenge is that there isn't a deeper discussion of the applications of artificial intelligence in advertising and design via social media in the study, which is why the research isn't generalizable due to a lack of information about the apps used. Future studies can concentrate more intently on a subset of AI applications, particularly via social media, while examining users' trust in these apps. Apart from emphasizing that these applications can be accessed with little financial or effort.

6   Conclusion

Artificial intelligence applications have made design and advertising for social media marketing simple and quick to develop. Furthermore, other facets of marketing are similarly affected by this. not limited to marketing on social media. The primary driver behind consumers' high demand for businesses' services and goods on social networking sites is artificial intelligence. These applications are also regarded as essential components of marketing campaigns because they enable the collection of personal data and the effective use of that data in campaigns, as well as the provision of methods for advertising and promotion that entice consumers to purchase. Beauty centers now have great opportunities to develop and compete by focusing on applications or developing them with artificial intelligence to identify potential customers, analyze their behavior, follow their habits, determine their motivations, and publish content in a harmonious manner that mimics their desires, etc., to provide a product or service that meets their needs and expectations.

Through a three-dimensional display that realistically depicts the entire process, artificial intelligence can create a more personalized experience for the brands used by beauty centers, whether they are associated with medical devices, cosmetic and medical products, or materials used to highlight and define different aesthetic features. This facilitates the development of user loyalty and participation. AI applications with an advertising focus are used by marketers as certified tools, payment processors, and sales tools to improve their financial and operational status. Social media marketers can now depend on AI-powered design software to handle the purchase process instead of having to figure it out on their own.

References

Abbas, S., et al.: Antecedents of trustworthiness of social commerce platforms: a case of rural communities using multi group SEM & MCDM methods. Electron. Commer. Res. Appl. **62**, 101322 (2023)

Abbott, M.L., McKinney, J.: Understanding and Applying Research Design. Wiley, Hoboken (2013)

Ahmed, A.D., Salih, M.M., Muhsen, Y.R.: Opinion weight criteria method (OWCM): a new method for weighting criteria with zero inconsistency. IEEE Access (2024)

Albinali, E.A., Hamdan, A.: The implementation of artificial intelligence in social media marketing and its impact on consumer behavior: evidence from Bahrain. In: Alareeni, B., Hamdan, A., Elgedawy, I. (eds.) ICBT 2020. LNNS, vol. 194, pp. 767–774. Springer, Cham (2021). https://doi.org/10.1007/978-3-030-69221-6_58

Al-Enzi, S.H.Z., Abbas, S., Abbood, A.A., Muhsen, Y.R., Al-Hchaimi, A.A.J., Almosawi, Z.: Exploring research trends of metaverse: a bibliometric analysis. In: Al-Emran, M., Ali, J.H., Valeri, M., Alnoor, A., Hussien, Z.A. (eds.) IMDC-IST 2024. LNNS, vol. 895, pp. 21–34. Springer, Cham (2023). https://doi.org/10.1007/978-3-031-51716-7_2

Ali, J., Hussain, K.N., Alnoor, A., Muhsen, Y.R., Atiyah, A.G.: Benchmarking methodology of banks based on financial sustainability using CRITIC and RAFSI techniques. Decis. Making: Appl. Manag. Eng. **7**(1), 315–341 (2024)

Alnoor, A., Atiyah, A.G., Abbas, S.: Toward digitalization strategic perspective in the European food industry: non-linear nexuses analysis. Asia-Paci. J. Bus. Adm. (2023)

Alnoor, A., Atiyah, A.G., Abbas, S.: Unveiling the determinants of digital strategy from the perspective of entrepreneurial orientation theory: a two-stage SEM-ANN approach. Global J. Flex. Syst. Manag. **25**(2), 1–18 (2024)

Alnoor, A., Chew, X., Khaw, K.W., Muhsen, Y.R., Sadaa, A.M.: Benchmarking of circular economy behaviors for Iraqi energy companies based on engagement modes with green technology and environmental, social, and governance rating. Environ. Sci. Pollut. Res. **31**(4), 5762–5783 (2024)

Alnoor, A., et al.: How positive and negative electronic word of mouth (eWOM) affects customers' intention to use social commerce? A dual-stage multi group-SEM and ANN analysis. Int. J. Hum. Comput. Interact. **40**(3), 808–837 (2024)

Arasu, B.S., Seelan, B.J.B., Thamaraiselvan, N.: A machine learning-based approach to enhancing social media marketing. Comput. Electr. Eng. **86**, 106723 (2020)

Atiyah, A.G.: Unveiling the quality perception of productivity from the senses of real-time multisensory social interactions strategies in metaverse. In: Al-Emran, M., Ali, J.H., Valeri, M., Alnoor, A., Hussien, Z.A. (eds.) IMDC-IST 2024. LNNS, vol. 876, pp. 83–93. Springer, Cham (2023). https://doi.org/10.1007/978-3-031-51300-8_6

Atiyah, A.G., Zaidan, R.A.: Barriers to using social commerce. In: Alnoor, A., Wah, K.K., Hassan, A. (eds.) Artificial Neural Networks and Structural Equation modeling, pp. 115–130. Springer, Singapore (2022). https://doi.org/10.1007/978-981-19-6509-8_7

Atiyah, A.G., Alhasnawi, M., Almasoodi, M.F.: Understanding metaverse adoption strategy from perspective of social presence and support theories: the moderating role of privacy risks. In: Al-Emran, M., Ali, J.H., Valeri, M., Alnoor, A., Hussien, Z.A. (eds.) IMDC-IST 2024. LNNS, vol. 876, pp. 144–158. Springer, Cham (2023). https://doi.org/10.1007/978-3-031-51300-8_10

Atiyah, A.G., All, N.D.A., Zaidan, A.S., Bayram, G.E.: Understating the social sustainability of metaverse by integrating adoption properties with users' satisfaction. In: Al-Emran, M., Ali, J.H., Valeri, M., Alnoor, A., Hussien, Z.A. (eds.) IMDC-IST 2024. LNNS, vol. 895, pp. 95–107. Springer, Cham (2023). https://doi.org/10.1007/978-3-031-51716-7_7

Atiyah, A.G., Faris, N.N., Rexhepi, G., Qasim, A.J.: Integrating ideal characteristics of chat-GPT mechanisms into the metaverse: knowledge, transparency, and ethics. In: Al-Emran, M., Ali, J.H., Valeri, M., Alnoor, A., Hussien, Z.A. (eds.) IMDC-IST 2024. Lecture Notes in Networks and Systems, vol. 895, pp. 131–141. Springer, Cham (2023). https://doi.org/10.1007/978-3-031-51716-7_9

Basri, W.: Examining the impact of artificial intelligence (AI)-assisted social media marketing on the performance of small and medium enterprises: toward effective business management in the Saudi Arabian context. Int. J. Comput. Intell. Syst. **13**(1), 142–152 (2020)

Benabdelouahed, R., Dakouan, C.: The use of artificial intelligence in social media: opportunities and perspectives. Exp. J. Mark. **8**(1), 82–87 (2020)

Berman, P.S., Jones, J., Udry, J.R., National Longitudinal Study of Adolescent Health.: Research design (2000)

Bjola, C.: AI for development: Implications for theory and practice. Oxf. Dev. Stud. **50**(1), 78–90 (2022)

Capatina, A., et al.: Matching the future capabilities of an artificial intelligence-based software for social media marketing with potential users' expectations. Technol. Forecast. Soc. Chang. **151**, 119794 (2020)

Capatina, A., Kachour, M., Lichy, J., Micu, A., Micu, A.E., Codignola, F.: Matching the future capabilities of an artificial intelligence-based software for social media marketing with potential users' expectations. Technol. Forecast. Soc. Chang. **151**, 119794 (2020)

Chen, D., et al.: The impact of artificial intelligence on firm performance: an application of the resource-based view to e-commerce firms. Front. Psychol. **13**, 884830 (2022)

Cheng, Y., Jiang, H.: How do AI-driven chatbots impact user experience? Examining gratifications, perceived privacy risk, satisfaction, loyalty, and continued use. J. Broadcast. Electron. Media **64**(4), 592–614 (2020)

Donaldson, S.I., Crano, W.D.: Theory-driven evaluation science and applied social psychology. Soc. Psychol. Eval. **141** (2011)

Duan, Y., et al.: Artificial intelligence for decision making in the era of big data–evolution, challenges and research agenda. Int. J. Inf. Manage. **48**, 63–71 (2019)

Dwivedi, Y.K., et al.: Setting the future of digital and social media marketing research: Perspectives and research propositions. Int. J. Inf. Manage. **59**, 102168 (2021)

Esmaeilzadeh, P.: Use of AI-based tools for healthcare purposes: a survey study from consumers' perspectives. BMC Med. Inform. Decis. Mak. **20**, 1–19 (2020)

Fathahillah, F.: et al.: Analysis of artificial intelligence literacy in the blended learning model in higher education. EduLine: J. Educ. Learn. Innov. **3**(4), 566–575 (2023)

Garg, P., Pahuja, S.: Social media: Concept, role, categories, trends, social media and AI, impact on youth, careers, recommendations. Managing social media practices in the digital economy, IGI Global: 172–192 (2020)

Gkikas, D.C., Theodoridis, P.K.: Artificial intelligence (AI) impact on digital marketing research. In: Kavoura, A., Kefallonitis, E., Giovanis, A. (eds.) Strategic Innovative Marketing and Tourism. SPBE, pp. 1251–1259. Springer, Cham (2019). https://doi.org/10.1007/978-3-030-12453-3_143

Groenewald, T.: A phenomenological research design illustrated. Int J Qual Methods **3**(1), 42–55 (2004)

Haleem, A., Javaid, M., Qadri, M.A., Singh, R.P., Suman, R.: Artificial intelligence (AI) applications for marketing: a literature-based study. Int. J. Intell. Netw. **3**, 119–132 (2022)

Hopcan, S., et al.: Artificial intelligence in special education: a systematic review. Interact. Learn. Environ. **31**(10), 7335–7353 (2023)

Husin, N.A., Abdulsaeed, A.A., Muhsen, Y.R., Zaidan, A.S., Alnoor, A., Al-mawla, Z.R.: Evaluation of metaverse tools based on privacy model using fuzzy MCDM approach. In: Al-Emran, M., Ali, J.H., Valeri, M., Alnoor, A., Hussien, Z.A. (eds.) IMDC-IST 2024. Lecture Notes in Networks and Systems, vol. 895, pp. 1–20. Springer, Cham (2023)

Imran, M., et al.: Using AI and social media multimodal content for disaster response and management: Opportunities, challenges, and future directions. Inf. Process. Manag. **57**(5), 102261 (2020)

Jha, A.: AI-Driven algorithms for optimizing social media advertising: prospects and challenges. Cases on social media and entrepreneurship, pp. 63–84 (2024)

Li, H.: Special section introduction: artificial intelligence and advertising. J. Advert. **48**(4), 333–337 (2019)

Liu, R., Gupta, S., Patel, P.: The application of the principles of responsible AI on social media marketing for digital health. Inf. Syst. Front. **25**(6), 2275–2299 (2023)

Lyu, Y., et al.: Communication in human–AI co-creation: perceptual analysis of paintings generated by text-to-image system. Appl. Sci. **12**(22), 11312 (2022)

Marczyk, G.R., DeMatteo, D., Festinger, D.: Essentials of Research Design and Methodology, vol. 2. Wiley, Hoboken (2010)

Meyer, S.B., Lunnay, B.: The application of abductive and retroductive inference for the design and analysis of theory-driven sociological research. Sociol. Res. Online **18**(1), 86–96 (2013)

Micu, A., et al.: Exploring artificial intelligence techniques' applicability in social media marketing. J. Emerg. Trends Mark. Manag. **1**(1), 156–165 (2018)

Mishra, S., et al.: Artificial intelligence focus and firm performance. J. Acad. Mark. Sci. **50**(6), 1176–1197 (2022)

Mitrović, A., Holland, J.: Effect of non-mandatory use of an intelligent tutoring system on students' learning. In: Bittencourt, I.I., Cukurova, M., Muldner, K., Luckin, R., Millán, E. (eds.) AIED 2020. LNCS (LNAI), vol. 12163, pp. 386–397. Springer, Cham (2020). https://doi.org/10.1007/978-3-030-52237-7_31

Mökander, J., Schroeder, R.: AI and social theory. AI & Soc. **37**(4), 1337–1351 (2022)

Morandé, S., Amini, M.: Digital persona: reflection on the power of generative ai for customer profiling in social media marketing (2023)

Muhsen, Y.R., Husin, N.A., Zolkepli, M.B., Manshor, N.: A systematic literature review of fuzzy-weighted zero-inconsistency and fuzzy-decision-by-opinion-score-methods: assessment of the past to inform the future. J. Intell. Fuzzy Syst. **45**(3), 4617–4638 (2023)

Nair, K., Gupta, R.: Application of AI technology in modern digital marketing environment. World J. Entrepreneurship, Manag. Sustain. Dev. **17**(3), 318–328 (2021)

Nazir, S., et al.: Exploring the influence of artificial intelligence technology on consumer repurchase intention: the mediation and moderation approach. Technol. Soc. **72**, 102190 (2023)

Nixon, L., et al.: AI and data-driven media analysis of TV content for optimised digital content marketing. Multimedia Syst. **30**(1), 25 (2024)

Perkins, M.B., et al.: Applying theory-driven approaches to understanding and modifying clinicians' behavior: what do we know? Psychiatr. Serv. **58**(3), 342–348 (2007)

Shaik Vadla, M.K., et al.: Enhancing product design through AI-driven sentiment analysis of Amazon reviews using BERT. Algorithms **17**(2), 59 (2024)

Stone, M., et al.: Artificial intelligence (AI) in strategic marketing decision-making: a research agenda. Bottom Line **33**(2), 183–200 (2020)

Terranova, C., et al.: AI and professional liability assessment in healthcare. A revolution in legal medicine? Front. Med. **10**, 1337335 (2024)

Theodoridis, P.K., Gkikas, D.C.: How artificial intelligence affects digital marketing. In: Kavoura, A., Kefallonitis, E., Giovanis, A. (eds.) Strategic Innovative Marketing and Tourism. SPBE, pp. 1319–1327. Springer, Cham (2019). https://doi.org/10.1007/978-3-030-12453-3_151

Wamba-Taguimdje, S.L., Wamba, S.F., Kamdjoug, J.R.K., Wanko, C.E.T.: Impact of artificial intelligence on firm performance: exploring the mediating effect of process-oriented dynamic capabilities. In: Agrifoglio, R., Lamboglia, R., Mancini, D., Ricciardi, F. (eds.) Digital Business Transformation. LNNS, vol. 38, pp. 3–18. Springer, Cham (2020). https://doi.org/10.1007/978-3-030-47355-6_1

Yu, Y.: The role and influence of artificial intelligence on advertising industry. In: 2021 International Conference on Social Development and Media Communication (SDMC 2021). Atlantis Press (2022)

Zhang, H., Song, M.: How Big data analytics, AI, and social media marketing research boost market orientation. Res. Technol. Manag. **65**(2), 64–70 (2022)

Artificial Intelligence and Environmental, Social and Governance: A Bibliometric Analysis Review

Mushtaq Yousif Alhasnawi[1,2]([✉]) [iD], Sajead Mowafaq Alshdaifat[3] [iD],
Noor Hidayah Ab Aziz[4,5] [iD], and Muthana Faaeq Almasoodi[6]

[1] School of Business and Economics, Universiti Putra Malaysia, Seri Kembangan, Selangor,
Malaysia
alhasnawi78@utq.edu.iq, aidi@uum.edu.my
[2] Faculty of Administration and Economics, University of Thi-Qar, Nasiriyah, Iraq
[3] School of Business, Al Al-Bayt University, Mafraq, Jordan
salshdaifat@aabu.edu.jo
[4] Faculty of Accountancy, Universiti Teknologi MARA, Johor Branch, Segamat Campus,
Masai, Johor, Malaysia
noorh469@uitm.edu.my
[5] Putra Business School, Serdang, Malaysia
[6] College of Tourism Science, Department of Tourism Studies, University of Kerbala, Karbala,
Iraq

Abstract. Studies on sustainability have been given more attention in recent
years. Although there is an increasing body of literature on sustainability, there
has been little bibliometric analysis about sustainability and the environmental-
social, governance (ESG). This study analyse the literature on sustainability and
environmental, social, and governance from (2015–2024). Based on the search
in the Scopus database, 148 publications have been analysed, including publication trends, most active institutions and countries, most highly cited documents,
and author's keywords. VOSviewer software was used to analyse the data. The
data analysis showed an increase in publications during the analysis period, which
indicate this topic evolving in the future. The results show that most papers were
published in the English language and most publications were the article type.
The findings illustrate that the most active institution is the Jadara University,
while the most active country is United States. This review contributes to existing
knowledge on sustainability and ESG research field by providing insights into publication trends, citation patterns, and collaboration networks between institutions
and countries.

Keywords: Robotic process automation · Audit · Enterprise resource planning ·
Artificial intelligence

© The Author(s), under exclusive license to Springer Nature Switzerland AG 2024
A. Alnoor et al. (Eds.): AIRDS 2024, LNNS 1033, pp. 123–143, 2024.
https://doi.org/10.1007/978-3-031-63717-9_8

1 Introduction

Over the last few years, there has been extensive growth in the use of concepts such as sustainability and ESG, not only as a research area in academia but also in business. These concepts extent represent a giant factual change in the consciousness of organizations and the roles they play in the society and environment alike (Punj et al., 2023; Ali et al., 2024; Zhang et al., 2024). The sustainability is defined by the Brundtland Commission in 1987, namely, meeting the needs of the present without shortening the durability of the Earth's capital in terms of resources for future generations to meet their own needs (Imperiale et al., 2023; Alnoor et al., 2024; Atiyah et al., 2023;). On the other hand, ESG is the sub components of sustainability that focus on how firms operate to achieve sustainability at the same time financial performance (Ab Aziz et al., 2023; Atiyah et al., 2023; Abate et al., 2021). There has been a mounting connection between sustainability and ESG as companies understand their responsibility to bare their environmental and social issues while still maintaining the required governance elements. The ESG framework has emerged as a holistic approach for evaluating companies' performance in three key areas: In relation to ESG which can be considered as the key factors (Atiyah et al., 2023; Alnoor et al., 2023; Ng et al., 2020). Firms with ESG integration tend to be more profitable to the extent that they can manage risks better, attract investment easier, and produce long-term value (Ab Aziz et al., 2023). This research investigates past literature produced on the concepts of sustainability and ESG, that identify the link between these two, and how these affect each other.

Currently, there is a general understanding that what corporations do for a profit does not simply affect the organizational activities and society but also affects the whole ecosystem (Oprean-Stan et al., 2020; Alnoor et al., 2024; Zaidan et al., 2023). Thus, the sustainability and ESG area research focal item to investigate the connection of ESG with financial results and general corporate sustainability (Alhasnawi et al., 2023; Atiyah and Zaidan, 2022; Atiyah, 2022; Nousheen & Kalsoom, 2022). Therefore, a significant gap in the literature on the spectrum of sustainability and the implications that the ESG framework can generate is important for investigation. The problem is to understand the complex interaction between sustainability and ESG issues and their influence on organizational performance, risk management, and sustainability. On the other hand, with system identity being one of the crucial elements in stakeholder management, organizations are presented with the issue of involving ESG aspects into their strategic decision-making successfully. This study will strive to fill the void that currently exists in the literature on the link between sustainability and ESG by conducting a thorough study, sharing the main tendencies, and identifying possible subjects to be examined in future studies.

The current literature on sustainability and ESG studies demonstrates that a body of evidence that proves ESG issues as part of business operations is crucial (Siao et al., 2022; Atiyah, 2022). Table 1 shows some research studies conducted about sustainability and ESG, which emphasized specific issues in this relationship. Yet, there is a lack of literature that links ESG with sustainability (M. Y. Alhasnawi et al., 2023; Atiyah et al., 2023; Steblianskaia et al., 2023). For instance, many articles are dedicated to specific issues and are not complete in terms of the full understanding of the link between ESG and sustainability (Gao et al., 2021; Atiyah, 2022).

Table 1. Summary of Review Papers on Sustainability and ESG

Author	Domain/Search Strategy	Data Source & Scope	TDE	Bibliometric Attributes Examined
(Steblianskaia et al., 2023)	Analyze articles concerning ESG	Web of Science and Scopus	100	Building a development model for ESG
(Siao et al., 2022)	Analyze articles regarding ESG (2002 to 2021)	Web of Science	3599	Research trends, keyword analysis, research fields, authors, and national institutions
(Gao et al., 2021)	Analysis of the literature on sustainability, which published between 1987 and 2022	Scopus	500	Conducting a literature review on ESG to provide a research agenda for future research
(Nobanee et al., 2021)	Analyse and apply the bibliometric approach to review literature on sustainability practices and management of risk (1989–2020)	Scopus	1233	Sustainability development and topics related to risk management
(Yadav & Saini, 2023)	ESG in the domain of business and management	Scopus	460	Research trends, keyword analysis, research fields, authors, and national institutions
(Zhang et al., 2024)	Analysis of the literature on ESG, which was published between 1991 and 2023	Web of Science	400	Research directions, analysis of inter-country and partnerships research fields, authors, and Research frontier identification

TDE = Total documents examined

As shown in Table 1, some of the prior review papers have focused on the ESG concept (e.g., Gao et al., 2021; Steblianskaia et al., 2023; Abbas et al., 2023; Yadav & Saini, 2023). On the other hand, the second stream focuses on sustainability issues (e.g., Nobanee et al., 2021). As a result, there has been little bibliometric analysis linking ESG and sustainability. Therefore, the present study attempts to fill this gap through a review of research conducted on ESG and sustainability by answering the following research questions:

1. What are the publication trends in the field of sustainability and ESG, and how have they changed over time?
2. Who are the most active institutions and countries in the field of sustainability and ESG?
3. What are the most highly cited documents in the field of sustainability and ESG?
4. What are the most common keywords in the literature on sustainability and ESG?

2 Methods

The bibliometric mapping approach can present how keywords, citations, and author names have connections. This is a useful tool to analyse complicated bibliometric information (Abdullah, 2022; Muhsen et al., 2023). A bibliographic review was conducted by identification of different dimensions of sustainability and environmental-social-governance research. Although there are many peer-reviewed publication databases, Scopus is possibly one of those thus having the highest number of records. Scopus has a wide range of areas of interest taking them in such technical, medical, and social sciences topic spheres that this online resource is the most effective for the analysis of the literature on this topic.

2.1 Search Strategy and Data Collection

The current investigation used papers published from 2015 to April 1, 2024, to collect the largest body of literature. It includes many dimensions, including document type, source type, language, subject area, publication trends, the number of authors per document, institutional contributions to publications, the distribution of publications across countries, and common keywords. The article title was used as the primary search field to attempt precise and relevant results on sustainability and environmental social-governance research by using the following query ("sustainability AND "Environment social governance" OR "ESG"). Based on this query, 151 papers have been collected; however, due to some missing data, 3 papers have been deleted. The final total of 148 documents, addressing the link between ESG and sustainability factors.

2.2 Data Cleaning and Harmonisation

Data cleaning and harmonization constitute determinative features of bibliometric analysis, meaning accuracy and reliability are the most important measures. This study was used via the application of two special tools- OpenRefine (Ahmi, 2023; Ahmed et al., 2024; Alnoor et al., 2024; Husin et al., 2023), which were specifically designed to assist in finding and merging data related to author names, affiliations, keywords, and essential bibliographic information. As a starting point, we collected the data from the Scopus database in the CSV format and then we proceeded to pick the suitable files for cleaning. My clustering utilities were employed to improve the values of certain pieces of information like some keywords, author names and data and locations. The BiblioMagika platform also provided additional means of describing holes in the data, resulting in the consecutive cleansing and harmonizing phases being more easily carried out better to the automatic identification and filling of the missing data.

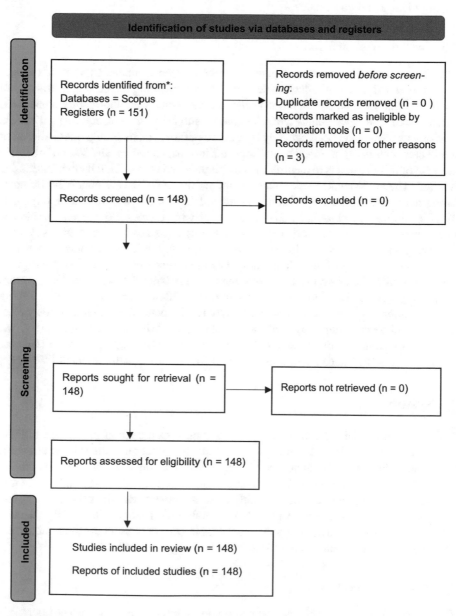

The next stage following cleaning of data is scrutiny of all keywords which have been prepared and edited. In this phase, careful analysis of these keywords is conducted to ensure their correctness. Through these stages, we could ensure the accuracy of following procedures minimizing possibilities to reduce bias and aiding in establishing convincing and trusting results. As a direct consequence of the harmonizing of the dataset and cleaning, the quality of the dataset and its clarity were improved significantly. It further provided a robust basis to inquire and analyze the complex issue of sustainability and environment social governance.

2.3 Data Analysis and Tools

The results of the analysis were organized in the manner to enable answer the research questions stated in the introduction section. By documenting the current status of sustainability and ESG studies seven microlevel dimensions were thoroughly considered, such as document types, source types, languages, subject areas, and citation metrics. The disclosures were made systematically by various parameters, including publications and intellectual contributions made by authors, affiliations, countries, and source titles. The purpose of this method is to identify the key players and overarching themes in this curriculum. The bibliMagika was a foundational instrument in the systematic bibliometric assessment, ranging from h-index and g-index, h-core citation sum, among others. To provide additional details and visual evidence of what themes and concepts are in that field, the co-occurrence network, thematic mapping, and statistical analysis of keywords belonging to that author are shown. A comprehensive bibliometric analysis was conducted using a variety of tools. The initial data cleansing and organisation phases were carried out using Microsoft Excel, providing the foundation for analyses. Furthermore, BiblioMagika contributed to streamlining author, affiliation, and country data cleaning, harmonisation, and standardisation. This study used OpenRefine to meticulously clean and harmonise the author's keywords, and Biblioshiny (Aria and Cuccurullo, 2017) was used to provide informative visual representations of our findings. Together, these tools and techniques facilitated a thorough and robust analysis of unemployment litretare.

3 Results

This segment provides the results of a bibliometric examination performed utilising Biblioshiny in order to tackle fundamental inquiries concerning environmental social governance (ESG) and sustainability. Through the utilisation of customisable co-citation analyses, this study investigates various aspects of the field: publication trends, influential authors, institutions, and countries, highly cited documents and their themes, common keywords and themes, as well as patterns of collaboration and co-citation. By means of these examinations, the research provides valuable perspectives on the dynamic field of sustainability studies and its interdisciplinary convergences with ESG.

3.1 Documents Profiles

The dataset's documents were profiling based on their types and origins. The documents were categorised by type in Table 1, and their sources were classified into distinct groups in Table 2. Seven distinct types of documents were identified through the analysis: conference papers, book chapters, editorials, notes, books, and reviews. Articles constituted the most prevalent type, accounting for 67.33% of the overall publications (101 articles out of 148). Five distinct categories of sources were discerned, namely trade journals, conferences proceedings, books, and book series. The most prevalent source of documents was journals, which accounted for 71.33% of the total publications (107 articles out of 148). The substantial academic attention given to sustainability and environmental social governance (ESG) is supported by the 148 scholarly articles that have been compiled thus far on this topic.

Table 2. Document Type

Document Type	Total Publications	Percentage (%)
Article	101	67.33%
Conference Paper	23	15.33%
Book Chapter	16	10.67%
Editorial	3	2.00%
Note	3	2.00%
Book	2	1.33%
Review	2	1.33%
Total	**148**	**100.00**

Table 3. Source Type

Source Type	Total Publications	Percentage (%)
Journal	107	71.33%
Conference Proceeding	18	12.00%
Book	13	8.67%
Book Series	10	6.67%
Trade Journal	2	1.33%
Total	**148**	**100.00**

In addition, documents are classified according to their respective academic disciplines, as illustrated in Table 3. 72 documents (48%) of the total research on sustainability and ESG are classified as journal articles under the heading Business, Management, and Accounting. Following this are the Social Sciences (65 documents, 36.7%), Economics, Econometrics, and Finance (61 documents, 40.7%), Environmental Science (48 documents, 32%). Furthermore, the results suggest that scholarly articles have been published in numerous other fields, including energy, computer science, engineering, and others, that examine sustainability and ESG.

Table 4. Subject Area

Subject Area	Total Publications	Percentage (%)
Business, Management and Accounting	72	48.0%
Economics, Econometrics and Finance	61	40.7%
Social Sciences	55	36.7%
Environmental Science	48	32.0%
Energy	39	26.0%
Computer Science	20	13.3%
Engineering	19	12.7%
Decision Sciences	8	5.3%
Arts and Humanities	6	4.0%
Earth and Planetary Sciences	6	4.0%
Mathematics	5	3.3%
Chemical Engineering	3	2.0%
Materials Science	3	2.0%
Medicine	2	1.3%
Pharmacology and Pharmaceutics	2	1.3%
Psychology	2	1.3%
Multidisciplinary	1	0.7%

3.2 Publication Trends

Table 4 provides a detailed summary of annual publications on sustainability and environmental social governance (ESG) from 2015 to 2024. According to the table, the highest number of publications occurred in 2023, totalling 61 documents out of 148. Based on the table it can be seen that year 2020 hold the highest numbers of total citation of 743 with 92.88 average number of citations. On the contrary, the least number citation is in 2015 with 1 citation on average. From year to year there is a growing trend on the citation as sustainability and ESG become more importance in various sectors particularly in business, finance and policy making (Patil et al., 2020; Sharma et al., 2020). It is further intensified when covid-19 pandemic hit the world and the rising concern on climate change which makes research on sustainability and ESG increase during this period (Nousheen & Kalsoom, 2022; Al-Enzi et al., 2023). While lower citation in 2015 indicate that the topic is still at infancy stage where the scholarly interest on this topic were still growing during that period.

The graph shown in Fig. 1 displays the yearly changes in publications related to sustainability and ESG starting from 2015. There has been a significant increase in the number of articles on this topic from 2019 to 2024, with a significant increase from 2022 to 2023. As mentioned above, the field of sustainability and ESG keep on

Table 5. Year of Publication

Year	TP	NCP	TC	C/P	C/CP	h	g
2015	1	1	1	1.00	1.00	1	1
2018	1	1	38	38.00	38.00	1	1
2019	4	4	549	137.25	137.25	4	4
2020	8	7	743	92.88	106.14	6	8
2021	8	5	77	9.63	15.40	4	8
2022	40	25	504	12.60	20.16	12	22
2023	61	33	270	4.43	8.18	10	14
2024	25	3	5	0.20	1.67	1	2
Total	**148**						

Notes: TP = total number of publications; NCP = number of cited publications; TC = total citations; C/P = average citations per publication; C/CP = average citations per cited publication; h = h-index; and g = g-index

growing due to awareness of global environmental issues, corporate governance, and social responsibility (Oliver Yébenes, 2024).

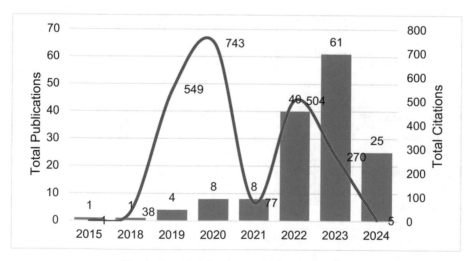

Fig. 1. Total Publications and Citations by Year

Besides, there is a pressure from market and regulators that require companies to consider ESG factors in their operations. Additionally, investors are increasingly focused on ESG issues for their investment decision (Baid & Jayaraman, 2022). Lastly due to the nature of sustainability and ESG studies which involve cross-discipline that encourages collaboration among scholars from various backgrounds, helping to further expand research in this area (Nunhes et al., 2021).

3.3 Publications By Institutions

Moving to RQ2, this study examining the top academic institutions that actively involved in sustainability and ESG research. Table 5 showed the top 14 universities globally which are most active. From the table, it can be seen that Jadara University has the highest number of publications in the field, followed by Universitas Padjadjaran and Bina Nusantara University in second and third place respectively. There are several factors contribute to the active involvement of the universities in this field of research. Firstly, due to the culture of the university that focus on addressing the sustainability and global issues (Aithal & Aithal, 2023). Secondly due to having scholars that specializing in sustainability topics (Sady et al., 2019). Lastly when the universities have good resources and facilities as well as good collaboration and funding will also help to increase the research activities in sustainability and ESG.

Table 6. Most active institutions

Institution	TP	TC	NCP	C/P	C/CP	h-index	g-index
Jadara University	8	82	5	10.25	16.40	4	8
Universitas Padjadjaran	5	2	1	0.40	2.00	1	1
Bina Nusantara University	4	0	0	0.00	0.00	0	0
Universiti Kebangsaan Malaysia	3	15	3	5.00	5.00	2	3
University of Antwerp	3	5	3	1.67	1.67	1	2
Universiti Malaysia Terengganu	3	64	3	21.33	21.33	3	3
Macquarie University	2	3	2	1.50	1.50	1	1
Amman Arab University	2	9	1	4.50	9.00	1	2
Universitas Sumatera Utara	2	3	1	1.50	3.00	1	1
University of Muhammadiyah	2	0	0	0.00	0.00	0	0
Irbid National University	2	4	1	2.00	4.00	1	2
Dar Alhekma University	2	5	1	2.50	5.00	1	2
Victoria University of Wellington	2	2	1	1.00	2.00	1	1
Universiti Sains Malaysia	2	4	1	2.00	4.00	1	2

3.4   Publications By Countries

This section discusses the most active country that involve in sustainability and ESG research. Table 6 and Fig. 2 highlight the top 20 countries that contribute the most publications in this field. United states (US) is the top country with 33 total publications and 467 citations. From the table it can be concluded that both developed and developing country have contribute to the significant amount of research in sustainability and ESG (Singhania & Saini, 2023). There are factors that contribute to US as leading country in sustainability and ESG research. Firstly, due to its strong economy and industrial base that put emphasize on investment in sustainability and ESG from both the public and private sectors (Scheyvens et al., 2016). In addition, there is a strong group of scholars who are pioneering in research on sustainability and ESG, helping to establish the earliest research. After the US is China with a total of 23 published works, followed by Australia with 18 publications, and Jordan with 16 publications. When refer to the

Table 7. Top 20 Countries contributed to the publications.

Country	TP	TC	NCP	C/P	C/CP	h-index	g-index
United States	33	467.00	26	14.15	17.96	14	21
China	23	43.00	16	1.87	2.69	4	6
Australia	18	126.00	12	7.00	10.50	6	11
Jordan	16	102.00	9	6.38	11.33	5	10
Malaysia	15	96.00	12	6.40	8.00	5	9
United Kingdom	11	252.00	11	22.91	22.91	8	11
Romania	10	60.00	6	6.00	10.00	4	7
Italy	8	23.00	6	2.88	3.83	3	4
Turkey	7	14.00	3	2.00	4.67	2	3
Saudi Arabia	7	33.00	6	4.71	5.50	4	5
New Zealand	4	0.00	0	0.00	0.00	0	0
South Korea	4	174.00	4	43.50	43.50	3	4
United Arab Emirates	3	12.00	2	4.00	6.00	1	3
Belgium	3	17.00	3	5.67	5.67	2	3
Pakistan	3	21.00	3	7.00	7.00	3	3
Bangladesh	3	13.00	3	4.33	4.33	3	3
Spain	3	19.00	3	6.33	6.33	3	3
Portugal	3	29.00	2	9.67	14.50	2	3
Russian Federation	3	4.00	3	1.33	1.33	1	2
Bahrain	3	11.00	1	3.67	11.00	1	3

number of citations, the United Kingdom (UK) took second place after the US, with a total of 252 citations, followed by South Korea with 174 citations. Thus it can be seen that papers from UK and South Korea give more impact due to its quality even though these two countries among the fewer contributors of publications.

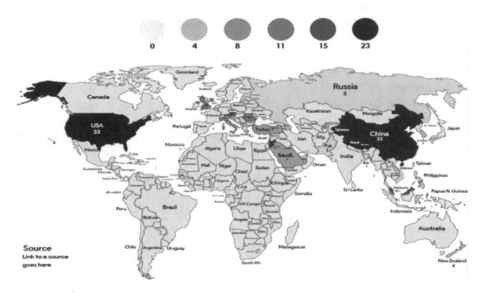

Fig. 2. The publications distribution based on Countries' contributions

3.5 Highly Cited Documents

Moving to RQ3, Table 7 showcases the top 20 most referenced papers in the field of sustainability and ESG. These papers have garnered a significant amount of recognition based on the number of times they have been cited. The two most referenced papers are authored by Drempetic et al. in 2020, titled "The Influence of Firm Size on the ESG Score: Corporate Sustainability Ratings Under Review," and Buallay A. in 2019, titled "Is sustainability reporting (ESG) associated with performance? Evidence from the European banking sector." These studies have gained widespread recognition for their pioneering contributions to the field of sustainability and ESG research. The high citation counts for articles are probably attributable to a number of factors. Firstly, the subjects they tackle are pertinent and current matters concerning ESG and sustainability. These concerns hold considerable importance for policymakers, stakeholders, practitioners, and scholars. Furthermore, both investigations provide fresh perspectives and empirical data that enhance the discipline, thereby advancing our comprehension of the correlation between organisational sustainability initiatives and the financial performance of businesses. Furthermore, the researchers' analytical approaches, data sources, and methodologies are rigorous and robust, thereby bolstering the credibility and dependability of their results. Furthermore, as early investigations in the discipline, these papers may have established fundamental understandings and conceptual frameworks that have been expanded upon by subsequent research, thereby augmenting their influence and citation counts. Furthermore, it is possible that the high citation counts of these publications were influenced by the reputation and stature of the authors and their affiliations (Table 8).

Table 8. Top 20 highly cited articles.

No	Authors	Title	Cites	Cites per Year
1	(Drempetic et al., 2020)	The Influence of Firm Size on the ESG Score: Corporate Sustainability Ratings Under Review	Journal of Business Ethics	388
2	(Buallay, 2019)	Is sustainability reporting (ESG) associated with performance? Evidence from the European banking sector	Management of Environmental Quality: An International Journal	305
3	(Alsayegh et al., 2020)	Corporate economic, environmental, and social sustainability performance transformation through ESG disclosure	Sustainability (Switzerland)	202
4	(Escrig-Olmedo et al., 2019)	Rating the raters: Evaluating how ESG rating agencies integrate sustainability principles	Sustainability (Switzerland)	182
5	(Adams & Abhayawansa, 2022)	Connecting the COVID-19 pandemic, environmental, social and governance (ESG) investing and calls for 'harmonisation' of sustainability reporting	Critical Perspectives on Accounting	154
6	(Abdi et al., 2022)	Exploring the impact of sustainability (ESG) disclosure on firm value and financial performance (FP) in airline industry: the moderating role of size and age	Environment, Development and Sustainability	86

(continued)

Table 8. (*continued*)

No	Authors	Title	Cites	Cites per Year
7	(Hübel & Scholz, 2020)	Integrating sustainability risks in asset management: the role of ESG exposures and ESG ratings	Journal of Asset Management	59
8	(Garcia et al., 2019)	Corporate sustainability, capital markets, and ESG performance	Individual Behaviors and Technologies for Financial Innovations	38
9	(Ng et al., 2020)	Sustainability in Asia: The Roles of Financial Development in Environmental, Social and Governance (ESG) Performance	Social Indicators Research	36
10	(Oprean-Stan et al., 2020)	Impact of sustainability reporting and inadequate management of esg factors on corporate performance and sustainable growth	Sustainability (Switzerland)	34
11	(Jebe, 2019)	The Convergence of Financial and ESG Materiality: Taking Sustainability Mainstream	American Business Law Journal	32
12	(Abate et al., 2021)	The level of sustainability and mutual fund performance in Europe: An empirical analysis using ESG ratings	Corporate Social Responsibility and Environmental Management	32
13	(Imperiale et al., 2023)	Sustainability reporting and ESG performance in the utilities sector	Utilities Policy	31
14	(Ismail & Latiff, 2019)	Board diversity and corporate sustainability practices: Evidence on environmental, social and governance (ESG) reporting	International Journal of Financial Research	30

(*continued*)

Table 8. (*continued*)

No	Authors	Title	Cites	Cites per Year
15	(Rahman et al., 2023)	ESG and firm performance: The rarely explored moderation of sustainability strategy and top management commitment	Journal of Cleaner Production	29

3.6 Co-authorship Analysis

Co-authorship By Organizations

In order to address the last study question, 30 nodes were used to examine organisational collaboration. The interconnected organisations are depicted in Fig. 3; characteristics like colour, text size, circle size, and line thickness show how strongly related the organisations are to one another. Organisations that are connected are shown in the same colour, which connects them. Insights into the collaborative landscape within the field are provided by Fig. 3, which emphasises the closely linked organisations engaging in research activities.

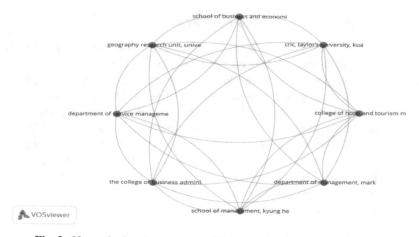

Fig. 3. Network visualisation map of the co-authorship by organisations

3.7 Co-authorship by Countries

Figure 4 displays a network visualization of author affiliations by country. The analysis focuses on countries that have cited more than three articles and received more than five citations. According to the scoring method used, the results highlight the significant role of the US in collaborating with other countries. The US is closely connected

with Spain, Germany, and Sweden in terms of cooperation. Similarly, the UK collabo-
rates closely with Italy. Additionally, Turkey is found to collaborate with India, France,
and other countries. These findings suggest strong patterns of international cooperation
among these countries in the field of sustainability and environmental social governance
research.

Fig. 4. Network visualization map of the co-authorship by countries

3.8   Co-occurrence Analysis of Author's Keywords

Utilising the VOS viewer software, keyword and co-occurrence analyses were performed
in order to address the final research question. By means of this application, bibliometric networks can be generated and displayed, precisely outlining the keywords that are
linked to individual documents. The map visualisation of authors' keywords produced
by VOS viewer is illustrated in Fig. 5. The intensity of connections between keywords
is demonstrated through the use of colour, line thickness, circle size, and font. When
keywords are closely related, they are frequently depicted in the same colour grouping.
Based on author keywords, the analysis reveals the formation of clusters in sustainability and environmental social governance (ESG) research. The diagram illustrates, for
instance, that blue-shaded keywords including sustainability, CSR, and ESG ratings are
highly interconnected and appear frequently together in the literature.

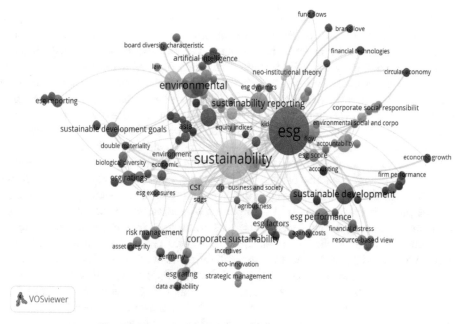

Fig. 5. Network visualisation of the author's keywords

4 Discussion and Conclusion

Given the significance of the sustainability and ESG field, our objective is to offer a thorough review of the existing literature in this domain using bibliometric analysis. Bibliometric analysis offers a multifaceted approach to understanding research dynamics within a field, providing quantitative insights into publication trends, citation patterns, and collaboration networks. This analysis highlights the rise in scholarly publications happened in 2023, totaling 61 out of 148 documents, indicating the expansion of the body of knowledge relevant to the field. It is worth mentioning that in 2020, citations increased substantially, totalling 743; this trend underscores the growing acknowledgement of sustainability and ESG research. Consistent with the observed expansion of the literature from 2019 to 2024, this pattern persisted. The institution that achieved the highest number of publications was Jadara University, which was succeeded by Universitas Padjadjaran and Bina Nusantara University. The United States maintained its leading position in sustainability and ESG research, as evidenced by its overwhelming number of publications (33) and citations (467). Key facets of corporate sustainability and performance were the subject of the most frequently cited articles, which were written by Drempetic et al. (2020) and Buallay (2019). Close relations existed between countries, as evidenced by the United States' partnerships with Spain, Germany, and Sweden, the United Kingdom's collaborate with Italy, and Turkey's close collaborations with France, India, and other countries. Moreover, the presence of interrelated terms like sustainability, CSR, and ESG ratings indicated shared research interests and focal points.

This study yields significant insights with implications for knowledge advancement in sustainability and ESG. By delving into publication trends, active institutions, countries, most-cited publications, and key topics and keywords, it offers a comprehensive synthesis of existing research in the field. The study highlights an imbalance number of publications on sustainability and ESG across countries, emphasizing the importance of understanding their impact in various contexts. To address this issue, future research could explore developing and emerging markets to create a more diverse and comprehensive knowledge base. Collaboration between scholars from different countries is essential for enhancing research engagement and reducing this imbalance. In the end, the importance of more research on sustainability and ESG highlights the ongoing necessity for a deep understanding of these crucial matters and offers helpful direction for upcoming researchers as they navigate this ever-changing field. Although bibliometric analysis provides useful insights, it is important to address some limitations for future improvement. Firstly, due to this analysis focus on limited keywords which might result in overlook of certain information and leading to bias result. Thus, it is important to carefully filter and clean the data to ensure reliability of the data. Besides this study utilize Scopus as the powerful database to search for the articles. Thus, this will limit other publication from database such as google scholar and WoS. Future researcher might want to consider using more than one database to get more sample. Nevertheless, this research makes a contribution to our understanding by summarizing current trends in sustainability and ESG literature, providing a new direction for future research in the field.

References

Ab Aziz, N.H., Abdul Latiff, A.R., Alshdaifat, S.M., Osman, M.N.H., Azmi, N.A.: ESG disclosure and firm performance: evidence after the revision of Malaysian code of corporate governance. Int. J. Acad. Res. Bus. Soc. Sci. **13**(12) (2023). https://doi.org/10.6007/ijarbss/v13-i12/19140

Abate, G., Basile, I., Ferrari, P.: The level of sustainability and mutual fund performance in Europe: an empirical analysis using ESG ratings. Corp. Soc. Responsib. Environ. Manag. **28**(5), 1446–1455 (2021). https://doi.org/10.1002/csr.2175

Abbas, S., et al.: Antecedents of trustworthiness of social commerce platforms: a case of rural communities using multi group SEM & MCDM methods. Electron. Commer. Res. Appl. **62**, 101322 (2023)

Abdi, Y., Li, X., Càmara-Turull, X.: Exploring the impact of sustainability (ESG) disclosure on firm value and financial performance (FP) in airline industry: the moderating role of size and age. Environ. Dev. Sustain. **24**(4), 5052–5079 (2022). https://doi.org/10.1007/s10668-021-016 49-w

Adams, C.A., Abhayawansa, S.: Connecting the COVID-19 pandemic, environmental, social and governance (ESG) investing and calls for 'harmonisation' of sustainability reporting. Crit. Perspect. Account. **82**, 102309 (2022). https://doi.org/10.1016/j.cpa.2021.102309

Ahmed, A.D., Salih, M.M., Muhsen, Y.R.: Opinion weight criteria method (OWCM): a new method for weighting criteria with zero inconsistency. IEEE Access (2024)

Aithal, P.S., Aithal, S.: How to increase emotional infrastructure of higher education institutions. Int. J. Manag. Technol. Soc. Sci. (IJMTS), **8**(3), 356–394 (2023). https://doi.org/10.47992/ijmts.2581.6012.0307

Al-Enzi, S.H.Z., Abbas, S., Abbood, A.A., Muhsen, Y.R., Al-Hchaimi, A.A.J., Almosawi, Z.: Exploring research trends of metaverse: a bibliometric analysis. In: Al-Emran, M., Ali, J.H., Valeri, M., Alnoor, A., Hussien, Z.A. (eds.) IMDC-IST 2024. LNNS, vol. 895, pp. 21–34. Springer, Cham (2023). https://doi.org/10.1007/978-3-031-51716-7_2

Alhasnawi, M.Y., Mohd Said, R., Mat Daud, Z., Muhammad, H.: Enhancing managerial performance through budget participation: insights from a two-stage A PLS-SEM and artificial neural network approach (ANN). J. Open Innovation: Techn. Market, Complexity 9(4), 100161 (2023). https://doi.org/10.1016/j.joitmc.2023.100161

Alhasnawi, M., Said, R.M., Daud, Z.M., Muhamad, H.: Budget participation and managerial performance: bridging the gap through budget goal clarity. Adv. Soc. Sci. Res. J. 10(9), 187–200 (2023). https://doi.org/10.14738/assrj.109.15539

Ali, J., Hussain, K.N., Alnoor, A., Muhsen, Y.R., Atiyah, A.G.: Benchmarking methodology of banks based on financial sustainability using CRITIC and RAFSI techniques. Decis. Making: Appl. Manag. Eng. 7(1), 315–341 (2024)

Alnoor, A., Atiyah, A.G., Abbas, S.: Toward digitalization strategic perspective in the European food industry: non-linear nexuses analysis. Asia-Pac. J. Bus. Adm. (2023)

Alnoor, A., Atiyah, A.G., Abbas, S.: Unveiling the determinants of digital strategy from the perspective of entrepreneurial orientation theory: a two-stage SEM-ANN approach. Glob. J. Flex. Syst. Manag. 25, 1–18 (2024)

Alnoor, A., Chew, X., Khaw, K.W., Muhsen, Y.R., Sadaa, A.M.: Benchmarking of circular economy behaviors for Iraqi energy companies based on engagement modes with green technology and environmental, social, and governance rating. Environ. Sci. Pollut. Res. 31(4), 5762–5783 (2024)

Alnoor, A., et al.: How positive and negative electronic word of mouth (eWOM) affects customers' intention to use social commerce? A dual-stage multi group-SEM and ANN analysis. Int. J. Hum. Comput. Interact. 40(3), 808–837 (2024)

Alsayegh, M.F., Abdul Rahman, R., Homayoun, S.: Corporate economic, environmental, and social sustainability performance transformation through ESG disclosure. Sustain. (Switzerland) 12(9), 3910 (2020). https://doi.org/10.3390/su12093910

Atiyah, A.G.: Impact of knowledge workers characteristics in promoting organizational creativity: an applied study in a sample of Smart organizations. PalArch's J. Archaeol. Egypt/Egyptology 17(6), 16626–16637 (2020)

Atiyah, A.G.: Effect of temporal and spatial myopia on managerial performance. J. La Bisecoman 3(4), 140–150 (2022)

Atiyah, A.G.: Strategic network and psychological contract breach: the mediating effect of role ambiguity. Int. J. Res. Manag. Stud. (IJRMS), 13(1) (2023)

Atiyah, A.G.: Unveiling the quality perception of productivity from the senses of real-time multisensory social interactions strategies in metaverse. In: Al-Emran, M., Ali, J.H., Valeri, M., Alnoor, A., Hussien, Z.A. (eds.) IMDC-IST 2024. LNNS, vol. 876, pp. 83–93. Springer, Cham (2023). https://doi.org/10.1007/978-3-031-51300-8_6

Atiyah, A.G., Zaidan, R.A.: Barriers to using social commerce. In: Artificial Neural Networks and Structural Equation Modeling: Marketing and Consumer Research Applications, pp. 115–130. Springer, Singapore (2022)

Atiyah, A.G., Alhasnawi, M., Almasoodi, M.F.: Understanding metaverse adoption strategy from perspective of social presence and support theories: the moderating role of privacy risks. In: Al-Emran, M., Ali, J.H., Valeri, M., Alnoor, A., Hussien, Z.A. (eds.) IMDC-IST 2024. LNNS, vol. 876, pp. 144–158. Springer, Cham (2023)

Atiyah, A.G., All, N.D.A., Zaidan, A.S., Bayram, G.E.: Understating the social sustainability of metaverse by integrating adoption properties with users' satisfaction. In: Al-Emran, M., Ali, J.H., Valeri, M., Alnoor, A., Hussien, Z.A. (eds.) IMDC-IST 2024. LNNS, vol. 895, pp. 95–107. Springer, Cham (2023). https://doi.org/10.1007/978-3-031-51716-7_7

Atiyah, A.G., Faris, N.N., Rexhepi, G., Qasim, A.J.: Integrating ideal characteristics of Chat-GPT mechanisms into the metaverse: knowledge, transparency, and ethics. In: Al-Emran, M., Ali, J.H., Valeri, M., Alnoor, A., Hussien, Z.A. (eds.) IMDC-IST 2024. LNNS, vol. 895, pp. 131–141. Springer, Cham (2023). https://doi.org/10.1007/978-3-031-51716-7_9

Baid, V., Jayaraman, V.: Amplifying and promoting the "S" in ESG investing: the case for social responsibility in supply chain financing. Manag. Finan. **48**(8), 1279–1297 (2022). https://doi.org/10.1108/MF-12-2021-0588

Buallay, A.: Is sustainability reporting (ESG) associated with performance? Evidence from the European banking sector. Manag. Environ. Qual. Int. J. **30**(1), 98–115 (2019). https://doi.org/10.1108/MEQ-12-2017-0149

Drempetic, S., Klein, C., Zwergel, B.: The influence of firm size on the ESG Score: corporate sustainability ratings under review. J. Bus. Ethics **167**(2), 333–360 (2020). https://doi.org/10.1007/s10551-019-04164-1

Escrig-Olmedo, E., Fernández-Izquierdo, M.Á., Ferrero-Ferrero, I., Rivera-Lirio, J.M., Muñoz-Torres, M.J.: Rating the raters: evaluating how ESG rating agencies integrate sustainability principles. Sustainability (Switzerland) **11**(3), 915 (2019). https://doi.org/10.3390/su11030915

Gao, S., Meng, F., Gu, Z., Liu, Z., Farrukh, M.: Mapping and clustering analysis on environmental, social and governance field a bibliometric analysis using scopus. Sustainability (Switzerland) **13**(13), 7304 (2021). https://doi.org/10.3390/su13137304

Garcia, A.S., Mendes-Da-Silva, W., Orsato, R.J.: Corporate sustainability, capital markets, and ESG performance. In: Mendes-Da-Silva, W. (ed.) Individual Behaviors and Technologies for Financial Innovations, pp. 287–309. Springer, Cham (2019). https://doi.org/10.1007/978-3-319-91911-9_13

Hübel, B., Scholz, H.: Integrating sustainability risks in asset management: the role of ESG exposures and ESG ratings. J. Asset Manag. **21**(1), 52–69 (2020). https://doi.org/10.1057/s41260-019-00139-z

Husin, N.A., Abdulsaeed, A.A., Muhsen, Y.R., Zaidan, A.S., Alnoor, A., Al-mawla, Z.R.: Evaluation of metaverse tools based on privacy model using fuzzy MCDM approach. In: Al-Emran, M., Ali, J.H., Valeri, M., Alnoor, A., Hussien, Z.A. (eds.) IMDC-IST 2024. LNNS, vol. 895, pp. 1–20. Springer, Cham (2023). https://doi.org/10.1007/978-3-031-51716-7_1

Imperiale, F., Pizzi, S., Lippolis, S.: Sustainability reporting and ESG performance in the utilities sector. Utilities Policy **80**, 101468 (2023). https://doi.org/10.1016/j.jup.2022.101468

Ismail, A.M., Latiff, I.H.M.: Board diversity and corporate sustainability practices: evidence on environmental, social and governance (ESG) reporting. Int. J. Finan. Res. **10**(3), 31–50 (2019). https://doi.org/10.5430/ijfr.v10n3p31

Jebe, R.: The convergence of financial and ESG materiality: taking sustainability mainstream. Am. Bus. Law J. **56**(3), 645–702 (2019). https://doi.org/10.1111/ablj.12148

Muhsen, Y.R., Husin, N.A., Zolkepli, M.B., Manshor, N.: A systematic literature review of fuzzy-weighted zero-inconsistency and fuzzy-decision-by-opinion-score-methods: assessment of the past to inform the future. J. Intell. Fuzzy Syst. **45**(3), 4617–4638 (2023)

Ng, T.H., Lye, C.T., Chan, K.H., Lim, Y.Z., Lim, Y.S.: Sustainability in Asia: the roles of financial development in environmental, social and governance (ESG) performance. Soc. Indic. Res. **150**(1), 17–44 (2020). https://doi.org/10.1007/s11205-020-02288-w

Nobanee, H., et al.: A bibliometric analysis of sustainability and risk management. Sustainability (Switzerland) **13**(6), 3277 (2021). https://doi.org/10.3390/su13063277

Nousheen, A., Kalsoom, Q.: Education for sustainable development amidst COVID-19 pandemic: role of sustainability pedagogies in developing students' sustainability consciousness. Int. J. Sustain. High. Educ. **23**(6), 1386–1403 (2022). https://doi.org/10.1108/IJSHE-04-2021-0154

Nunhes, T.V., Garcia, E.V., Espuny, M., Santos, V.H. de M., Isaksson, R., de Oliveira, O.J.: Where to go with corporate sustainability? Opening paths for sustainable businesses through the collaboration between universities, governments, and organizations. Sustainability (Switzerland) **13**(3), 1429 (2021).https://doi.org/10.3390/su13031429

Oliver Yébenes, M.: Climate change, ESG criteria and recent regulation: challenges and opportunities. Eurasian Econ. Rev. (2024). https://doi.org/10.1007/s40822-023-00251-x

Oprean-Stan, C., Oncioiu, I., Iuga, I.C., Stan, S.: Impact of sustainability reporting and inadequate management of ESG factors on corporate performance and sustainable growth. Sustainability (Switzerland) **12**(20), 1–31 (2020). https://doi.org/10.3390/su12208536

Patil, R.A., Ghisellini, P., Ramakrishna, S.: Towards sustainable business strategies for a circular economy: environmental, social and governance (ESG) performance and evaluation. In: Liu, L., Ramakrishna, S. (eds.) An Introduction to Circular Economy, pp. 527–554. Springer, Singapore (2020). https://doi.org/10.1007/978-981-15-8510-4_26

Punj, N., Ahmi, A., Tanwar, A., Abdul, S.: Mapping the field of green manufacturing : a bibliometric review of the literature and research frontiers. J. Clean. Prod. **423**(August), 138729 (2023). https://doi.org/10.1016/j.jclepro.2023.138729

Rahman, H.U., Zahid, M., Al-Faryan, M.A.S.: ESG and firm performance: The rarely explored moderation of sustainability strategy and top management commitment. J. Clean. Prod. **404**, 136859 (2023). https://doi.org/10.1016/j.jclepro.2023.136859

Sady, M., Żak, A., Rzepka, K.: The role of universities in sustainability-oriented competencies development: Insights from an empirical study on polish universities. Adm. Sci. **9**(3), 62 (2019). https://doi.org/10.3390/admsci9030062

Scheyvens, R., Banks, G., Hughes, E.: The private sector and the SDGs: The need to move beyond 'business as usual.' Sustain. Dev. **24**(6), 371–382 (2016). https://doi.org/10.1002/sd.1623

Sharma, P., Panday, P., Dangwal, R.C.: Determinants of environmental, social and corporate governance (ESG) disclosure: a study of Indian companies. Int. J. Discl. Gov. **17**(4), 208–217 (2020). https://doi.org/10.1057/s41310-020-00085-y

Siao, H.J., Gau, S.H., Kuo, J.H., Li, M.G., Sun, C.J.: Bibliometric analysis of environmental, social, and governance management research from 2002 to 2021. Sustainability (Switzerland) **14**(23), 16121 (2022). https://doi.org/10.3390/su142316121

Singhania, M., Saini, N.: Institutional framework of ESG disclosures: comparative analysis of developed and developing countries. J. Sustain. Finan. Investment **13**(1), 516–559 (2023). https://doi.org/10.1080/20430795.2021.1964810

Steblianskaia, E., Vasiev, M., Denisov, A., Bocharnikov, V., Steblyanskaya, A., Wang, Q.: Environmental-social-governance concept bibliometric analysis and systematic literature review: do investors becoming more environmentally conscious? Environ. Sustain. Ind. **17**, 100218 (2023)

Yadav, M., Saini, M.: Environmental, social and governance literature: a bibliometric analysis. Int. J. Manag. Financ. Account. **15**(2), 231–254 (2023). https://doi.org/10.1504/IJMFA.2023.129844

Zaidan, A.S., Alshammary, K.M., Khaw, K.W., Yousif, M., Chew, X.: Investigating behavior of using metaverse by integrating UTAUT2 and self-efficacy. In: Al-Emran, M., Ali, J.H., Valeri, M., Alnoor, A., Hussien, Z.A. (eds.) IMDC-IST 2024. LNNS, vol. 895, pp. 81–94. Springer, Cham (2023). https://doi.org/10.1007/978-3-031-51716-7_6

Zhang, N., Yang, C., Wang, S.: Research progress and prospect of environmental, social and governance: a systematic literature review and bibliometric analysis. J. Clean. Prod. **447**, 141489 (2024). https://doi.org/10.1016/j.jclepro.2024.141489

The Mediating Influence of Energy Reduction on the Relationship Between Green Production and Digital Sustainability: Insights from Iraqi Oil Companies

Abbas Gatea Atiyah[1], Mushtaq Yousif Alhasnawi[1], Sajead Mowafaq Alshdaifat[2], Mohammed Basendwah[3], and Ridzwana Mohd Said[4(✉)]

[1] Faculty of Administration and Economics, University of Thi-Qar, Nasiriyah, Iraq
{abbas-al-khalidi,alhasnawi78}@utq.edu.iq
[2] School of Business, Al al-Bayt University, Mafraq, Jordan
salshdaifat@aabu.edu.jo
[3] Department of Tourism and Hotel Management, Umm Al-Qura University, Makkah, Saudi Arabia
[4] School of Business and Economics, Universiti Putra Malaysia (UPM), 43400 Serdang, Selangor, Malaysia
ridzwana@upm.edu.my

Abstract. Over the past three decades, sustainability issues have become essential goals worldwide; therefore, new concepts, strategies, frameworks, and systems have been developed to address this issue. The study explores the relationship between green production and sustainability and how energy reduction mediates this relationship. Data from Iraqi oil companies were analyzed using Partial Least Squares Structural Equation Modeling (PLS-SEM). The research findings indicate that green production and energy reduction positively impact sustainability, and energy reduction acts as a positive mediator in the relationship between green production and sustainability. The study findings highlight the significance of green production and energy reduction as key factors driving sustainability. This study provides practical and managerial implications for oil companies, offering a framework managers and policymakers can utilize to ensure sustainability within their operations.

Keywords: Green Production · Sustainability · Energy Reduction · Oil Companies · Iraq

1 Introduction

Sustainability issues in recent years have been a growing concern worldwide, particularly in highly sensitive industries like oil production companies due to their significant environmental impact. The recent trend where companies embrace green production is a sign of a significant step toward meeting sustainability goals (Li et al., 2023). These measures include diminishing air pollution and saving energy to protect employees and society. In an effort to enhance sustainability, green production ensures lower costs and provides significant market competitiveness (Afum et al., 2020). Besides, it involves the creation of environmentally friendly products aimed at conserving resources and minimizing pollution and waste (Almoussawi et al., 2022). However, Iraqi oil companies face huge challenges in aligning their operations with sustainable goals due to the complexities of their industry and environmental conditions. Reducing carbon emissions, sustaining water, recycling waste, and protecting wildlife in sensitive ecological zones are among the concerns of oil companies in Iraq (Saeed et al., 2016). Moreover, the Iraqi oil sector is the key to the country's economic stability that can be regarded as the cornerstone of the national economy.

Managing the optimal use of resources, particularly energy, becomes crucial to maintain sustainability. Energy reduction initiatives do not only promote sustainability but also bring a broader dimension of long-term economic growth (Ali et al., 2021; Raihan et al., 2022). Energy reduction function as a basic and the most important principle of green productions and sustainability by protecting the environment, conserve the resources and reduce the cost (Abdul Kareem et al., 2024; Al-Wattar et al., 2019). Moreover this moves also promote technological advancement, increase compliance to regulation, and enhance corporate reputation (Bhardwaj et al., 2020). Through taking the lead in energy efficiency, businesses are signaling that they can provide financial stability while not compromising on nature conservation to achieve the sustainable future (Chang-xin et al., 2021; Renna & Materi, 2021). Henceforth, corporations need to devote efforts into the field of resource use optimization while safeguarding their economic interests. This calls for a comprehensive approach that integrates sustainable practices into their operations while maintaining financial stability (Li et al., 2023).

As green production practice is intensified, the recent research has focused on the impact of green production on sustainability. For instance, Zhang et al. (2019) studied how the adoption of green production improve society in terms of social effect, whereas Guo et al. (2020) investigated the potential of green technology for sustainable development. Nevertheless, there is a lack of understanding concerning the mediating role of energy reduction and it is questionable how energy reduction mediating the relationship between green production and sustainability (Li et al., 2023; Muhsen et al., 2023). Thus, the main focus of this analysis is to investigate the link between green production, energy reduction, and sustainability drawing insights from Iraqi oil companies.

This study attempts to fill the gap in previous studies. It is designed to understand how green production and energy reduction affect sustainability through empirical research carried out within the context of Iraqi oil companies. It also explores energy reduction's mediating role in this relationship, which is essential to achieving sustainable environmental goals. This paper contributes to existing knowledge and encourages dialogues within academia. The study offers significant insights that industry practitioners, policymakers, and stakeholders can utilize to promote sustainable development and environmental stewardship in the oil industry.

2 Literature Review and Hypotheses Development

2.1 Green Production and Energy Reduction

Green production is the process of improving manufacturing or production environmentally (Akhtar et al., 2023). Green production decreases negative environmental influences through environmentally friendly practices and conserves natural resources (Chen et al., 2006). This procedure includes the entire production, from raw resources to the end product, utilizing renewable resources, energy-efficient processes, and waste recycling (Borowski & Barwicki, 2023). Green production aims to decrease the environmental harm that manufacturing causes at a lower cost and higher profit (Akhtar et al., 2023). For example, GP practices may use new materials obtained from recycled waste (Akhtar et al., 2023), manufacture long-lasting products (Sun et al., 2023), and adopt methods that allow limiting electric conductivity during production. A body of literature has shown that green production is one of the key environmental agents since it reduces waste during the manufacturing process and reduces vehicle pollutants while also introducing recycling or reuse of waste or products and renewable resources for product production. Based on the above argument, we suppose the following hypothesis.

H1: Green Production Has a Positive Impact on Energy Reduction.

2.2 Energy Reduction and Sustainability

Energy reduction and environmental sustainability are critical parts of modern society; thus, our main goal is to relieve efficiency, reduce the environmental load, and promote sustainable development. Sustainable energy storage green practices have a critical role in the energy conversion and storage of proper conductivity needed for carbon dioxide reduction, which greatly influences the environment (Li et al., 2023). Energy management is realized as a necessity for sustainable development, and systematic energy-saving approaches are supported to accomplish sustainable development goals (Chang-xin et al., 2021). The literature offers insights into factors influencing energy intensity, consumption, and sustainability, providing a basis for policy development and efficiency assessment (Aboagye, 2019). Green practices play a significant role in sustainable energy storage, with processes like bifunctional oxygen reduction and evolution essential for energy conversion and storage (Chang-xin et al., 2021). Energy management is vital for sustainable development, stressing the need for systematic approaches to achieve energy savings and contribute to sustainable development goals (Awan et al., 2014). The environmental standards for the certification of ISO 14000, green energy, and resource efficiency are vital for a sustainable environment (Karaeva et al., 2023). As a result, green technology should be added by production industries to fulfill their production process, which is green and move sustainability ahead [10]. Based on the above argument, we suppose the following hypothesis.

H2: Energy Reduction Has a Positive Impact on Sustainability.

2.3 Green Production and Sustainability

In recent years, the energy shortage has been one of the leading global issues faced by most countries (Afum et al., 2021). It runs industries, and development would have been impossible without the energy. Diffusion of Innovation theory emphasizes the social effect and perceived advantage of adopting green protection practices and technologies in industries. The use of energy-saving technologies in production and manufacturing, as well as reliance on renewable energy sources, is one of the ways in which green production can help achieve sustainability (Li et al., 2023). Energy production is the energy consumed during manufacturing or production (Sardana et al., 2020). Effective energy production consists of environmentally friendly practices that decrease the environmental effects of manufacturing (Aboagye, 2019). Manufacturing companies should decrease electrical conductivity to the level possible to enhance and maintain sustainability. It also includes optimization of the manufacturing process and the lowest possible consumption of resources [90, 91]. With the help of these practices, industries can reduce environmental effects (He et al., 2019). Based on the above argument, we suppose the following hypothesis.

H3: Green Production Has a Positive Impact on Sustainability.

2.4   The Mediating Role of Energy Reduction

Based on the diffusion of innovation theory, engaging the public in the positive side of innovative actions, such as environmental production, can lead to easier adoption of "energy reduction" in conservation (Guo et al., 2020). As a result, there can be a great reduction in energy consumption, which is simultaneously the essential requirement of those customers who prefer a green or sustainable choice. Energy reduction is a factor of adapting environmentally friendly manufacturing techniques. Besides the issue, renewable energy utilization not only manages to save energy but also evolves into a traditional norm for enterprises, which guarantees its promotion (Khoshnava et al., 2019). Therefore, higher environmental reduction efforts correlate directly with enhanced sustainability, hence trending towards a conciliating role between efficient production and sustainability.

Based on the literature on green production processes and sustainability, green production technologies are a method to improve sustainability outcomes (Afum et al., 2021). Besides resource efficiency and waste reduction, green practices like producing lower energy consumptions within the manufacturing process are anticipated to rise (He et al., 2019). Anticipated energy usage will potentially comprise more sustainability goals, i.e., conservation of the environment, the efficiency of the economy, and social responsibility (Afum et al., 2021). Also, Green manufacturing encompasses the production of environmentally friendly materials and processes, among other practices, and the primary intention is to decrease the environmental effect, consequently deploying green manufacturing practices across organizations (Afum et al., 2021). Energy conservation influences this equation by simplifying the shift towards the use of renewable energy, as energy efficiency and sustainable practices have a relationship with environmental impact and the conservation of resources (Aboagye, 2019). This hypothesis aligns with the broader discourse on sustainable energy technologies and the imperative of reducing energy consumption for environmental preservation and long-term sustainability (Wang et al., 2017).

H4: Energy Reduction Positively Mediates the Relationship Between Green Production and Sustainability.

3 Methodology

3.1 Our Sample Size and Procedure

The present study is a quantitative study, adopting a deductive approach and using questionnaire-based survey approaches, which were widely used in prior research (e.g., Alhasnawi et al., 2023a, b; Atiyah et al., 2023; Hirst et al., 2011). It enables sample comparison and the cost- and time-effective collection of standardised data; therefore, it was used by many respondents (Sekaran & Bougie, 2016). The population of this study is individuals in oil companies in Iraq, which include both private and public sectors of Iraqi oil companies. Iraq has many companies that work in the oil sector. However, the present study has focused on companies that work in Basra. The reason for choosing these companies is that the companies in Basra produce about 60% of the total Iraqi oil production, which has an environmental effect on this city. G*Power (Hair et al., 2017) was utilized to determine the minimal sample size, and it showed that the sample size required is 98 to achieve a power of 0.8 for the proposed framework. The number of respondents in the present study is 373, which achieves the minimum sample size requirement. Items pertaining to each variable in the study instrument were adapted from many existing studies.

3.2 Instrument Development and Measurement

The items have been adopted from prior studies and were adapted and modified to suit the study's specific needs while maintaining the content and accuracy of the measurements. Significant effort was invested in developing the questionnaire, focusing on the language used and the organization of the questions, ensuring that the questionnaire was clear and easy to understand (Alhasnawi et al., 2023a, b; Podsakoff et al., 2012). Additionally, specialists in both languages evaluated the English and Arabic grammar; the terminology used in the questionnaire was chosen to suit the comprehension level of managers in Iraqi oil companies. The validity of the adapted or adopted items was examined by seeking the opinions of three practitioners and five academic experts from different companies. Energy reduction (4 items), green production (5 items), and sustainability (6 items) were measured using the scale adapted from (Chen et al., 2006), (Li et al., 2023), and (Sardana et al., 2020), respectively. All constructs were measured using a 5-point Likert scale with responses ranging from 1 (strongly disagree) to 5 (strongly agree).

4 Results

In the present research, data analysis has been divided into two stages. Firstly, SPSS software was used to analyze and process data. These processes include entering data, screening the data, detecting outliers, computing descriptive statistics, and examining normality via skewness and kurtosis. The next stage is partial least squares structural equation modeling (PLS-SEM) using SmartPLS 4, which includes an assessment of the measurement model and an assessment of the structural model.

4.1 Descriptive Analysis of Respondents

The demographic characteristics of the participants indicate that a significant proportion of the respondents were male (71%), between 30 and 40 years old, and possessed doctoral degrees (see Table 1). The participants' average experience in their current job was more than six years, with most holding the post of college department head.

Table 1. Demographic Information

Description of n = 373	Demographic characteristics	Frequency	Percentage (%)
Gender	Mele	265	71.05%
	Female	108	28.95%
Age	Below 30	6	1.61%
	30–40	151	40.48%
	40–50	132	35.39%
	Above 50	84	22.52%
Education	Diploma	39	10.46%
	Bachelor	79	21.18%
	Master	22	5.90%
	Ph.D.	15	4.02%
	Others	218	58.45%
Work duration	Less than two years	6	1.61%
	2–6 years	76	20.38%
	6–10 years	145	38.87%
	10–15 years	108	28.95%
	More than 15 years	38	10.19%

4.2 Measurement Model

Convergent and discriminant validity are two important aspects of measuring the validity of a model. Convergent validity is typically evaluated using factor loadings (FL > 0.7), composite reliability (CR > 0.7), and average variance extracted (AVE > 0.5) (Hair et al., 2017). Table 2 shows that the model has no issues regarding convergent validity, CR exceeded 0.7, AVE values exceeded 0.5, and all item loadings exceeded 0.7.

The test results confirm that all constructs have sufficient discriminant validity, as evidenced by the HTMT values below 0.85 (Henseler et al., 2015). As shown in Table 3 no multicollinearity problem exists between the constructs, implying that each construct measures a unique and distinct concept and has no solid link to other constructs.

Table 2. Measurement Model Summary

Construct/Items	FL	CR	AVE
Green Production (Chen et al., 2006)			
GP1	0.841	0.918	0.658
GP2	0.833		
GP3	0.878		
GP4	0.853		
Energy Reduction (Li et al., 2023)			
ER1	0.846	0.839	0.640
ER2	0.785		
ER3	0.871		
ER4	0.837		
ER5	0.792		
Sustainability (Sardana et al., 2020)			
SU1	0.828	0.813	0.641
SU2	0.786		
SU3	0.872		
SU4	0.877		
SU5	0.825		
SU6	0.840		

AVE, average variance extracted; FL, factor loading; CR, composite reliability

Table 3. Validity of Constructs Using HTMT Method

No.	Constructs	1	2	3
1	Energy Reduction			
2	Green Production	0.198		
3	Sustainability	0.456	0.113	

4.3 Structural Model

The structural model aims to clarify the causal relationships between the exogenous and endogenous variables. Table 4 and Fig. 1 show that both direct hypotheses have been accepted, where green production positively impacts energy reduction ($\beta = 0.174$, $t = 3.330$, $p < 0.01$, $f^2 = 0.153$). In addition, energy reduction also positively impacts sustainability ($\beta = 0.417$, $t = 7.919$, $p < 0.01$, $f^2 = 0.210$). Furthermore, the result shows that energy reduction positively mediates the relationship between green production and sustainability ($\beta = 0.072$, $t = 3.151$, $p < 0.01$, $f^2 = 0.123$), which means accepting the indirect hypothesis.

Table 4. Structural Model and Hypotheses Testing

Path coefficient	St. Beta	St. Dv	T	P	2.50% LB	97.50% U B	f^2	Support
Green Production - > Energy Reduction	0.173	0.053	3.259	0.001	0.082	0.282	0.153	Yes
Energy Reduction - > Sustainability	0.409	0.054	7.516	0.000	0.314	0.522	0.210	Yes
Green Production - > Sustainability	0.724	0.019	37.648	0.000	0.314	0.522	0.030	Yes
Green Production - > Energy Reduction - > Sustainability	0.071	0.023	3.103	0.002	0.035	0.123		Yes

The path coefficient findings show that Sustainability yielded the highest coefficient determination ($R^2 = 18.1\%$) by green production and energy reduction. Furthermore, the relationship between green production and energy reduction produced an R^2 of 13.0% (see Fig. 1). The value of Q^2 refers to the model having predictive relevance. The Q^2 value in the present study was more than zero (ER = 0.153 SU = 0.077), which indicates the present model has predictive relevance.

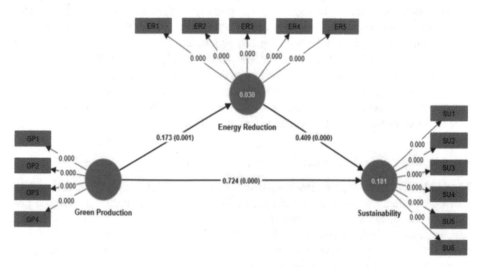

Fig. 1. Structural Model

5 Discussion

The first hypothesis in this study assumes that Green Production positively impacts energy reduction. Based on the results of the analysis, this hypothesis has been accepted. This means companies implementing green production practices experience decreased energy feasting in their manufacturing procedures. This result aligns with prior studies highlighting the role of green production in enhancing resource efficacy and environmental sustainability (Tariq et al., 2022). The second hypothesis assumes that energy reduction positively affects sustainability. The analysis shows compelling evidence to accept this hypothesis, which indicates that companies that successfully decrease their energy consumption display higher levels of sustainability across several dimensions. The outcome aligns with past studies like Li et al. (2023). The third hypothesis proposes that green production positively affects sustainability. Based on the findings, the hypothesis has been supported. These results are consistent with those of other studies and suggest that the companies implementing green production exhibit higher levels of sustainability. This means by using eco-friendly measures like renewable energy utilization and waste reduction; companies can lower their environmental effect and contribute to the larger goal of sustainability. Moreover, the expanding green production sustaining element goes beyond industrial and systemic changes. With sustainability becoming a mainstream topic in companies by recognizing that it is both a catalyst for innovation, market differentiation, and creating value for stakeholders, green production practices will become more prevalent across various sectors (González-Benito & González-Benito, 2006). This transforming shift to sustainable manufacturing seems to address key diversities that are in place, such as climatic change, resource scarcity, and social inequality, as well as to promote the economy while the people's well-being is enhanced (Sarkis et al., 2011). The last hypothesis stated in this research is that energy reduction is a positive mediator in the relationship between green production and sustainability. Our analysis strongly supports this hypothesis, demonstrating that energy reduction is critical in improving the sustainability of organizations that implement green production practices. This discovery highlights energy efficiency initiatives as an influential mechanism by which green production supports more comprehensive sustainability objectives (Dangelico & Pontrandolfo, 2015). Through decreasing energy use and reducing environmental pollution, organizations will attain considerable success in their sustainability performance in terms of economic, environmental, and social aspects. These findings are consistent with those of other studies, such as Sarkis et al. (2011), Roselló-Soto et al. (2016), and Li et al. (2023).

5.1 Theoretical Contribution

The study results have provided insight into the theory background in environmental protection, energy conservation, and organizational studies. Energy efficiency is a prerequisite for sustainable development. Through real measurement of the relationship between energy savings and sustainability, the research underpins the role played by energy efficiency in attaining environmental, economic, and social sustainability goals. In addition, the energy diminution action as an intermediary in the relationship between eco-friendly production operations and sustainability signifies a key role in instigating organizations to adopt sustainable production strategies. It is highlighted that it is vital to implement energy saving principles to ensure the success of clean production schemes. Inclusion of energy saving approach will ensure that aspect of sustainability is universally incorporated in various fields. Furthermore, the elements of these impressions become the ingredient for the apprehension of insights on the dynamics of sustainable production underpinning resource-based sectors, and thus provide a new level of understanding in sustainability economics in the resource-based economy complex. This study underscores the main issues concerning ecological productivity accounting and aims to enrich the theoretical framework by elucidating the complex interplay between energy preservation, "green" production and sustainability of a specific industrial context contributing to the enrichment of both academia and practice.

5.2 Practical and Policy Contribution

The results of this study suggest that the sustainability policies and regulations in the oil sector will contribute substantially to greater environmental friendliness. The proof of a positive correlation between energy production and sustainability is clear and emphasizes the need to set a policy and provide incentives to encourage energy efficiency practices and add more renewable energy. Therefore, policymakers can utilize these data to avail relevant programs like tax incentives, subsidies, and proper regulation to provide a roadmap and strategies on how organizations can ditch old technologies by using new ones and clean energy resources. Moreover, it should be noted that energy reduction's role in sustainability is among the factors that make green production successful. Policymakers can also focus on pushing action that makes energy management a part of strategies in general.

Additionally, this research provides decision-makers, such as managers within Iraq's petroleum organizations, with actionable information that helps improve sustainability performance. By recognizing the significance of energy effectiveness in sustainability and its role in the bond between Iraq's green production and sustainability, decision-makers can put their investments in the energy efficiency process first. These measures can be undertaken by implementing energy-efficient production procedures, purchasing renewable power resources, and placing energy management systems to enhance technology utilization and diminish negative impacts. For the same reason, a complex involving both green production and energy saving will better guide toward broad-based sustainability strategies that address environmental issues and productivity improvement, among others.

6   Limitations and Future Directions Research

This study contributes a great deal to our understanding of the mediating effect of energy reduction consumption in the midst of the connection between green production and the sustainability of oil companies within Iraq; nevertheless, some shortcomings need to be considered. Firstly, the data from this research was collected at a single point in time (a cross-sectional study), giving no chance to analyze causal relationships among the variables. A future study could apply a more dynamic, longitudinal model that can tackle the temporal nature of relationship complexity. Besides that, the study mainly concentrated on Iraqi oil corporations, which was inadequate for generalizing the findings to other cases in different fields or regions. In future research, the researchers might develop a study that covers other industries and/or country-based companies to validate the original results more comprehensively. Additionally, the study was prone to bias due to self-reported data, which can cause common method bias. Selecting a multi-source, multi-method, and longitudinal data approach can be a helpful means to control the common method bias. Another possible limitation is partial least squares structural equation modeling (PLS-SEM), which is the unique analytical approach used. Although PLS-SEM can be utilized to model complex structures and analyze small samples, future investigations using alternative statistical methods that meet the same goals, for example, covariance-based structural equation modeling (CB-SEM), could validate these findings. Finally, although this research is considered helpful for oil company managers, future studies may take into account the views of other stakeholders, such as state authorities, non-governmental organizations, and local people, in an attempt to establish a more general understanding of the sustainability measures in the oil sector. Scientists can strive to expand the fields of research, rectify the limitations found, enhance the robustness of the findings, and contribute to the discourse on sustainable development and environmental stewardship.

7 Conclusion

The study explores the relationship between green production, energy reduction, and sustainability. The results show that green production and energy reduction positively and significantly impact sustainability. Moreover, the results confirmed the mediating role of energy reduction in the relationship between green production and sustainability. Thus, the study's findings emphasize that industries should invest in green production and energy reduction to enhance sustainability. The present study has theoretical, policy, and managerial contributions. It helps researchers further promote studies on other industrial factors to achieve more sustainability. Furthermore, it enhances investments in green practices and innovation, which promotes sustainability. Finally, adopting green production and energy reduction will guide managers in promoting sustainable practices.

References

Abdul Kareem, B.L. Zubaidi, S., Al-Ansari, N., Raad Muhsen, Y.: Review of recent trends in the hybridisation of preprocessing-based and parameter optimisation-based hybrid models to forecast univariate streamflow. Comput. Model. Eng. Sci. **138**(1), 1–41 (2024). https://doi.org/10.32604/cmes.2023.027954

Aboagye, S.: What drives energy intensity in Ghana? A test of the Environmental Kuznets hypothesis. In: OPEC Energy Review, vol. 43, no. 3, pp. 259–276 (2019). https://doi.org/10.1111/opec.12155

Afum, E., Agyabeng-Mensah, Y., Sun, Z., Frimpong, B., Kusi, L.Y., Acquah, I.S.K.: Exploring the link between green manufacturing, operational competitiveness, firm reputation and sustainable performance dimensions: a mediated approach. J. Manuf. Technol. Manag. **31**(7), 1417–1438 (2020). https://doi.org/10.1108/JMTM-02-2020-0036

Afum, E., Zhang, R., Agyabeng-Mensah, Y., Sun, Z.: Sustainability excellence: the interactions of lean production, internal green practices and green product innovation. Int. J. Lean Six Sigma **12**(6), 1089–1114 (2021). https://doi.org/10.1108/IJLSS-07-2020-0109

Akhtar, M.T., et al.: Sustainable production of biodiesel from novel non-edible oil seeds (Descurainia sophia L.) via green nano CeO2 catalyst. In: Energies, vol. 16, no. 3 (2023). https://doi.org/10.3390/en16031534

Al-Wattar, Y.M.A., Almagtome, A.H., Al-Shafeay, K.M.: The role of integrating hotel sustainability reporting practices into an accounting information system to enhance hotel financial performance: evidence from Iraq. Afr. J. Hospit. Tour. Leisure **8**(5), 1–16 (2019)

Alhasnawi, M., Said, R.M., Daud, Z.M., Muhamad, H.: Budget participation and managerial performance: bridging the gap through budget goal clarity. Adv. Soc. Sci. Res. J. **10**(9), 187–200 (2023). https://doi.org/10.14738/assrj.109.15539

Alhasnawi, M.Y., Mohd Said, R., Mat Daud, Z., Muhammad, H.: Enhancing managerial performance through budget participation: insights from a two-stage a PLS-SEM and artificial neural network approach (ANN). J. Open Innov.: Technol. Mark. Complexity **9**(4), 100161 (2023). https://doi.org/10.1016/j.joitmc.2023.100161

Ali, M.T., Muhsen, Y.R., Chisab, R.F., Abed, S.N.: Evaluation study of radio frequency radiation effects from cell phone towers on human health. Radioelectron. Commun. Syst. **64**(3), 155–164 (2021). https://doi.org/10.3103/S0735272721030055

Almoussawi, Z.A., Sarhed, J. N., Saeed, M., Ali, M.H., Wafqan, H.M., Alhasan, S.A.A.: Moderating the role of green trust in the relationship of green brand positioning, green marketing, green production, and green consumer value on green purchase intention of university students in Iraq. Transnational Mark. J. **10**(3), 738–750 (2022). https://doi.org/10.33182/tmj.v10i3.2177

Atiyah, A.G., Alhasnawi, M., Almasoodi, M.F.: Understanding metaverse adoption strategy from perspective of social presence and support theories: the moderating role of privacy risks. In: Al-Emran, M., Ali, J.H., Valeri, M., Alnoor, A., Hussien, Z.A. (eds.) IMDC-IST 2024. LNNS, vol. 876, pp. 144–158. Springer, Cham (2023). https://doi.org/10.1007/978-3-031-51300-8_10

Bhardwaj, A.K., Garg, A., Ram, S., Gajpal, Y., Zheng, C.: Research trends in green product for environment: a bibliometric perspective. Int. J. Environ. Res. Public Health **17**(22), 1–21 (2020). https://doi.org/10.3390/ijerph17228469

Borowski, P.F., Barwicki, J.: Efficiency of utilization of wastes for green energy production and reduction of pollution in rural areas. Energies **16**(1), 13 (2023). https://doi.org/10.3390/en16010013

Chen, Y.-S., Lai, S.-B., Wen, C.-T.: The influence of green innovation performance on corporate advantage in Taiwan. J. Bus. Ethics **67**, 331–339 (2006)

González-Benito, J., González-Benito, Ó.: The role of stakeholder pressure and managerial values in the implementation of environmental logistics practices. Int. J. Prod. Res. **44**(7), 1353–1373 (2006). https://doi.org/10.1080/00207540500435199

Guo, M., Nowakowska-Grunt, J., Gorbanyov, V., Egorova, M.: Green technology and sustainable development: assessment and green growth frameworks. Sustainability **12**(16), 6571 (2020)

Hair, J.F., Hult, G.T.M., Ringle, C.M., Sarstedt, M.: A Primer on Partial Least Squares Structural Equation Modeling (PLS-SEM). Sage (2017)

He, Y., Guo, S., Chen, K., Li, S., Zhang, L., Yin, S.: Sustainable green production: a review of recent development on rare earths extraction and separation using microreactors. ACS Sustain. Chem. Eng. **7**(21), 17616–17626 (2019). https://doi.org/10.1021/acssuschemeng.9b03384

Henseler, J., Ringle, C.M., Sarstedt, M.: A new criterion for assessing discriminant validity in variance-based structural equation modeling. J. Acad. Mark. Sci. **43**(1), 115–135 (2015). https://doi.org/10.1007/s11747-014-0403-8

Hirst, G., Van Knippenberg, D., Chen, C.H., Sacramento, C.A.: How does bureaucracy impact individual creativity? A cross-level investigation of team contextual influences on goal orientation-creativity relationships. Acad. Manag. J. **54**(3), 624–641 (2011). https://doi.org/10.5465/AMJ.2011.61968124

Khoshnava, S.M., et al.: Aligning the criteria of green economy (GE) and sustainable development goals (SDGs) to implement sustainable development. Sustainability **11**(17), 4615 (2019)

Li, C., et al.: Green production and green technology for sustainability: the mediating role of waste reduction and energy use. Heliyon **9**(12), e22496 (2023). https://doi.org/10.1016/j.heliyon.2023.e22496

Muhsen, Y.R., Husin, N.A., Zolkepli, M.B., Manshor, N., Al-Hchaimi, A.A.J., Ridha, H.M.: Enhancing NoC-based MPSoC performance: a predictive approach with ANN and guaranteed convergence arithmetic optimization algorithm. IEEE Access **11**, 90143–90157 (2023). https://doi.org/10.1109/ACCESS.2023.3305669

Podsakoff, P.M., MacKenzie, S.B., Podsakoff, N.P.: Sources of method bias in social science research and recommendations on how to control it. Annu. Rev. Psychol. **63**, 539–569 (2012). https://doi.org/10.1146/annurev-psych-120710-100452

Raihan, A., et al.: Nexus between carbon emissions, economic growth, renewable energy use, urbanization, industrialization, technological innovation, and forest area towards achieving environmental sustainability in Bangladesh. Energy and Climate Change **3**(August), 100080 (2022). https://doi.org/10.1016/j.egycc.2022.100080

Renna, P., Materi, S.: A literature review of energy efficiency and sustainability in manufacturing systems. Appl. Sci. (Switz.) **11**(16) (2021). https://doi.org/10.3390/app11167366

Roselló-Soto, E., et al.: Application of non-conventional extraction methods: toward a sustainable and green production of valuable compounds from mushrooms. Food Eng. Rev. **8**, 214–234 (2016)

Saeed, I.M., Ramli, A.T., Saleh, M.A.: Assessment of sustainability in energy of Iraq, and achievable opportunities in the long run. Renew. Sustain. Energy Rev. **58**, 1207–1215 (2016). https://doi.org/10.1016/j.rser.2015.12.302

Sardana, D., Gupta, N., Kumar, V., Terziovski, M.: CSR 'sustainability' practices and firm performance in an emerging economy. J. Clean. Prod. **258**, 120766 (2020). https://doi.org/10.1016/j.jclepro.2020.120766

Sarkis, J., Zhu, Q., Lai, K.: An organizational theoretic review of green supply chain management literature. Int. J. Prod. Econ. **130**(1), 1–15 (2011)

Sun, S., Wu, Q., Tian, X.: How does sharing economy advance cleaner production? Evidence from the product life cycle design perspective. Environ. Impact Assess. Rev. **99**, 107016 (2023). https://doi.org/10.1016/j.eiar.2022.107016

Tariq, G., et al.: Influence of green technology, green energy consumption, energy efficiency, trade, economic development and FDI on climate change in South Asia. Sci. Rep. **12**(1), 16376 (2022)

Exploring University Faculty Members' Sustainable Innovative Behavior of Work with Artificial Intelligence: A Review of the Literature

Hafiza Saadia Sharif[1], Al-Amin Bin Mydin[1], Israa M. Hayder[2], and Hussain A. Younis[3,4(✉)]

[1] School of Educational Studies, University Sains Malaysia, Gelugor, Malaysia
saadiasharif@lgu.edu.pk, alamin@usm.my
[2] Department of Computer Systems Techniques, Qurna Technique Institute, Southern Technical University, Basrah 61016, Iraq
israa.mh@stu.edu.iq
[3] College of Education for Women, University of Basrah, 61004 Basrah, Iraq
hussain.younis@uobasrah.edu.iq
[4] School of Computer Sciences, Universiti Sains Malaysia, 11800 Gelugor, Penang, Malaysia

Abstract. Sustainability and artificial intelligence intervention extensively in the field of education. In order to fabricate innovations, faculty members must be trained. However, the weak productivity growth shows that mostly countries passively require creativity and innovation, as evidenced by the inactive contribution of the entire aspect of productivity and education to the increase in output. This is despite the creation of a superb, creative human capital-producing education system on a global level. Innovative job behavior introduction, the deliberate invention, and application of novel ideas in the workplace with the aim of boosting both individual and organizational performance. As a result, lecturers who exhibit innovative behavior of work are capable of working creatively, contributing ideas, and producing fruitful results for the institution where they are employed. The factors or antecedents of innovative behavior of work. According to earlier research investigations; these indicators include commitment to the job, workplace facilities, autonomy, job instability, rewards, and job design. However, the majority of earlier studies held that job autonomy and commitment had a significant impact on innovative behavior of work. To close a knowledge gap regarding innovative behavior of work in educational institutions tasked with supplying faculty members with valuable and useful knowledge, this study aimed to create a review of literature for what factors university faculty members' innovative behavior of work. This material was researched by a thorough study of the pertinent literature. The study concludes that job autonomy and job dedication have a positive impact on creating the conditions that encourage university professors to engage in creative behavior of work. This article is expected to provide significant information about inventive behavior of work that will help the government develop work educational reform that will benefit government and its citizens.

Keywords: Job Commitment · Innovative Behavior · Job Autonomy · University teachers

A. Alnoor et al. (Eds.): AIRDS 2024, LNNS 1033, pp. 157–167, 2024.
https://doi.org/10.1007/978-3-031-63717-9_10

1 Introduction

Modern reconsideration has changed a number of factors that influence workforce competence and skills to become more imaginative. Adopting a work ethic can promote organization, management, and production competence, enable quality improvement, and foster creativity (Singh et al., 2020). It can also increase effectiveness, increase efficiency, and reduce costs. In the intervening time, business uprisings have given educational change new momentum (Xing & Marwala, 2017). Educationists and professionals have recently come to understand the significant impact that a wide range of technical advancements in in rank and make contact with technology (ICT) are having on education (Haseeb et al., 2019; Bilal et al., 2018). It is clear that education will be shaped by innovations and that faculty members will need to be taught how to create them (Ali, 2020). However, the weak productivity growth shows that most countries still lack creativity and innovation, as evidenced by the stagnant contribution of entire aspects of productivity and education to output growth (Senouci et al., 2010). This is despite the development of a world-class education system that produces excellent and creative human capital. Additionally, it is claimed in the 2010 publication of the New Economic Model that attempts to innovate and create are still insufficient. However, findings from earlier studies indicate that the educational system is not sufficient (Tok, 2020). Furthermore, innovative people are essential in the field of education, especially for university faculty members who take part in various teaching and learning processes (Chou et al., 2019). But most of the faculty members lack the motivation to innovate in their employment. Teachers prefer to stick to the tried-and-true conventional learning strategies in universities because they feel at ease there and don't see the need to develop or implement new learning strategies (Sutarto et al., 2020). This ignores the fact that faculty are diverse and need various teaching strategies. From a human perspective, creative behavior of work is defined as the creation, application, and spread of novel notes inside the group to encourage (Sciarelli et al., 2020).

The introduction, deliberate development, Sustainability and application of novel notes with the intention of improving one's own or an organization's appearance is innovative behavior of work. University teachers who behave in an innovative behavior at work will therefore be more equipped to think creatively, provide ideas, and result in positive outcomes for their employers. (Van Zyl et al., 2021). Numerous factors have been found to have an impact on innovative behavior of work.The components or reasons for innovative behavior of work differ according to the results of previous studies. Workplace commitment, facilities, autonomy, employment instability, rewards, and job designs are a few of these components. But the majority of earlier investigations point in a different direction (Singh, 2022). Job autonomy and dedication are strongly correlated with innovative behavior of work of work (Bawuro et al., 2019). The majority of these research studies on innovative behavior are concentrated in the benefit, up-to-date, and industrial regions. The field of education has not yet given research studies on innovative behavior of work the same level of attention (Parthasarathy & Premalatha, 2017). The researcher made very slight mention of research on the ways in which faculty members engage in innovation-related behavior of work s and the ways in which their active contributions might be encouraged and supported (Messmann et al., 2018). Many research's made a proposal to close a knowledge break regarding innovative behavior of work in educational organizations, whose goal is to give faculty members relevant and practical knowledge to combat this phenomenon. This study was to explore the factors that affect university faculty members innovative behavior of work. It is anticipated that this study will offer pertinent data on creative behavior of work that will help the government create an efficient educational reform that will benefit its university faculty members (Tosheva & Abdullaeva, 2022; (Younis, et al., 2020; Younis, et al., 2021; Noor, et al., 2024). Artificial Intelligence (AI) in education refers to the application of AI technologies, such as natural language processing (NLP), robotics, to enhance learning processes, improve educational outcomes machine learning (ML), and streamline administrative tasks within the education sector (Younis, et al., 2024; Noor, et al., 2024).

2 Literature Review Behavior of Work Innovative

2.1 Behavior of Work Innovative

Since the theory of "innovative behavior of work" (IWB) was confirmed in those two years, the literature utilizing innovative behavior of work has continuously grown. Any employee behavior that has to do with implementing, developing, presenting, and finding creative perceptions in the organization to improve innovative performance is engaging in creative behavior of work (Scott & Bruce, 1994). Another definition provided by other academics is the deliberate adoption and usage inside an organization of ideas, processes, goods (Battistelli et al., 2022). It is the deliberate expansion, propagation, and use of new concepts inside a work role, group, or organization with the goal of enhancing performance. Innovative behavior of work in the educational setting refers to alterations and enhancements made to the learning environment for the advantage of the students, such as the introduction of novel teaching methods, resources, technologies, and materials that benefit faculty members and foster their creative potential (Önhon, 2019). Similar to how new and helpful ideas, processes, goods, or procedures can be initiated and intentionally introduced; innovative behavior of work is typically described in terms of how individuals can do so (Baharuddin et al., 2019). Innovative behavior of work refers to employee actions that help the development and acceptance of innovations at work, both directly and indirectly (Setyawasih & Hamidah 2022; Efandi & Syuhada, 2021). Modern behavior of work can give an organization a good advantage (Hakimian et al., 2016). Innovation, on the other hand, is typically seen as a one-time endeavor (Ghani et al., 2009). Innovation is also considered a multi-stage procedure requiring varied actions and innovative behavior of work s at various stages. A human being is able to engage in any combination of these behaviors. An individual can therefore be expected to engage in various combinations of these actions at any given time (Spanuth & Wald, 2017). Help the development and acceptance of innovations at job, directly and indirectly. In the modern workplace, innovative behavior of work is one of the key elements for logistic growth and growth in both the public or private sectors. Innovative behavior of work can give an organization a competitive advantage, claim Hakimian et al. (2016). Innovation, on the other hand, is typically seen as a one-time endeavor. This pattern inspired organizational researchers to investigate the organizational variables that significantly influence workers' innovative behavior of work s (Muhamad et al., 2023). According to a survey of related literature, most earlier investigations into workers' innovative behavior of work work were carried out at the organizational level. There are stimulating reasons to promote educational innovation to maximize the return on public investment. According to earlier research, there are three primary, important reasons why universities require faculty members to exhibit innovative behavior of work s. First, to keep up with society's rapid development, innovative behavior of work is crucial. Their faculty are facing more demands in our knowledge-based society (Ausat et al., 2022; Sharif, et al., 2023).

Second, because faculty members and their methods of instruction cover biggest impact on learners' motivation and self-determination for learning, new breakthroughs and insights into teaching are requiring fresh behavior of work. (Yusof et al., 2018). Inventive behavior of work s in order for society to remain competitive. Creativity is a major force behind societal and economic advancement. Additionally, innovation is thought a tool to improve any company's capacity to adjust to shifting surroundings (Javed et al., 2019). Education is necessary to encourage faculty members' original and creative thinking, after all other words, innovative behavior is essential for the long-term success of educational organizations, professions, and the knowledge society. Few studies have specifically looked at innovative behavior of work and its determinants in university faculty personnel. Studies have demonstrated that a variety of characteristics (Baharuddin et al., 2019). Despite this, experiments suggest that a variety of motivational factors may, in the workplace, promote innovative work. Numerous studies demonstrate that job autonomy and dedication, which foster creative work, improve job performance across all including education and industries. On the other hand, internal rewards at work and the organization's external rewards (such as pay, positions, degrees, etc.) are significant elements in encouraging employees to embrace the concept of innovative behavior at work.

Opportunities exploration, knowledge generation, knowledge promotion, and idea realization (implementation) have previously been identified as the dimensions of innovative behavior of work (Kaur et al. 2020).

Clarify that opportunity discovery stage of the novelty process involves identifying opportunities to try to develop something new out of existing problems. It is characterized as dynamic creation and association process, growth in representations and opportunity categories, and exchange of ideas that can be expressed as tangible, abstract, or visual information. Exploring opportunities helps with idea generation. It is acknowledged that this is the stage where brand-new suggestions for goods, procedures, or services are generated with the intention of improvement. At the idea generation stage, novel concepts are created, and they are later promoted. Idea promotion involves introducing and spreading these concepts within the workplace by winning over significant figures or key leadership assembling of proponents of innovation (Messmann et al., 2018). Additionally, it can be used to represent someone who does not occupy a formal job but is capable of advancing a new service by exceeding all expectations (Leong & Rasli, 2014). The goal of concept promotion, often referred to as the implementation stage, is to obtain the resources necessary to realize the idea as well as the group's acceptance. When an organization decides to implement an idea, the idea promotion stage, also known as the implementation stage, tries to gain the support of the group and the funding necessary to make the idea a reality. When a company decides to create, evaluate, and market an idea, concept realization or implementation begins to occur. Innovation subsequently integrates into the workings of the organization (Fatoki, 2021; Alnoor et al., 2022; Alnoor et al., 2024a, b; XinYing et al., 2024).

2.2 Job Autonomy

Job autonomy is one of these perceived environmental influences. According to Sazandrishvili (2009), this idea has a key role in determining an employee's attitudes, motivation, and behavior. Job autonomy refers to how much control employees have over how their work is organized (Azim et al., 2012). Autonomy as the extent to which an employee has control over how to take out a line task. Workers can try out various work methodologies and techniques thanks to their autonomy. Additionally, by using these ideas on a small scale, it allows faculty members to discover new ideas and develop them further. According to Spiegelaere (2014), it is seen as a valuable job resource that is intrinsically motivation and prepares people to manage enormous workloads, according to Spiegelaere (2014). According to the literature, job autonomy is a crucial predecessor of an employee's creativity and innovative behavior of work (Usman et al., 2019; Werleman, 2016). Therefore, if job autonomy is having a significant impact, innovation managers should provide employees plenty of flexibility (Aizawl & Dhar, 2017). Hackman and others were among the first to discuss innovative behavior of work. They defined autonomy as the extent to which an employee has control over how to carry out a job assignment (Garg & Dhar, 2017). Furthermore, it has been proven that work autonomy has a significant influence on innovative behavior of work. For instance, work autonomy encourages employees to develop new working methods and approaches and assists them in putting those ideas into practice (Baig et al., 2022). Additionally, work autonomy enables staff members to come up with methods that are more effective and efficient for carrying out their duties. Due to the relationship between these two ideas, there is reason to think that the motivational factors that encourage active activity may also encourage innovative behavior of work (Messmann et al., 2018). Additionally, job autonomy plays a vital role in innovative behavior of work. It gives workers a sense of control over their work in the first place, and it is likely that job autonomy plays a vital role in innovative behavior of work. Additionally, it provides staff with the required room to test out different working methods and procedures that they may later offer as innovations. Third, having a high degree of job autonomy makes an employee feel more accountable for doing a good job, which should lead to more proactive activity like idea development and proposal submission. Therefore, workplace autonomy or independence may allow workers to practice "trial and error" and discover more effective and efficient methods for carrying out their duties (Orth & Volmer, 2017; Albahri et al., 2021; Khaw et al., 2022; Chew et al., 2023; Sandberg et al., 2023). Job autonomy gives employees a way to test out new ideas even if they fail because creativity entails trial and error, successes, and failures. Employees no longer have to adhere to a predetermined set of bureaucratic rules and regulations, thanks to autonomy. However, the concept of autonomy has been defined in a variety of disciplines, particularly in the discipline of education, and it has also been hotly debated in philosophy. The idea has also been challenged from several perspectives in educational research, as, for instance, scholars engaged in legal concerns were concentrating on creative work conduct and phenomena towards university teachers. Instead, highly controlling educators avoid developing students' internal motivational resources in favor of influencing behavior using external controls like rewards, limitations, and directives. Autonomy-supportive educators are those who try to reduce external controls and consider students' internal frames of reference regarding issues, concepts, and activities (Baharuddin et al., 2019).

2.3 Job Commitment

Knowledge workers who had been expressly chosen by the entrepreneur did not provide suggestions for innovation. Ideas originated from those who were sincerely dedicated and convinced, instead that, in addition, other people would embrace their thoughts (Hamzah & Hussain, 2018; Suhaimi & Panatik). Organizational research is now very interested in commitment. According to this theory. Additionally, challenging business environment forces organizations to rely on human resources. Organizations must take into account fundamental aspects of employees' commitment. For commercial organizations to succeed and operate well, committed people are a crucial and fundamental resource (Hakimian, 2016). Like events also take place in the educational environment, particularly in universities. Any university would consider having a faculty with a above average level of charge to be advantageous (Yeap et al., 2021). According to earlier research (Orindah, 2014), 57% of the university teachers agreed that their lack of commitment is a result of their low pay, and typically, feelings of dissatisfaction arise because they are not included in the decision-making process, particularly when it comes to teaching and there is no teamwork or collaborative work (Klnç et al., 2022). Numerous studies on organizational commitment were found in a review of the literature (Baharuddin et al., 2019). Meyer and Allen later created the three versions of the commitment paradigm in 1987: emotional commitment, continuation commitment, and normative commitment (Hakimian et al., 2016). These attitudinal commitment components (affective, continuation, and normative) were conceptually and experimentally separate but were not necessarily associated, according to the study they conducted. Organizations with devoted personnel enjoy various advantages, such as lower costs, greater performance, lower absenteeism, and improved production. There is a clear connection between teachers' dedication to their workplaces, their work groups, and their profession as teachers (Cetin et al., 2022; Eneizan et al., 2019; Abbas et al., 2023; Alnoor et al., 2024a, b).

Organizational commitment and correct feelings of pride in being a part of such a university and work group were found to be directly related. The dedication of the professors to their university, the workplace, and the teaching profession were all closely related. The components of university teachers' commitment, according to (Thien et al., 2021). In the meantime, commitment is predicted to have a favorable effect on innovative behavior of work. Teachers' dedication helps them be innovative and incorporate new concepts into their practices (Werleman, 2016).

3 Proposed Structure and Conclusion

Teaching in the digital age is a challenging undertaking since university faculty' responsibilities are shifting from being the primary knowledge providers to helping students become self-assured in environments that value creativity. As a result, the university teacher needs to possess the knowledge, abilities, competencies, and attitudes necessary to alter education in order to meet the changing requirements of society. With the addition of new competences, teachers' creative behavior of work will improve, resulting in a successful outcome.

As a result, in addition to filling in knowledge spaces, this study wants to contribute to the solution by advancing our understanding of the factors that influence university teachers' innovative behavior of work today. This will enable us to strategically prepare for future advancements that will benefit the entire country. To assist university instructors, this paper suggests a framework for the factors that determine innovative behavior of work. The connection between job autonomy, job commitment, and innovative behavior of work among university faculty members is shown. The framework demonstrates the clear relationship between workplace commitment and influencing innovative behavior of work. The idea was in line with early conceptions of innovative behavior of work from earlier investigations. Job autonomy has been noted as one of the most cited benefits (Juntunen & Martiskainen, 2021).

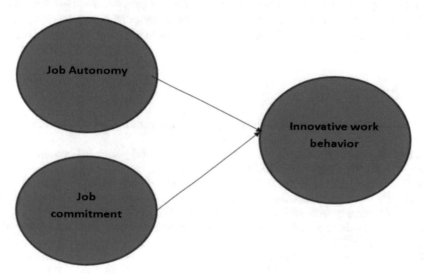

Fig. 1. Proposed framework illustrating the link between university faculty members' innovative behavior of work, commitment to their jobs, and job autonomy.

The association between job autonomy, job comment, and creative behavior of work among university professors is shown in Fig. 1. The framework demonstrates how job autonomy and commitment have a direct result on innovative behavior of work. This claim was uniform with the original theories on inventive behavior of work from earlier studies. According to Spiegel (2014), job autonomy is generally cited as one of the key factors influencing employee creativity and upgrades the possibility that proposals for developments will be made during the development of any innovation. Ramamurthy and others, 2005. The primary catalyst for employee creativity is the submission of proposals for enhancements in the development of some novelty. As per Ramamurthy et al. (2005). Increases a worker's sense of responsibility for his work, which should encourage proactive conduct like idea generation and suggestion. Stated that the importance of work autonomy and creative requirements—where instructors should be in command of their careers but also need to be challenged to be innovative—was crucial. Positively influenced idea development. In fact, one of Workplace devotion is one of the most significant variables affecting innovative behavior of work. Bawuro (2019), an employee's commitment to their employer inclination affects their capacity for innovation, problem-solving at work and information sharing. Thien et al. (2021). University faculty members' ability to create and incorporate new concepts into their own practice is supported by faculty' commitment. To assist university faculty members, this paper suggests a framework for the factors that determine innovative behavior of work. The framework illustrates the connection between job autonomy, commitment to the task, and innovative behavior of work. Utilizing the study's instrument, the proposed conceptual framework can assist educational institutions, particularly universities, in identifying existing faculty fellows' innovative behavior of work s because the bond between these aspects can be looked at. This study aims to support the government's efforts to create community with exemplary value by identifying a change that is required to enhance innovation in universities. a top-notch educational system that places special emphasis on the part that dynamic, creative human capital shows in boosting spirit and resourceful success.

References

Ali, W.: Online and remote learning in higher education institutes: a necessity in light of COVID-19 pandemic. High. Educ. Stud. **10**(3), 16–25 (2020)

Cetin, S., Davarci, M., Karakas, A.: The impact of organizational justice and trust on knowledge sharing behaviour. Управленец **13**(3), 30–45 (2022)

Chou, C.-M., Shen, C.-H., Hsiao, H.-C., Shen, T.-C.: Factors influencing teachers' innovative teaching behaviour with information and communication technology (ICT): the mediator role of organisational innovation climate. Educ. Psychol. **39**(1), 65–85 (2019)

Efandi, S.E., Syuhada, M.N.: Innovative behavior of work and influencing factors. J. Ipteks Terapan **15**(3), 241–250 (2021)

Fatoki, O.: Entrepreneurial leadership and employees' innovative work behaviour in small firms: mediating role of creative self-efficacy. Acad. Strateg. Manag. J. **20**, 1–9 (2021)

Garg, S., Dhar, R.: Employee service innovative behavior: the roles of leader-member exchange (LMX), work engagement, and job autonomy. Int. J. Manpower (2017)

Ghani, N.A.A., Hussin, T.A.B.S., b. R., & Jusoff, K.: Antecedents of psychological empowerment in the malaysian private higher education institutions. Int. Educ. Stud. **2**(3), 161–165 (2009)

Hakimian, F., Farid, H., Ismail, M.N., Nair, P.K.: Importance of commitment in encouraging employees' innovative behaviour. Asia-Pac. J. Bus. Adm. **8**(1), 70–83 (2016)

Hamzah, M.F.B., Hussain, M.N.B.M.: Effect of competency on the job performance of frontline employees in Islamic banks: The moderator effect of religiosity. In: Issues and Trends in Interdisciplinary Behavior and Social Science, pp. 321–327. CRC Press (2018)

Haseeb, M., Hussain, H.I., Ślusarczyk, B., Jermsittiparsert, K.: Industry 4.0: a solution towards technology challenges of sustainable business performance. Soc. Sci. **8**(5), 154 (2019)

Ausat, A.M.A., Widayani, A., Rachmawati, I., Latifah, N., Suherlan, S.: The effect of intellectual capital and innovative behavior of work on business performance. J. Econ. Bus. Account. Ventura **24**(3), 363–378 (2022)

Javed, B., Abdullah, I., Zaffar, M.A., ul Haque, A., Rubab, U.: Inclusive leadership and innovative behavior of work: the role of psychological empowerment. J. Manage. Organ. **25**(4), 554–571 (2019)

Juntunen, J.K., Martiskainen, M.: Improving understanding of energy autonomy: a systematic review. Renew. Sustain. Energy Rev. **141**, 110797 (2021)

Kılınç, A.Ç., Polatcan, M., Savaş, G., Er, E.: How transformational leadership influences teachers' commitment and innovative practices: understanding the moderating role of trust in principal. Educ. Manage. Adm. Leadersh. 17411432221082803 (2022)

Leong, C.T., Rasli, A.: The Relationship between innovative behavior of work on work role performance: an empirical study. Procedia Soc. Behav. Sci. **129**, 592–600 (2014)

Messmann, G., Mulder, R.H., Palonen, T.: Vocational education teachers' personal network at school as a resource for innovative work behaviour. J. Work. Learn. **30**(3), 174–185 (2018)

Muhamad, L.F., Bakti, R., Febriyantoro, M.T., Kraugusteeliana, K., Ausat, A.M.A.: Do innovative behavior of work and organizational commitment create business performance: a literature review. Commun. Dev. J.: Jurnal Pengabdian Masyarakat **4**(1), 713–717 (2023)

Noor, A.A., et al.: An analytical review of CHATGPT influence on healthcare, media, and education advancements. J. AL-Turath Univ. Coll. **2**(38) (2024)

Orindah, F.A.: Influence of organizational culture on teachers' job Commitment in public primary schools in Ndhiwa Subcounty, Kenya University of Nairobi (2014)

Orth, M., Volmer, J.: Daily within-person effects of job autonomy and work engagement on innovative behaviour: the cross-level moderating role of creative self-efficacy. Eur. J. Work Organ. Psy. **26**(4), 601–612 (2017)

Azim, A.M.M., Ahmad, A., Omar, Z., Silong, A.D.: Work-family psychological contract, job autonomy and organizational commitment. Am. J. Appl. Sci. **9**(5), 740 (2012)

Parthasarathy, J., Premalatha, T.: Impact of collective-efficacy and self-efficacy on the innovative work behaviour of teachers in the Nilgiris District, Tamil Nadu. Int. J. Indian Psychol. **5**(1) (2017)

Sciarelli, M., Gheith, M.H., Tani, M.: The relationship between soft and hard quality management practices, innovation and organizational performance in higher education. TQM J. **32**(6), 1349–1372 (2020)

Scott, S.G., Bruce, R.A.: Determinants of innovative behavior: a path model of individual innovation in the workplace. Acad. Manag. J. **37**(3), 580–607 (1994)

Senouci, K., et al.: The supporting independent immunization and vaccine advisory committees (SIVAC) initiative: a country-driven, multi-partner program to support evidence-based decision making. Vaccine **28**, A26–A30 (2010)

Setyawasih, R., Hamidah, A.D.B.: Organizational culture and innovative behavior of work in manufacturing company: the role of employee engagement as a mediator (2022)

Singh, A.: The mediating role of employee commitment between quality of work-life and job performance of the faculty. Ind. Commer. Train. **54**(2), 250–266 (2022)

Singh, C.K.S., et al.: Teaching strategies to develop higher-order thinking skills in English literature. Int. J. Innov. Creat. Change **11**(80), 211–231 (2020)

Spanuth, T., Wald, A.: How to unleash the innovative behavior of work of project staff? The role of affective and performance-based factors. Int. J. Proj. Manage. **35**(7), 1302–1311 (2017)

Suhaimi, I.W., Panatik, S.A.: Relationship between leader-member exchange, innovative work behaviour, and work engagement: a literature review. Global Business and Social Entrepreneurship Resources (GBSE) Taman Melawati Kuala Lumpur, 1

Sutarto, S., Sari, D.P., Fathurrochman, I.: Teacher strategies in online learning to increase students' interest in learning during COVID-19 pandemic. J. Konseling Pendidikan (JKP) **8**(3), 129–137 (2020)

Baharuddin, M.F., Masrek, M.N., Shuhidan, S.M.: Innovative work behaviour of school teachers: a conceptual framework. Int. E-J. Adv. Educ. **5**(14), 213–221 (2019)

Sharif, H.S., Bin Mydin, A.A., Younis, H.A.: Influence of authentic leadership practices on innovative work behaviour in higher educational institutions: a virtual reality perspective. In: Al-Emran, M., Ali, J.H., Valeri, M., Alnoor, A., Hussien, Z.A. (eds.) IMDC-IST 2024. LNNS, vol. 876, pp. 175–187. Springer, Cham (2023). https://doi.org/10.1007/978-3-031-51300-8_12

Tok, E.: The incentives and efforts for innovation and entrepreneurship in a resource-based economy: a survey on perspective of Qatari residents. Sustainability 12(2), 626 (2020)

Tosheva, N., Abdullaeva, G.: The concept of "innovation" and types of innovative technologies. Sci. Progr. 3(3), 586–589 (2022)

Van Zyl, L.E., Van Oort, A., Rispens, S., Olckers, C.: Work engagement and task performance within a global Dutch ICT-consulting firm: the mediating role of innovative behavior of work s. Curr. Psychol. 40, 4012–4023 (2021)

Werleman, I.: The two faces of corporate brands: the impact of product brands' contradictory CSR perceptions on the relationship with the brands and corporate brand University of Twente (2016)

Xing, B., Marwala, T.: Implications of the fourth industrial age on higher education. arXiv preprint arXiv:1703.09643 (2017)

Younis, H.A., Jamaludin, R., Wahab, M.N.A., Mohamed, A.S.A.: The review of NAO robotics in educational 2014–2020 in COVID-19 virus (pandemic era): technologies, type of application, advantage, disadvantage and motivation. In: IOP Conference Series: Materials Science and Engineering, vol. 928, no. 3, p. 032014. IOP Publishing (2020)

Younis, H.A., Mohamed, A.S.A., Jamaludin, R., Wahab, M.N.A.: Survey of robotics in education, taxonomy, applications, and platforms during covid-19. Comput. Mater. Continua 67(1), 687–707 (2021)

Younis, H.A., et al.: A systematic review and meta-analysis of artificial intelligence tools in medicine and healthcare: applications, considerations, limitations motivation and challenges. Diagnostics 14(1), 109 (2024)

Yeap, S.B., Abdullah, A.G.K., Thien, L.M.: Lecturers' commitment to teaching entrepreneurship: do transformational leadership, mindfulness and readiness for change matter? J. Appl. Res. High. Educ. 13(1), 164–179 (2021)

Baig, L.D., Azeem, M.F., Paracha, A.: Cognitive appraisal of job autonomy by nurses: a cross-sectional study. SAGE Open Nurs. 8, 23779608221127824 (2022)

Yusof, N., Hashim, R.A., Valdez, N.P., Yaacob, A.: Managing diversity in higher education: a strategic communication approach. J. Asian Pac. Commun. 28(1), 41–60 (2018)

Eneizan, B., Mohammed, A.G., Alnoor, A., Alabboodi, A.S., Enaizan, O.: Customer acceptance of mobile marketing in Jordan: an extended UTAUT2 model with trust and risk factors. Int. J. Eng. Bus. Manage. 11, 1847979019889484 (2019)

Abbas, S., et al.: Antecedents of trustworthiness of social commerce platforms: a case of rural communities using multi group SEM & MCDM methods. Electron. Commer. Res. Appl. 62, 101322 (2023)

Albahri, A.S., et al.: Based on the multi-assessment model: towards a new context of combining the artificial neural network and structural equation modelling: a review. Chaos Solitons Fractals 153, 111445 (2021)

Khaw, K.W., et al.: Modelling and evaluating trust in mobile commerce: a hybrid three stage fuzzy Delphi, structural equation modeling, and neural network approach. Int. J. Hum.-Comput. Interact. 38(16), 1529–1545 (2022)

Chew, X., Khaw, K.W., Alnoor, A., Ferasso, M., Al Halbusi, H., Muhsen, Y.R.: Circular economy of medical waste: novel intelligent medical waste management framework based on extension linear Diophantine fuzzy FDOSM and neural network approach. Environ. Sci. Pollut. Res. 30(21), 60473–60499 (2023)

Sandberg, H., Alnoor, A., Tiberius, V.: Environmental, social, and governance ratings and financial performance: evidence from the European food industry. Bus. Strateg. Environ. **32**(4), 2471–2489 (2023)

Alnoor, A., et al.: Uncovering the antecedents of trust in social commerce: an application of the non-linear artificial neural network approach. Competit. Rev.: Int. Bus. J. **32**(3), 492–523 (2022)

Alnoor, A., et al.: How positive and negative electronic word of mouth (eWOM) affects customers' intention to use social commerce? A dual-stage multi group-SEM and ANN analysis. Int. J. Hum.-Comput. Interact. **40**(3), 808–837 (2024)

XinYing, C., Tiberius, V., Alnoor, A., Camilleri, M., Khaw, K.W.: The dark side of metaverse: a multi-perspective of deviant behaviors from PLS-SEM and fsQCA findings. Int. J. Hum.–Comput. Interact. 1–21 (2024)

Battistelli, A., Odoardi, C., Cangialosi, N., Di Napoli, G., Piccione, L.: The role of image expectations in linking organizational climate and innovative work behaviour. Eur. J. Innov. Manag. **25**(6), 204–222 (2022)

Alnoor, A., Atiyah, A.G., Abbas, S.: Unveiling the determinants of digital strategy from the perspective of entrepreneurial orientation theory: a two-stage SEM-ANN approach. Glob. J. Flexible Syst. Manag. 1–18 (2024)

Bawuro, F.A., Shamsuddin, A., Wahab, E., Usman, H.: Mediating role of meaningful work in the relationship between intrinsic motivation and innovative work behaviour. Int. J. Sci. Technol. Res. **8**(9), 2076–2084 (2019)

Bilal, M., et al.: Knowledge, attitudes, and practices among nurses in Pakistan towards diabetic foot. Cureus **10**(7) (2018)

Önhon, Ö.: The relationship between organizational climate for innovation and employees innovative work behavior: ICT sector in Turkey. Vezetéstudomány-Bp. Manag. Rev. **50**(11), 53–64 (2019)

Exploring Customer Engagement Intentions with Interactive Smart Tables of AI for Full-Service Restaurants Sustainability

Ghada Taher Al-Lami[1]([✉]) [ID], Hadi AL-Abrrow[2] [ID], Hasan Oudah Abdullah[3] [ID], and Alhamzah Alnoor[4] [ID]

[1] Faculty of Management, Finance, and Economics, University of Sciences and Arts in Lebanon (USAL), Beirut, Lebanon
tahir.ghada45@gmail.com
[2] Organization Studies, Business Administration Department, University of Basrah, Basrah, Iraq
[3] Business Administration Department, Basrah University College of Science and Technology, College of Administration and Economics/Qurna, University of Basrah, Basrah, Iraq
[4] Southern Technical University, Basrah, Iraq
Alhamzah.malik@stu.edu.iq

Abstract. The study seeks to explore the most important elements which may impact customers' intention or desire to engage with smart tables in full-service restaurants utilizing the Unified Theory of Acceptance and Use of Technology 2 (UTAUT2) model. As applications of technology and artificial intelligence continue to develop, comprehending in what way customers perceive and engage with smart systems is crucial for restaurants looking to achieve the best possible customer experience, as well as achieve sustainability. The research adopts a quantitative approach utilizing a questionnaire to gather and examine data from 306 participants. The research focus is identifying elements like as hedonic motivation, societal impact, facilitating conditions, and performance expectation on interacting intention smart tables, in addition to, moderator role for personal innovativeness with the five dimensions. The findings supported all hypotheses except for effort expectancy which did not receive backing. The outcomes of this research add insights to the field of technology adoption and consumer behavior within the restaurant industry offering guidance, for restaurant managers and policymakers in effectively implementing interactive technologies to enrich customer engagement and satisfaction.

Keywords: Artificial intelligence · UTAUT2 · Smart tables · Restaurants · Restaurant sustainability · Personal innovativeness

1 Introduction

In recent times, there trend has been a growing fascination in emerging technologies like artificial intelligence (AI) among customers, society and companies. Gansser & Reich, (2021) characterize AI as a technology designed to enhance human existence and provide

© The Author(s), under exclusive license to Springer Nature Switzerland AG 2024
A. Alnoor et al. (Eds.): AIRDS 2024, LNNS 1033, pp. 168–186, 2024.
https://doi.org/10.1007/978-3-031-63717-9_11

support in certain situations. Implementing AI can result in increased productivity and enhanced customer service procedures (Berezina et al., 2019). AI technologies have a broad scope, including several fields including economics (Belanche et al., 2020), the education sector (Atshan et al., 2023), health sector (Fan et al., 2020), transportation sector (Ribeiro et al., 2022), and tourism sector (Go et al., 2020). Restaurants have a significant role as a central point for social interactions in nearly all nations and cultures worldwide. Restaurants not only offer meals and beverages to individuals, but they also serve as a gathering spot for socializing, connecting, and creating memories (Sandberg et al., 2023; Chew et al., 2023).

Generally, the hospitality sector in general, particularly restaurants, is highly attractive for applying artificial intelligence. AI facilitates various aspects, including improving quality, reducing faults, regulating cost planes, and increasing response speed (Berezina et al., 2019). AI and robotics play a large and increasing role in many aspects of the restaurant industry, such as planning and arranging the kitchen, controlling food quality, training and developing workers, and improving customer service. The process of transformation or interest in applying AI technologies does not take place in one stage, as it needs a gradual planning process for application and service improvement (Abdullah et al., 2022). Moreover, the restaurant service has not witnessed many changes in its operational activities over the past century. Despite the tremendous technological developments, the restaurant sector still needs to exploit the technical revolution to develop and improve restaurant services. AI applications can contribute to overcoming some of the various challenges resulting from human limitations through the interaction between robots and humans (Achan et al., 2022).

Technology used in restaurant service activities is moderately novel (Berezina et al., 2019; Mathath & Fernando, 2015). Organizations still face challenges in the arena of relying on artificial intelligence and robotics applications to create optimal economic value (Hoffman et al. 2020). Previous literature has shown increasing interest in this field (i.e., Cain et al., 2019; Maier & Edwards, 2020; Yang et al., 2020). The restaurant industry has attracted the attention of many study areas (Rosete et al. 2020) including studying the acceptance of artificial intelligence applications with respect to subjective principles (Belanche et al., 2020), social influences (Sindermann et al., 2021), technical aptitude and its outcome on human jobs (Vu & Lim, 2022), and enjoyment of utilization (Rese et al., 2020). Consequently, the need to evaluate and identify attidiutes of individuals towards AI has augmented recently. Individuals may have optimistic expectations about AI, while at the same time they may have concerns about this technology (Alnoor et al., 2023; Khaw et al., 2023; Faris et al., 2023).

This study seeks to research customers' intention and attitude regarding the use of AI, specifically smart tables, in restaurants in order to enhance performance. The smart table was established to improve waiting durations for customers, as well as increase restaurant service efficiency. In this framework, it is important for restaurant owners to comprehend the degree which customers accept the use of smart tables. Hence, this study aims to deliver essential understandings and commendations to restaurant owners, developers, operators, and servers regarding any decision to integrate smart tables within full-service restaurants, in Iraq. To do so, this research utilized UTAUT 2 as a tool for its research. UTAUT is a prevailing theoretical tool which delivers a framework to estimate the degree and probability of accepting AI technology usage in different situations. Thus, revision of literature will take place, followed by building the hypothesis model, determining the details of methodology used, reviewing the results, and finally, the discussing the results along with clarifying their theoretical and implied limitation.

2 Literature Review and Study Hypothesis Development

2.1 Theoretical Foundation

Amongst a myriad of popular technology models, the current study relies on the UTAUT2 model. UTAUT2 is connected to current technical developments, thus relying on it to comprehend the efficacy of employing ai systems in the workplace attains important results (Cabrera-Sánchez et al., 2021). The UTAUT model, established by Venkatesh et al., clarifies prognostic performance valuation of plausible behavior to novel applications and technology. UTAUT2, the advanced model of UTAUT, UTAUT2, is extensively utilized in place of an extensive framework which recognizes numerous facets that impact individual's intent to accept technologies.

The original UTAUT model, powerful and broadly used, is constructed on an amalgamation of eight diverse theories (Williams et al., 2015; Meet et al., 2022). The eight theories are: the Theory of Reasoned Action (Fishbein & Ajzen, 1977), the Theory of Planned Behavior (Ajzen, 1991), the Technology and Acceptance Model (Davis, 1989), the Motivational Model (Davis et al., 1989), the Social Cognitive Theory (Compeau & Higgins, 1995), the Personal Computer Utilization Model (Thompson et al., 1991), the combination of TAM and TPB (C-TAM-TPB) (Taylor & Todd, 1995), and the Diffusion of Innovation Theory (Moore and Benbasat, 1991). Founded on the aforementioned, a customer's intent considering utilizing novel technology is predisposed based on various facets which include exertion expectation, social impact, facilitating conditions, and performance expectation. Applying UTAUT in customer-service frameworks, through accenting the hedonistic value (intrinsic motivation) of technological handlers has developed importance. With the addition of three new constructs – hedonic motivation, price value, and habit, an updated form also referred to as UTAUT2 (Tamilmani et al., 2021) outperforms the original UTAUT (Chu et al., 2022). UTAUT 2 has been extensively utilized in many fields to understand the integral elements that may impact customers' behavioral intent to accept and engage with new information systems and technological innovations. UTAUT has been utilized as a study model in various fields, including mobile payment in hotels (Morosan & DeFranco, 2016), travel and tour guides (Lai, 2015), and diet restaurants (Okumus et al., 2018). Similarly, Zhu et al. (2022) relied on UTAUT to explore the elements impacting the embracing of smart elevators by Taiwanese customers. Additionally, Magsamin-Cunard et al. (2015) institute age and gender yielding a moderating outcome, whereas performance expectation, exertion expectation, facilitating conditions, and societal impacts yield a direct outcome on the intent smart devices. Lian & Yen (2014) examined how age and gender influence user intention to engage in online shopping in Taiwan.

Despite the model's accuracy, numerous researchers believe it is essential to expand it to provide a more comprehensive explanation for specific research subjects (i.g., Merhi et al., 2019; Sun et al., 2023; Gansser and Reich, 2021). This study examined how personal innovativeness effect the behavioral intention to use smart tables in restaurants. Furthermore, the initial assumptions proposed in UTAUT 2 have been modified in this study to ensure that the model applies to Smart Tables. The UTAUT2 framework of this study identifies five structures that are performance expectation, exertion expectation, societal impact, facilitating conditions, and hedonic motivation. For this research, habit and price-value factors have been omitted as they do not align with the objectives of the study.

2.2 Attitude Towards AI

Artificial Intelligence (AI) is utilized in several advantageous domains, such as disease diagnosis, environmental conservation, natural catastrophe prediction, education enhancement, violence prevention, and workplace risk reduction (Brooks, 2019). Hence, the advantages of AI might potentially result in individuals having more leisure time to engage in learning, experimenting, and exploring, eventually yielding development in human creativity and life quality (Kayaa et al., 2024). However, Neudert et al. (2020) contend that there has been a thorough debate surrounding the possible ethical, social, and economic hazards. According to Huang & Rust (2018), AI poses a danger to human-provided services. Frey & Osborne (2017) highlighted that almost half of the American workforce, specifically 47%, faces the possibility of job displacement in the near future as a result of computerization, which includes advancements in artificial intelligence and robots. In addition, Bossman (2016) asserted that the reduction in employment opportunities will result in a more limited distribution of income, exacerbating global inequality. Furthermore, the use of AI technology may potentially give rise to a range of security concerns. Issues such as legal and organizational complications that arise while utilizing such advancements, as well as the societal skepticism and lack of confidence towards AI (Dwivedi et al., 2019). These are the detrimental elements of artificial intelligence, towards which individuals might exhibit an adverse attitude. The notion of attitudes towards artificial intelligence has augmented in significance in current years.

3 Hypothesis Development

3.1 Hedonic Motivation

Hedonic motivation refers to the internal factor related to the pleasure an individual has when using a particular technology. This concept may greatly affect the acceptance or rejection and use of novel innovations. Prior research indicated that the element of enjoyment entails an important part in determining consumers' intent to use novel technology. Many researches have argued that customers with higher levels of hedonic motivation reflect positive behavioral intents toward technology, including recurring purchase and continual use (i.g., Chiu et al., 2014; Zhang et al., 2012; Juaneda-Ayensa et al., 2016). Based on this, hedonic motives are an important factor affecting the intent of end users to accept new technology. Palas et al. (2022) proposed that Hedonic Motivation displayed a noteworthy impact on the behavioral intent of older individuals to utilize health services, within the same framework. Based on observed and discussed evidence aforementioned, the following was hypothesized:

H1a. Hedonic motivation is positively correlated to interacting intention.

3.2 Social Influence

Prior studies indicate that when customers make a choice to adopt or reject new technology, they also take into account how this choice would impact their relationships with others. As a result, customers tend to adhere to the decisions of the mass (Thajil et al., 2022). Conformity, in this case, can be understood as the individual's acceptance of certain beliefs or ideas, which then leads to a change in their personal attitudes. Societal impact, in the realm of technology, signifies to the extent to which consumers consider that influential individuals believe they ought to accept a specific technological innovation (Venkatesh et al., 2012). Prior research examining technology adoption (Tseng et al., 2019) has established the presence of societal impact, which has been recognized as a substantial influence in users' disposition to adopting technologies. The technology has been applied in various researches including mobile payments, the embrace of technology by older adults, electronic data in healthcare (Hoque & Sorwar, 2017), the embrace of e-commerce pharmacies by consumers (Mun et al., 2006), and Internet banking (Alalwan et al., 2018; Morosan & DeFranco, 2016). Current studies have established societal impact, social influence, as a prevailing aspect that impacts the intent of adopting a certain technology (Baudier et al., 2020; Cabrera-Sánchez et al., 2021). Based on observed and discussed evidence aforementioned, the following was hypothesized:

H1b. Social influence is positively related to interacting intent.

3.3 Facilitating Condition

Facilitating conditions is well-defined as "the degree to which an individual believes that an organizational and technical infrastructure exists to support use of the system" (Venkatesh et al., 2003), while also encompassing customers' perception of resource availability enhancing their effective technology utilization (Hossain et al., 2017). Based on the dissonance theory (Festinger, 1957), when the conditions that help or support something instead hinder or prevent it, individuals may change their attitudes in a negative way to align with that scenario. Conversely, when individuals have sufficient resources, they are more inclined to develop favorable views since less obstacles are preventing them from participating in the behavior. According to the UTAUT, enabling factors impact both behavioral intent and usage behavior (Venkatesh et al., 2003). Numerous researches (Okumus et al., 2018) have confirmed the elements which facilitate the adoption of technology for individuals. These studies have found that these factors have a substantial impact on the actual use of technology and the intent to use it. Several researches have revealed that facilitating conditions have a positive influence on intentions to engage in new innovations, such as mobile payment systems (Sobti, 2019), mobile wallets (Chawla & Joshi, 2019), and the use of tablets (Magsamen-Conrad et al., 2015), NFC mobile payments in hotels (Morosan & DeFranco, 2016) Based on the previous discussion, we assumed what follows:

H1c. Facilitating condition is positively related to interacting intention.

3.4 Effort Expectancy

Effort expectancy, or exertion expectation, refers to the level of comfort and ease that users experience when utilizing a system (Venkatesh et al., 2003; Boontarig et al., 2012).

Customers' willingness to adopt new innovations is impacted by both the apparent benefits and the ease of use, which includes the minimal effort required to operate it (Davis, 1989). Put simply, customers are more content with a self-service technology that is user-friendly (Meuter et al., 2005). Hence, the level of effort expected by clients could significantly influence their views towards novel technology. While certain users might view the additional exertion as insignificant, others might find it troublesome and hence develop a negative attitude toward smart retail technology (Roy et al., 2021). For instance, dealing with a sophisticated mobile application for smart retailing might be a significant source of difficulty (Grewal et al., 2020). In addition, clients may view the abundance of information as intricate. Consumer behavior is considered favorable when the process of adopting technology is simple and uncomplicated (Alalwan et al., 2018). According to Sun et al. (2013), the level of effort expected by users directly affects their intention to use a clinical mobile health monitoring system. Based on observed and discussed evidence aforementioned, the following was hypothesized:

H1d. Effort expectancy is positively related to interacting intention.

3.5 Performance Expectancies

Performance expectancy (PE) is the believed utility of implementing a framework and the notion that utilizing the framework will augment job performance (Venkatesh et al., 2003; Venkatesh et al., 2012; Schmitz et al., 2022). If customers realize the advantages and gains of using technology, their intentions towards use will be positive (Venkatesh et al., 2012). Many studies argue that performance expectation is an integralelement that influences behavioral intention (Shaw & Sergueeva, 2019; Baudier et al., 2020; Cabrera-Sánchez as al., 2021). In another context, Alkhwaldi et al. (2022) found that performance expectations have a positive relationship with willingness to accept technology in emerging countries throughout the COVID-19 epidemic. Similarly, Chauhan et al. (2018) obtained comparable results concerning voting machines. A follow-up study led by Dwivedi et al. (2019) further validated the predictive efficacy of PE and emphasized the pivotal influence of individual attitude or disposition. Nevertheless, while the impact of these components might differ depending on specific circumstances, user character-istics, or platform being examined, the overall data consistently confirms their role in promoting the adoption of a novel technological system (Dakduk et al., 2023). Based on observed and discussed evidence aforementioned, the following was hypothesized:

H1e. Performance expectancy is positively related to interacting intention.

3.6 Personal Innovativeness

Personal innovativeness is a trait that indicates an individual's inclination to explore and novel technologies (Slade et al., 2015). The adoption of technological innovations is contingent upon users' capacity to effectively navigate and engage with the invention (Belanche et al., 2020; Flavián et al., 2023). Peslak et al. (2011) posited that persons who possess a proclivity for innovation and curiosity towards novel systems are more inclined to form favorable intents and are more probable to embrace novel advance-ments. Conversely, those who are hesitant to devote their time and effort to exploring new technology are less likely to accept it. Empirical research has shown that personal innovativeness displayed substantial influence on how consumers perceive the value of and their desire to utilize new technologies, including e-commerce, mobile payment systems, and biometric payment systems. Hasan et al. (2021) have shown that within the realm of virtual assistants (VAs), the impact of novelty value on brand loyalty is more pronounced amongst customers who exhibit higher levels of innovation. Personal innovativeness behaves as a moderator amid technological features and behavioral intent, conceivably augmenting the professed pleasure of utilizing technology (Venkatesh et al., 2003; Jang & Lee, 2018). According to Tan et al. (2017), a client displaying a high degree of creativity is more inclined to choose new technology and expect more satisfaction from the newest services compared to a customer with only a moderate level of creativity, as shown in Fig. 1. Based on observed and discussed evidence aforementioned, the following was hypothesized:

H2a. Personal innovativeness moderates the hedonic motivation-interacting intention relationship.
H2b. Personal innovativeness moderates the social influence-interacting intention relationship.
H2c. Personal innovativeness moderates the facilitating conditions-interacting intention relationship.
H2d. Personal innovativeness moderates the effort expectancy-interacting intention relationship.
H2e. Personal innovativeness moderates the performance expectancy-interacting intention relationship.

Figure 1 shows the study model.

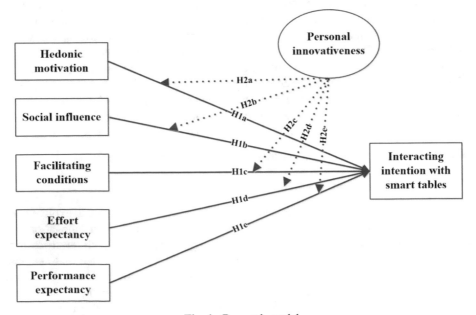

Fig. 1. Research model

4 Methods

4.1 Participants and Procedure

The survey questions were carefully crafted to make sure they were easy to understand. Participants were guaranteed that their participation would include confidentiality and anonymity. Data analysis included using methods, like regression analysis to investigate the connections between factors and evaluate how well the UTAUT2 model predicts customer engagement intentions with interactive smart tables in upscale restaurants. To address any biases in the data collection process we used the Harman single factor method, which helped us achieve a variance of 19%. We handed out questionnaires randomly to 350 customers who frequented high end restaurants and showed interest in using tables. Out of these we received 311 completed questionnaires resulting in a response rate of around 89%. After checking for accuracy and consistency we found that data from 306 respondents met our criteria for analysis.

4.2 Instrumentation

The survey tool included questions borrowed from research studies to measure factors related to the Unified Theory of Acceptance and Use of Technology 2 (UTAUT2) model. The questionnaire covered factors including hedonic motivation, societal impact, facilitating conditions, effort expectancy, and performance expectation, interacting intention, and personal innovativeness. We measured the five components of the UTAUT2 model by adapting items used in studies by Na et al. (2021). Venkatesh et al. (2012). In respect to assessing the desire, for interaction the questions used were taken from Halal, who borrowed them from Na et al. (2021). Venkatesh et al. (2012). Additionally the Personal Innovativeness scale was derived from Van Droogenbroeck and Van Hove (2021). This scale gauges how inclined participants are to try out information technologies in comparison to others, in their circle. Appendix A shows the scale items used in the study.

5 Results

5.1 Demographic Distribution

The study sample included 55% male clients and 45% female. The age group between 30–39 was the largest, at 30%, followed by the age group between 20–21, at 25%. Regarding the marital status of the sample, 55% were married and 35% were single. Concerning the samples degree of education, the majority (40%) of the sample members obtained a bachelors degree and 30% were high school diploma holders.

5.2 Assessment of Measurement Model

Assessing the measurement model's validity is empirical in the present research (Hair et al., 2010). As the present study pursues to comprehend the impact of rudiments of the UTAUT2 model on intents to utilize smart tables in a restaurant and to confirm the collaborative function of personal innovativeness, PLS-SEM was selected for its aptitude to work with multifaceted models (Hair et al., 2016). Based on Hair et al. (2014) and Henseler et al. (2009), the investigative method of PLS-SEM involves two phases: evaluating measurement and data structural model assessment. Primarily, to confirm the precision of the measurement model as well as verify convergent rationality and dependability, factor loading, average variance extracted (AVE), composite reliability, and Cronbach's alpha were assessed. Meanwhile, Heterotrait-Monotrait ratio (HTMT) will be relied on to verify discriminant rationality (Alnoor et al., 2024a, b; Ali et al., 2024; XinYing et al., 2024; Alnoor et al., 2024b).

5.3 Convergent Validity

Convergent validity valuation purposes to confirm that the scales contest the notions to be valued. Factor loading and AVE were evaluated to verify convergent validity. The results shown in Table 1 specify that most of the factor loading of the constructs surpassed the 0.500 minimum, and this specifies that acceptable convergent validity has been attained. In addition, the values of composite reliability (CR) and Cronbach's alpha were relied upon to evaluate reliability and internal consistency, as results indicated that all values exceeded 0.70, indicating high reliability.

Table 1. Convergent Validity

Construct	Range of Outer Loading	Cronbach's alpha	Composite reliability	AVE
Hedonic motivation	0.532–0.866	0.743	0.804	0.554
Social influence	0.511–0.843	0.764	0.798	0.523
Facilitating conditions	0.534–0.892	0.865	0.895	0.602
Effort expectancy	0.564–0.909	0.845	0.904	0.593
Performance expectancy	0.523–0.859	0.769	0.834	0.523
Interacting intention,	0.564–0.976	0.778	0.872	0.572
Personal innovativeness	0.578–0.923	0.803	0.904	0.554

5.4 Discriminant Validity

In addition, there is a need to verify discriminant validity in order to ensure that scales designed to measure concepts are actually different. This is a basic condition for evaluating the measurement model. Henseler et al. (2015) suggest the HTMT ratio as one of the approaches utilized to assess discriminant rationality. The HTMT ratio compares the average item correlation between constructs (heterogeneous correlations) to the geometric mean of the average correlations across items measuring individual constructs (univariate correlations) (Henseler et al., 2015). HTMT values exceeding 0.85 indicate potential concerns about the discriminant validity between the two concepts (Henseler et al., 2015). The results in Table 2 indicate that the HTMT values did not exceed the critical value, indicating that discriminant validity was achieved in an acceptable manner.

Table 2. Discriminant Validity

	1	2	3	4	5	6
1-Hedonic motivation						
2-Social influence	0.252					
3-Facilitating conditions	0.243	0.814				
4-Effort expectancy	0.846	0.397	0.315			
5-Performance expectancy	0.285	0.286	0.134	0.127		
6-Interacting intention,	0.393	0.638	0.693	0.356	0.313	
7-Personal innovativeness	0.354	0.329	0.148	0.112	0.594	0.284

5.5 Assessment of Structural Model

The second phase involves assessing the structural model, which refers to explaining the causal relations among variables (i.e., independent, moderator, and dependent variables). The integral strictures to evaluate the structural model in PLS-SEM inclide the beta coefficient (β), standard error (SE), t value, p value, F2, and R2 (Hair et al., 2017; Alnoor et al., 2024a; Hesselbarth et al., 2023; Abbas et al., 2023). The aptitude of the independent variables to clarify the difference in the dependent variable is indicated by the coefficient of determination (R2). According to Chin (1998), the coefficient of determination that is greater than 0.67 is strong, and between 0.33 to 0.67 is moderate, while 0.19 to 0.33 is weak, and less than 0.19 is considered rejected. While the Effect Size (F2) designates the degree to which each independent variable contributes to the prediction of the dependent variable. An effect size greater than 0.35 is strong, between 0.15 to 0.35 is medium, 0.02 to 0.15 is weak, and less than 0.02 is considered rejected (Hair et al., 2017).

Subsequently, Structural Equation Modeling (SEM) was employed to deduce the hypothesized model's estimations based on the given dataset. The significance threshold of a P-value at 0.05 and a t-value of 1.65 was embraced, acknowledging the statistical implication of the associations among the endogenous, exogenous, and mediating variables. It is clear from Table 4 that the first sub-hypotheses were reinforced at a significance level less than (0.05), with the exception of not receiving support for the relationship of hedonic motivation, societal impact, facilitating conditions, and performance expectancy on interacting intent, with the exception of the effort expectation

Table 3. Testing Hypotheses

Hypotheses		β	S.E	T Statistics	P Values	Results
H1a	Hedonic motivation	0.186	0.032	4.704	0.000	Supported
H1b	Social influence	0.139	0.020	4.000	0.000	Supported
H1c	Facilitating conditions	0.184	0.039	4.144	0.000	Supported
H1d	Effort expectancy	0.033	0.042	0.771	0.441	Unsupported
H1e	Performance expectancy	0.167	0.044	4.782	0.000	Supported
H2a	Hedonic motivation* Personal innovativeness	0.257	0.041	3.736	0.000	Supported
H2b	Social influence* Personal innovativeness	0.264	0.045	4.664	0.000	Supported
H2c	Facilitating conditions* Personal innovativeness	0.248	0.040	3.450	0.000	Supported
H2d	Effort expectancy* Personal innovativeness	0.274	0.037	4.412	0.000	Supported
H2e	Performance expectancy* Personal innovativeness	0.245	0.041	5.098	0.000	Supported

Table 4. Prediction quality indicators for the hypotheses model

Variable	R2	F2					
		Hedonic motivation	Social influence	Facilitating conditions	Effort expectancy	Performance expectancy	Personal innovativeness
Interacting intention	0.505	0.243	0.104	0.085	0.052	0.302	0.203

dimension. As for the second sub-hypotheses, they were all supported at a significance level less than (0.05), as shown in Table 3 and 4.

As shown in Table 4, the results indicate that the F2 values are acceptable and have a positive sign towards the strength of the model. The R2 value also indicates that the independent variables clarify 50.5% of the change in intentions to engage in smart tables.

6 Discussion

This study explores the factors influencing customers' intentions towards using smart tables in restaurants in the Iraqi context by relying on the UTAUT2 model. The results of the research additionally underscore the fact that customer engagement in technology adoption in restaurants is integral. The research results show that elements including pleasure, societal acceptance, ease of sure, and expectation of performance can impact a customers intent of using a smart table application in a restaurant. The findings come in line with those of other studies indicating the link of adopting a news technology to the idea of personal benefit and a pleasurable experience (i.g., Juaneda-Ayensa et al., 2016; Okumus et al., 2018; Tseng et al., 2019).

The facet of force, or exertion, expectations intriguingly did not indicate any positive impact on customers' intent to engage with smart tables. In turn, this finding proposes the possibility that customers to don't find smart tables easy to use. This proposition, in return highlights the urge to address any implications or concerns regarding usability and thus educate society via programs to encourage users to engage with novel technological innovations. Furthermore, the research findings elucidate personal innovation positively interacting with the UTAUT models five dimensions with the purpose of enhancing customer intent toward engaging with smart tables. This finding, further indicated the impact of personal traits on accepting the adopting novel technology.

The research finding insinuate suggestions for restaurant managers, owners, operators, and decision makers having ability, will, and desire to integrate smart tables in their operations. Comprehending the elements that impact a customer's intent to engage allows the aforementioned to modify their marketing strategies and customer/user experiences to enhance user willingness to engage with new technology.

6.1 Theoretical Implications

The research results of the study at hand elucidate the influence of customers' intent regarding technological adoption on the restaurant industry. Motivational pleasure, societal impact, performance and exertion expectation, according to UTAUT2, influence customers' perception of smart tables and their intent to engage with them (Venkatesh et al., 2012). It is noteworthy that exertion expectation showed no impact on customer intent to engage apart from its interface with personal innovation. The finding hence insinuates how customers vision and experience technological innovations. Customers may consider that smart applications are more intricate or complex that traditional dining methods, highlighting the need to explore methods to improve customer or user perception of usability of smart restaurant technological innovations, and doing so necessitates suitable training and support to enhance user intent toward new technological innovations (Cain et al., 2019; Chawla & Joshi, 2019).

Furthermore, individual characteristics impacting the adoption of new technological system is reflected in the collaborative role of personal innovativeness, specifically regarding ai. The intention to engage with smart tables was considerable influenced by extent of personal innovation. This means that individuals with a stronger inclination towards technology exhibit more likelihood to adopt and engage with restaurant technologies (Na et al., 2021). All in all, the theoretical understandings brought by this research, underscore the key elements influencing acceptation of ai technologies. By assessing the fitness of the UTAUT2 model to smart tables, this research yields pragmatic indications of the elements that influence customer intent to engage with smart technology and highlights the need to consider individual characteristics in the adoption process.

6.2 Practical Implications

The current study's research findings highlight understandings and recommendations for restaurant owners, operators, managers, and decision makers in the restaurant industry, aiming to enhance quality of service via ai-related technological innovations. First, focusing on improving employees' awareness of aspects of the dining experience can enhance customer interaction with smart tables. Secondly, restaurant managers can capitalize on the expected fun element by incorporating it into menus, games and entertainment options to create a fun and immersive dining atmosphere. By incorporating fun and entertainment elements into the dining experience, restaurants can ensure positive customer attitudes towards smart tables. Third, the study emphasizes the importance of social influence on customers' intentions toward smart tables, so restaurant managers can exploit this element by creating a crowd of social opinions with a positive outlook toward the change in service provision in restaurants. This can be done by promoting word-of-mouth referrals and sharing on platforms. Socialize and form partnerships with influencers to raise awareness to boost intentions towards smart table technology. Fourth, it is important to create conditions to support the use of smart tables in restaurants. This can be done by providing support before, during and after service, and guidance and training videos to illustrate the use of smart tables. Fifth, although no association with expected effort was found, it is still necessary for smart table designers to focus on simplifying the smart table interface, reducing errors to a minimum, and facilitating menu navigation and effort. Finally, there is a need to target innovative individuals as early adopters and advocates of interactive smart tables, and can also be used as promoters for these restaurants and motivate customers to explore this new technology.

6.3 Limitations and Future Studies

While this research provides insight into factors that may influence customers' interaction intentions with smart tables, it is necessary to review some limitations which could impact the generalizability and accuracy of the results. One limitation of this research is its focus on the restaurant sector in Iraq, and given the cultural and social differences with other countries, generalizing the results to other societies may be related to conducting studies with more comprehensive samples or making comparisons between diverse samples. Another limitation of the study is the dependance on self-reported data and the study's cross-sectional nature. Although preventive and remedial mechanisms are used to eliminate common bias, self-reported data can be affected by biases such as desirability and memory errors, which may impact the accuracy and reliability of the findings. In addition, because of the study's cross-sectional nature, the results of the current study do not allow causal inferences to be made about the relationships between variables. Accordingly, future research can address these limitations by using objective designs and measures to measure customers' intentions toward engaging with smart tables in full-service restaurants.

Furthermore, the current research focused principally on how customers view the use of smart tables, and then plan the shift from traditional to smart tables. However, the study did not take into account the perspective of restaurant employees or managers.

Accordingly, future studies could examine the opinions and attitudes of restaurant workers towards switching to smart tables, in addition to studying the organizational factors
that could affect the success of their integration and use. Additionally, although the study
identified factors influencing customers' engagement intentions with smart tables, there
may be other unexplored elements that could also be important. For instance, the research
did not examine how individual differences such as age, gender, and socioeconomic status may influence engagement with restaurant technologies. Therefore, future research
can examine how these factors influence customer behavior to gain an understanding.

6.4 Conclusion

The research highlights the importance of tables powered by artificial intelligence in
improving customer engagement at sit-down restaurants. Factors such as enjoyment
and peer influence have an effect on customers' willingness to interact. The ease of use
remains a hurdle. Additionally, individual openness to ideas significantly influences customer behavior. These results indicate that incorporating AI technologies into restaurant
operations can provide opportunities to deliver engaging dining experiences and meet
changing customer needs, setting the stage for progress in the hospitality sector.

References

Abbas, S., et al.: Antecedents of trustworthiness of social commerce platforms: a case of rural
communities using multi group SEM & MCDM methods. Electron. Commer. Res. Appl. **62**,
101322 (2023)

Abdullah, H.O., Al-Abrrow, H., Atshan, N.A., Abbas, S.: Determinants of customer intentions to
use social commerce. In: Alnoor, A., Wah, K.K., Hassan, A. (eds.) Artificial Neural Networks
and Structural Equation Modeling: Marketing and Consumer Research Applications, pp. 97–
114. Springer, Singapore (2022). https://doi.org/10.1007/978-981-19-6509-8_6

Ajzen, I.: The theory of planned behavior. Organ. Behav. Hum. Decis. Process. **50**(2) (1991)

Alalwan, A.A., Dwivedi, Y.K., Rana, N.P., Algharabat, R.: Examining factors influencing Jordanian customers' intentions and adoption of internet banking: extending UTAUT2 with risk. J.
Retail. Consum. Serv. **40**, 125–138 (2018)

Ali, J., Hussain, K.N., Alnoor, A., Muhsen, Y.R., Atiyah, A.G.: Benchmarking methodology of
banks based on financial sustainability using CRITIC and RAFSI techniques. Decis. Making
Appl. Manag. Eng. **7**(1), 315–341 (2024)

Alkhwaldi, A.F., Alharasis, E.E., Shehadeh, M., Abu-AlSondos, I.A., Oudat, M.S., Bani Atta,
A.A.: Towards an understanding of FinTech users' adoption: Intention and e-loyalty post-
COVID-19 from a developing country perspective. Sustainability **14**(19), 12616 (2022)

Alnoor, A., Chew, X., Khaw, K.W., Muhsen, Y.R., Sadaa, A.M.: Benchmarking of circular economy behaviors for Iraqi energy companies based on engagement modes with green technology
and environmental, social, and governance rating. Environ. Sci. Pollut. Res. **31**(4), 5762–5783
(2024a)

Alnoor, A., Khaw, K.W., Chew, X., Abbas, S., Khattak, Z.Z.: The influence of the barriers of
hybrid strategy on strategic competitive priorities: evidence from oil companies. Glob. J. Flex.
Syst. Manag. **24**(2), 179–198 (2023)

Alnoor, A., et al.: How positive and negative electronic word of mouth (eWOM) affects customers'
intention to use social commerce? A dual-stage multi group-SEM and ANN analysis. Int. J.
Hum.-Comput. Interact. **40**(3), 808–837 (2024b)

Atshan, N.A., Abdullah, H.O., AL-Abrrow, H., Abbas, S.: How are brand activity and purchase behavior affected by digital marketing in the metaverse universe? In: Al-Emran, M., Ali, J.H., Valeri, M., Alnoor, A., Hussien, Z.A. (eds.) IMDC-IST 2024, pp. 112–128. Springer, Cham (2023). https://doi.org/10.1007/978-3-031-51300-8_8

Atshan, N.A., Al-Abrrow, H., Abdullah, H.O., Al Halbusi, H.: Mobile commerce and social commerce with the development of Web 2.0 technology. In: Alnoor, A., Wah, K.K., Hassan, A. (eds.) Artificial Neural Networks and Structural Equation Modeling: Marketing and Consumer Research Applications, pp. 149–161. Springer, Singapore (2022). https://doi.org/10.1007/978-981-19-6509-8_9

Baudier, P., Ammi, C., Deboeuf-Rouchon, M.: Smart home: highly-educated students' acceptance. Technol. Forecast. Soc. Chang. 153, 119355 (2020)

Belanche, D., Casaló, L.V., Flavián, C., Schepers, J.: Service robot implementation: theoretical framework and research agenda. Serv. Ind. J. 40(3–4), 203–225 (2020)

Berezina, K., Ciftci, O., Cobanoglu, C.: Robots, artificial intelligence, and service automation in restaurants. In: Robots, Artificial Intelligence, and Service Automation in Travel, Tourism and Hospitality, pp. 185–219. Emerald Publishing Limited (2019)

Boontarig, W., Chutimaskul, W., Chongsuphajaisiddhi, V., Papasratorn, B.: Factors influencing the Thai elderly intention to use smartphone for e-Health services. In: 2012 IEEE Symposium on Humanities, Science and Engineering Research, pp. 479–483. IEEE (2012)

Bossman, J.: Top 9 ethical issues in artificial intelligence. World Economic Forum (2016). https://www.weforum.org/agenda/2016/10/top-10-ethical-issues-in-artificial-intelligence/

Brooks, A.: The benefits of AI: 6 societal advantages of automation. Rasmussen University (2019). https://www.Rasmussenedu/degrees/technology/blog/benefits-of-ai/. Accessed 15 Mar 2023

Cabrera-Sánchez, J.P., Villarejo-Ramos, Á.F., Liébana-Cabanillas, F., Shaikh, A.A.: Identifying relevant segments of AI applications adopters – expanding the UTAUT2's variables. Telemat. Inform. 58, 101529 (2021)

Cabrera-Sánchez, J.P., Villarejo-Ramos, A.F.: Factors affecting the adoption of big data analytics in companies. Revista de Administração de Empresas 59, 415–429 (2020)

Cain, L.N., Thomas, J.H., Alonso, M., Jr.: From sci-fi to sci-fact: the state of robotics and AI in the hospitality industry. J. Hospitality Tourism Technol. 10(4), 624–650 (2019)

Chauhan, S., Jaiswal, M., Kar, A.K.: The acceptance of electronic voting machines in India: a UTAUT approach. Electron. Gov. Int. J. 14(3), 255–275 (2018)

Chawla, D., Joshi, H.: Consumer attitude and intention to adopt mobile wallet in India–an empirical study. Int. J. Bank Mark. 37(7), 1590–1618 (2019)

Chew, X., Khaw, K.W., Alnoor, A., Ferasso, M., Al Halbusi, H., Muhsen, Y.R.: Circular economy of medical waste: novel intelligent medical waste management framework based on extension linear Diophantine fuzzy FDOSM and neural network approach. Environ. Sci. Pollut. Res. 30(21), 60473–60499 (2023)

Chin, W.W.: The partial least squares approach to structural equation modeling. Mod. Methods Bus. Res. 295(2), 295–336 (1998)

Chu, T.H., Chao, C.M., Liu, H.H., Chen, D.F.: Developing an extended theory of UTAUT 2 model to explore factors influencing Taiwanese consumer adoption of intelligent elevators. SAGE Open 12(4), 21582440221142209 (2022)

Compeau, D.R., Higgins, C.A.: Computer selfefficacy: development of a measure and initial test. MIS Q. 19(2), 189–211 (1995)

Dakduk, S., Van der Woude, D., Nieto, C.A.: Technological Adoption in Emerging Economies: Insights from Latin America and the Caribbean with a Focus on Low-Income Consumers (2023)

Davis, F.D.: Perceived usefulness, perceived ease of use, and user acceptance of information technology. MIS Q. 13(3), 319–339 (1989)

Davis, F.D., Bagozzi, R.P., Warshaw, P.R.: User acceptance of computer technology: a comparison of two theoretical models. Manag. Sci. **35**(8), 982–1003 (1989)

Dwivedi, Y.K., Rana, N.P., Jeyaraj, A., Clement, M., Williams, M.D.: Re-examining the unified theory of acceptance and use of technology (UTAUT): towards a revised theoretical model. Inf. Syst. Front. **21**, 719–734 (2019)

Fan, W., Liu, J., Zhu, S., Pardalos, P.M.: Investigating the impacting factors for the healthcare professionals to adopt artificial intelligence-based medical diagnosis support system (AIMDSS). Ann. Oper. Res. **294**(1), 567–592 (2020)

Faris, M., Mahmud, M.N., Salleh, M.F.M., Alnoor, A.: Wireless sensor network security: a recent review based on state-of-the-art works. Int. J. Eng. Bus. Manag. **15**, 18479790231157220 (2023)

Festinger, L.: Theory of Cognitive Dissonance. Stanford University Press, Stanford (1957)

Fishbein, M., Ajzen, I.: Belief, attitude, intention, and behavior: an introduction to theory and research (1977)

Flavián, C., Akdim, K., Casaló, L.V.: Effects of voice assistant recommendations on consumer behavior. Psychol. Mark. **40**(2), 328–346 (2023)

Frey, C.B., Osborne, M.A.: The future of employment: how susceptible are jobs to computerisation? Technol. Forecast. Soc. Chang. **114**, 254–280 (2017)

Gansser, O.A., Reich, C.S.: A new acceptance model for artificial intelligence with extensions to UTAUT2: an empirical study in three segments of application. Technol. Soc. **65**, 101535 (2021)

Go, H., Kang, M., Suh, S.C.: Machine learning of robots in tourism and hospitality: interactive technology acceptance model (iTAM)–cutting edge. Tourism Rev. **75**(4), 625–636 (2020)

Grewal, D., Hulland, J., Kopalle, P.K.: The future of technology and marketing: a multidisciplinary perspective. J. Acad. Mark. Sci. **48**, 1–8 (2020)

Hair, J.F., Jr., Sarstedt, M., Hopkins, L., Kuppelwieser, V.G.: Partial least squares structural equation modeling (PLS-SEM). Eur. Bus. Rev. (2014)

Hair, J.F., Anderson, R.E., Babin, B.J., Black, W.C.: Multivariate Data Analysis: A Global Perspective. Pearson, Upper Saddle River (2010)

Hair, J.F., Hult, G.T.M., Ringle, C.M., Sarstedt, M., Thiele, K.O.: Mirror, mirror on the wall: a comparative evaluation of composite-based structural equation modeling methods. J. Acad. Mark. Sci. **45**(5), 616–632 (2017)

Hair, J.F., Jr., Sarstedt, M., Matthews, L.M., Ringle, C.M.: Identifying and treating unobserved heterogeneity with FIMIX-PLS: part I–method. Eur. Bus. Rev. **28**(1), 63–76 (2016)

Hasan, R., Shams, R., Rahman, M.: Consumer trust and perceived risk for voice-controlled artificial intelligence: the case of Siri. J. Bus. Res. **131**, 591–597 (2021)

Henseler, J., Ringle, C.M., Sarstedt, M.: A new criterion for assessing discriminant validity in variance-based structural equation modeling. J. Acad. Mark. Sci. **43**, 115–135 (2015)

Henseler, J., Ringle, C.M., Sinkovics, R.R.: The use of partial least squares path modeling in international marketing. In: New Challenges to International Marketing. Emerald Group Publishing Limited (2009)

Hesselbarth, I., Alnoor, A., Tiberius, V.: Behavioral strategy: a systematic literature review and research framework. Manag. Decis. **61**(9), 2740–2756 (2023)

Hofmann, P., Jöhnk, J., Protschky, D., Urbach, N.: Developing purposeful AI use cases-a structured method and its application in project management. In: Wirtschaftsinformatik (Zentrale Tracks), pp. 33–49 (2020)

Hoque, R., Sorwar, G.: Understanding factors influencing the adoption of mHealth by the elderly: an extension of the UTAUT model. Int. J. Med. Inform. **101**, 75–84 (2017)

Hossain, M.A., Hasan, M.I., Chan, C., Ahmed, J.U.: Predicting user acceptance and continuance behaviour towards location-based services: the moderating effect of facilitating conditions on behavioural intention and actual use. Australas. J. Inf. Syst. **21**, 1–22 (2017)

Jang, S.H., Lee, C.W.: The impact of location-based service factors on usage intentions for technology acceptance: the moderating effect of innovativeness. Sustainability **10**(6), 1876 (2018)

Juaneda-Ayensa, E., Mosquera, A., Sierra Murillo, Y.: Omni channel customer behavior: key drivers of technology acceptance and use and their effects on purchase intention. Front. Psychol. **7**, 1–11 (2016)

Kaya, F., Aydin, F., Schepman, A., Rodway, P., Yetişensoy, O., Demir Kaya, M.: The roles of personality traits, AI anxiety, and demographic factors in attitudes toward artificial intelligence. Int. J. Hum.-Comput. Interact. **40**(2), 497–514 (2024)

Khaw, K.W., Alnoor, A., Al-Abrrow, H., Tiberius, V., Ganesan, Y., Atshan, N.A.: Reactions towards organizational change: a systematic literature review. Curr. Psychol. **42**(22), 19137–19160 (2023)

Lai, I.K.: Traveler acceptance of an app-based mobile tour guide. J. Hospitality Tourism Res. **39**(3), 401–432 (2015)

Lian, J.W., Yen, D.C.: Online shopping drivers and barriers for older adults: age and gender differences. Comput. Hum. Behav. **37**, 133–143 (2014)

Lu, Y., Yang, S., Chau, P.Y., Cao, Y.: Dynamics between the trust transfer process and intention to use mobile payment services: a cross-environment perspective. Inf. Manag. **48**(8), 393–403 (2011)

Magsamen-Conard, K., Upadhyaya, S., Joa, C.Y., Dowd, J.: Bridging the divide: Using UTAUT to predict multigenerational tablet adoption practices. Comput. Hum. Behav. **50**, 186–196 (2015)

Maier, T., Edwards, K.: Service system design and automation in the hospitality sector. J. Hospitality **2**(1–2), 1–14 (2020)

Mathath, A., Fernando, Y.: Robotic transformation and its business applications in food industry. In: Robotics, Automation, and Control in Industrial and Service Settings, pp. 281–305. IGI Global (2015)

Meet, R.K., Kala, D., Al-Adwan, A.S.: Exploring factors affecting the adoption of MOOC in Generation Z using extended UTAUT2 model. Educ. Inf. Technol. **27**(7), 10261–10283 (2022)

Merhi, M., Hone, K., Tarhini, A.: A cross-cultural study of the intention to use mobile banking between Lebanese and British consumers: extending UTAUT2 with security, privacy and trust. Technol. Soc. **59**, 101151 (2019)

Meuter, M.L., Bitner, M.J., Ostrom, A.L., Brown, S.W.: Choosing among alternative service delivery modes: an investigation of customer trial of self-service technologies. J. Mark. **69**, 61–83 (2005)

Moore, G.C., Benbasat, I.: Development of an instrument to measure the perceptions of adopting an information technology innovation. Inf. Syst. Res. **2**(3), 192–222 (1991)

Morosan, C., DeFranco, A.: Co-creating value in hotels using mobile devices: a conceptual model with empirical validation. Int. J. Hosp. Manag. **52**, 131–142 (2016)

Na, T.K., Yang, J.Y., Lee, S.H.: Determinants of behavioral intention of the use of self-order kiosks in fast-food restaurants: focus on the moderating effect of difference age. SAGE Open **11**(3), 21582440211031907 (2021)

Neudert, L.M., Knuutila, A., Howard, P.N.: Global attitudes towards AI, machine learning & automated decision making. Working paper 2020.10, Oxford Commission on AI & Good Governance (2020). https://oxcaigg.oii.ox.ac.uk

Palas, J.U., Sorwar, G., Hoque, M.R., Sivabalan, A.: Factors influencing the elderly's adoption of mHealth: an empirical study using extended UTAUT2 model. BMC Med. Inform. Decis. Mak. **22**(1), 191 (2022)

Peslak, A., Shannon, L.J., Ceccucci, W.: An empirical study of cell phone and smartphone usage. Issues Inf. Syst. **12**(1), 407–417 (2011)

Rese, A., Ganster, L., Baier, D.: Chatbots in retailers' customer communication: how to measure their acceptance? J. Retail. Consum. Serv. **56**, 102176 (2020)

Ribeiro, M.A., Gursoy, D., Chi, O.H.: Customer acceptance of autonomous vehicles in travel and tourism. J. Travel Res. **61**(3), 620–636 (2022)

Rosete, A., Soares, B., Salvadorinho, J., Reis, J., Amorim, M.: Service robots in the hospitality industry: an exploratory literature review. In: Nóvoa, H., Drǎgoicea, M., Kühl, N. (eds.) IESS 2020. LNBIP, vol. 377, pp. 174–186. Springer, Cham (2020). https://doi.org/10.1007/978-3-030-38724-2_13

Sandberg, H., Alnoor, A., Tiberius, V.: Environmental, social, and governance ratings and financial performance: evidence from the European food industry. Bus. Strateg. Environ. **32**(4), 2471–2489 (2023)

Schmitz, A., Díaz-Martín, A.M., Guillén, M.J.Y.: Modifying UTAUT2 for a cross-country comparison of telemedicine adoption. Comput. Hum. Behav. **130**, 107183 (2022)

Shaw, N., Sergueeva, K.: The non-monetary benefits of mobile commerce: extending UTAUT2 with perceived value. Int. J. Inf. Manag. **45**, 44–55 (2019)

Sindermann, C., et al.: Assessing the attitude towards artificial intelligence: introduction of a short measure in German, Chinese, and English language. KI-Künstliche intelligenz **35**(1), 109–118 (2021)

Slade, E.L., Dwivedi, Y.K., Piercy, N.C., Williams, M.D.: Modeling consumers' adoption intentions of remote mobile payments in the United Kingdom: extending UTAUT with innovativeness, risk, and trust. Psychol. Mark. **32**(8), 860–873 (2015)

Sobti, N.: Impact of demonetization on diffusion of mobile payment service in India: antecedents of behavioral intention and adoption using extended UTAUT model. J. Adv. Manag. Res. **16**(4), 472–497 (2019)

Sun, W., Dedahanov, A.T., Shin, H.Y., Kim, K.S.: Extending UTAUT theory to compare south Korean and Chinese institutional investors' investment decision behavior in Cambodia: a risk and asset model. Symmetry **11**(12), 1524 (2013)

Tan, G.W.H., Lee, V.H., Lin, B., Ooi, K.B.: Mobile applications in tourism: the future of the tourism industry? Ind. Manag. Data Syst. **117**(3), 560–581 (2017)

Taylor, S., Todd, P.: Assessing IT usage: the role of prior experience. MIS Q. **19**(4), 561–570 (1995)

Thajil, K.M., Al-Abrrow, H., Abdullah, H.O.: The role of blockchain adoption and supply chain practices on social commerce. In: Alnoor, A., Wah, K.K., Hassan, A. (eds.) Artificial Neural Networks and Structural Equation Modeling: Marketing and Consumer Research Applications, pp. 131–148. Springer, Singapore (2022). https://doi.org/10.1007/978-981-19-6509-8_8

Thompson, R.L., Higgins, C.A., Howell, J.M.: Personal computing: toward a conceptual model of utilization. MIS Q. **15**(1), 125–143 (1991)

Tseng, T.H., Lin, S., Wang, Y.S., Liu, H.X.: Investigating teachers' adoption of MOOCs: the perspective of UTAUT2. Interact. Learn. Environ. 1–16 (2019)

Van Droogenbroeck, E., Van Hove, L.: Adoption and usage of e-grocery shopping: a context-specific UTAUT2 model. Sustainability **13**(8), 4144 (2021)

Venkatesh, V., Thong, J.Y., Xu, X.: Consumer acceptance and use of information technology: extending the unified theory of acceptance and use of technology. MIS Q. **36**(1), 157–178 (2012)

Vu, H.T., Lim, J.: Effects of country and individual factors on public acceptance of artificial intelligence and robotics technologies: a multilevel SEM analysis of 28-country survey data. Behav. Inf. Technol. **41**(7), 1515–1528 (2022)

Williams, M.D., Rana, N.P., Dwivedi, Y.K.: The unified theory of acceptance and use of technology (UTAUT): a literature review. J. Enterp. Inf. Manag. **28**(3), 443–448 (2015)

XinYing, C., Tiberius, V., Alnoor, A., Camilleri, M., Khaw, K.W.: The dark side of metaverse: a multi-perspective of deviant behaviors from PLS-SEM and fsQCA findings. Int. J. Hum.–Comput. Interact. 1–21 (2024)

Yang, Li., Henthorne, T.L., George, B.: Artificial intelligence and robotics technology in the hospitality industry: current applications and future trends. In: George, B., Paul, J. (eds.) Digital transformation in business and society, pp. 211–228. Springer, Cham (2020). https://doi.org/10.1007/978-3-030-08277-2_13

Zhang, L., Zhu, J., Liu, Q.: A meta-analysis of mobile commerce adoption and the moderating effect of culture. Comput. Hum. Behav. **28**(5), 1902–1911 (2012)

The Role of Artificial Intelligence Trading Robots in Rationalizing Cryptocurrency Trading Decisions by Application to Bitcoin Currency

Muntather Fadel Saad Al-Batat[1], Ammar Shehab Ahmed Al-Ahmad[2(✉)], and Halla Sami Khudier Al-Yaseen[3]

[1] College of Administration and Economics, University of Basrah, Basrah, Iraq
muntader.saad@uobasrah.edu.iq
[2] General Directorate of Nineveh Education, Financial Affairs Department, Ministry of Education, Mosul, Iraq
ammarshhab81@gmail.com
[3] College of Administration and Economics, University of Mosul, Mosul, Iraq

Abstract. The subject of trading in electronic trading platforms based on artificial intelligence and through so-called trading robots has become the main concern of traders in the crypto currency markets, as the need for it has increased due to the continuation of trading operations for 24 h a day and 5 days a week. Which made it impossible for traders to keep up with it over time, which could cause them to miss important opportunities to enter trading at a low price or exit it at a high price, which motivated them to search for solutions to this large and important problem, to reach the appropriate mechanism that enables them to track the movement of currency prices on electronic trading platforms, and they found what they were looking for in artificial intelligence, which is represented by trading robots that can work on their behalf for 24 h and for 5 days a week. Which can contribute to maximizing their profits and minimizing their losses. The research was divided into two sections. The first dealt with the theoretical framework of artificial intelligence for trading robots and trading platforms. The second focused on reversing the working mechanism of artificial intelligence for trading robots and how to harness it in following up on electronic trading on the electronic trading platform Instead of Traders. The research concluded that artificial intelligence, represented by trading robots, can contribute to maximizing traders' profitability and reducing traders' losses, as well as reducing their efforts in managing and monitoring electronic trading platforms. A recommendation was made regarding the importance of traders training on trading robots and then adopting them in trading in order to take advantage of time. During periods when they are not in front of electronic trading platforms, in addition to eliminating the influences of the psychological state of trading decisions.

Keywords: Artificial Intelligence · Trading Robots · Trading Platforms · Crypto currencies · Technical analysis

1 Introduction

Traders in the field of cryptocurrencies through trading platforms to maximize their profits and grow their money, which they spent a lot of effort to obtain. Through their continuous monitoring of the price trend of cryptocurrencies to determine the Tops peaks and bottoms of cryptocurrencies. Which enables traders to make decisions about entering or exiting trading at the appropriate time. Thus achieving profits commensurate with their targeted goals. Despite the advantages offered by cryptocurrency markets, the most important of which is the ability to trade them 24 h a day, 7 days a week. However, at the same time, this advantage exceeds the ability of traders to keep up with trading operations all the time. Which may cause them to lose important trading opportunities that could have maximized their profits. Therefore, the need has increased to find something that enables them to benefit from trading operations on trading platforms during their periods of rest or absence from following them for any reason whatsoever. These efforts culminated in the innovation of trading robots that operate with artificial intelligence and have the ability to keep up with trading operations 24 h a day, 7 days a week. Which achieves many advantages for the trader, the most important of which is reducing the time required of him to follow the trading platforms and the possibility of exploiting it in other matters. In addition to allowing them to make the most of ongoing trading operations in order to maximize their profits with the least time and effort expended.

2 Literature Review

2.1 The Concept of Artificial Intelligence

Artificial intelligence is one of the most important topical topics in all sectors in general and the financial sector in particular. With the emergence of technology, the financial sector witnessed a revolution in its various branches, especially in cryptocurrency trading platforms (Atiyah, 2023; Alnoor et al., 2023). Artificial intelligence can contribute to facilitating the technical analysis process by analyzing trading data and strategies that have been framed within the powers granted to the trading robot (Wahid, 2013). Artificial Intelligence techniques also help in improving the shortcomings of traders' behavior such as greed, avarice, fear and other trader behaviours. Which negatively affects cryptocurrency trading decisions (Alnoor et al., 2024a, b; Atiyah et al., 2023a, b, c; Abbas et al., 2023). Taking it away from their behavior and making them more balanced and controlled in trading operations, such as buying and selling, at the appropriate time. In addition, it eliminates the need for traders to follow the cryptocurrency market during their absence, as it is a market that operates 24 h a day, 5 days a week, to replace them, trade in their place, and generate many additional profits for them (Wahid, 2013; Ali et al., 2024).

A cryptocurrency trading robot is an electronic program that relies on artificial intelligence in its work. It helps traders determine the appropriate time to buy or sell cryptocurrencies at any time (Atiyah et al., 2023a, b, c). And enter trading automatically based on a set of permissions granted to him. The robots are designed to remove the psychological factors of trading that can negatively affect their decisions. A trading robot can help them test and develop their strategies (Alnoor et al., 2024a, b). You can trade 24 h a day, 5 days a week in a logical, unemotional manner and always alert for profitable trading operations. It can automatically carry out trading operations in the cryptocurrency market on behalf of traders, depending on the settings that were programmed with it when the opportunity arises. Which gives trading robots the ability to implement specific trading strategies (Abednego, 2018; Atiyah and Zaidan, 2022).

2.2 Trading Platforms

A trading platform is a software system usually offered by a brokerage firm or other financial institution that allows a trader to trade online. The platform provides traders with an interface through which they can access various markets in general and cryptocurrency markets in particular. As well as conducting trading operations, monitoring their positions, and managing their accounts (Atiyah et al., 2023a, b, c). Trading platforms can also provide a number of features such as real-time quotes, live financial and trading news feeds, instant access to a wide range of streaming and historical financial data, technical analysis tools, investment research, and educational resources (Muhsen et al., 2023; Al-Enzi et al., 2023). Given the capabilities provided by the cryptocurrency market, such as trading 24 h a day and 5 days a week, it has become difficult for human traders to follow it around the clock, which has led to the search for what will help them control this problem (Husin et al., 2023). They found what they wanted to seek refuge and benefit from the great technological development and benefit from artificial intelligence, as it contributed to deputizing the trading operations for traders at all times or during their absence by linking the platform to robots called trading robots that are programmed and given powers by them (Ahmed et al., 2024). Therefore, trading operations have been automated and the trading platform has been managed with artificial intelligence (Chen, 2022).

2.3 Technical Analysis Indicators

Japanese candlesticks are considered one of the methods of technical analysis that reflects the psychological state and collective human feelings of traders in the cryptocurrency market. Analysis in this method leads to studying the price movement (opening price, highest price, lowest price, closing price) that constitutes. They are all in the form of the Japanese candle, which is reflected in a chart that resembles a candle, and it is not a requirement that the four prices of the Japanese candles reflect the real value of the cryptocurrencies, which may affect traders' trading decisions, as each type interprets a specific case of upward or downward, peak selling, or peak buying. From this standpoint, relying on trading robots has become an important matter because it is not affected by the psychological state and optimistic or pessimistic feelings, which improves trading decisions and profitability (Choudhuri, 2019). The Relative Strength Index was developed by J. Welles Wilder and was presented in his book New Concepts of Technical Trading Systems in 1978. It is distinguished by the simplicity of its interpretation and the quality of its signals, as the RSI fluctuates within a certain range between a maximum value (100) and a minimum value (0) (Moroşan, 2011), and the greatest effect of the indicator is observed when it is close to its maximum value. Between 70 to 100 is the overbought area, while from 0 to 30 is the oversold area (Choudhuri, 2019).

Figure 1 explain the technology used in the financial field in general and cryptocurrency trading markets in particular has become one of the most important topics that exist in order to address the problem facing traders, which is their inability to keep up with trading operations 24 h a day and 5 days a week, which affects the behavior of traders, in addition to requiring effort. A lot of effort is needed to keep up with it, and therefore the research assumes the following:

H1: The research assumes that a trading robot contributes to maximizing profits and minimizing losses, thus rationalizing trading decisions.
H2: The research assumes that the trading robot contributes to eliminating the psychological state from trading decisions.
H3: The research assumes that the trading robot contributes to reducing the effort expended in trading operations.

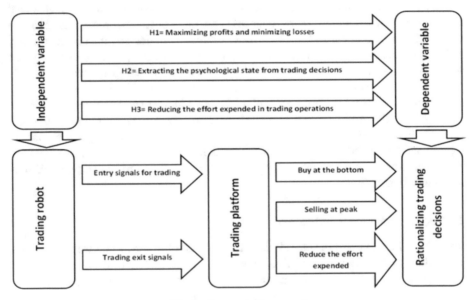

Fig. 1. Research Framework.

3 Methodology

In order to carry out trading operations in cryptocurrencies on trading platforms. It requires traders to expend a lot of time and effort in front of the trading platforms in an attempt to seize the best trading opportunities, represented by buying at low price levels and selling at high price levels. To make good profits for them. However, these trading operations collide with a number of problems, which are represented by the barrier of lack of time and lack of experience in managing trading operations that are affected by emotions and their psychological state. This may affect the quality of cryptocurrency trading decisions and negatively affect their investment goals. The research aims to shed light on the trading mechanism with trading robots that rely on artificial intelligence techniques and how to benefit from them in improving cryptocurrency trading operations on trading platforms. Artificial intelligence-based trading in cryptocurrencies by trading robots is of great importance to traders due to their ability to work tirelessly for long periods of time, exceeding the capabilities of human traders. Which may contribute to maximizing their profitability and achieving their goals faster. The research is based on the descriptive analytical method, as a number of research and studies published in periodicals and specialized scientific journals were reviewed. As for the applied aspect, the technical analysis method was adopted by programming trading robots with it, as shown in Fig. 2.

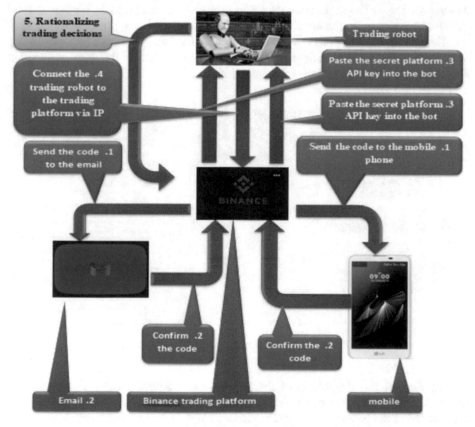

Fig. 2. Research model.

To test the proposed hypotheses and analyze data we relied on AMOS version 25 and SPSS version 25. The validity and reliability of the measures were confirmed, and then descriptive statistics and correlation coefficients were presented.

4   Results

The trader selects the robot application that he wishes to use to trade cryptocurrencies on its behalf using artificial intelligence, which is represented by the trading robot application. Register in the trading robot application by clicking on its registration link to open the application window that asks the trader to enter his email, then clicking on the verification code (Send Verification Code) to send a code (Code) to him which is copied and pasted into the Verification Code field of the application to verify the validity of his ownership of the trader, in addition to adding the Password to him. Then click on Register.

After entering the robot application, it must be loaded with the balance as it is a new account by looking at the (Asset) wallet, which indicates that the balance is (0) dollars, as shown in Fig. 3. The wallet contains three options, which are Deposit. Withdrawal and transfer of amounts. The trader charges the balance to the robot application by clicking on "Deposit" to open a window providing him with the link or bar code for transferring the USDT funds from the trader's money wallet (Trust Wallet) to the robot application for paying fees. Subscribe to the robot application, as shown in Fig. 4.

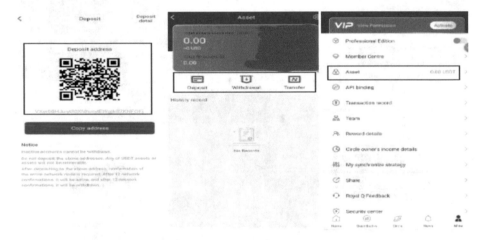

Fig. 3.   Charging the balance trading robot application.

After completing the process of charging the balance (USDT), go to the (Activats) box and click on it to activate the robot application. A window will open showing him the amount of the balance of (USDT) that was charged, in addition to clicking on the Conditions box to indicate the trader's agreement to the terms of applying the trading robot.

After opening the trading platform and in order to complete the process of linking the trading robot, click on the More icon to open a new window that includes several options. (API management) is selected, to open a window for creating a link. The (Create API) that can be linked to the trading robot by writing its name in the field designated for that, then clicking on (Create API), then a new window opens for the security clearance

Fig. 4. Activating the trading robot application.

to prevent account piracy, which requires inclusion The mobile number is in the (Phone Veritation Code) field, in addition to the email in the (E-Mail Veritation Code) field. After listing them, two codes are sent to them in order to rewrite them in their respective fields to verify the validity of the trader's ownership. For the account on the trading platform, then click on (Submit).

The trader returns to the robot application and copies its API and pastes it back into the Binance platform in order to identify it, then clicks on Save. After that, a new code is sent to the mobile phone. Copy it to complete the linking process on the part of the trading platform. Finally, the trader returns to the robot application to send the confirmation code to the mobile phone to complete the linking process on the part of the robot application by clicking on "Send." After receiving the code, he pastes it in the "Veritation Code" box and then clicks on "Bind" to activate the linking between the robot application. And the trading platform.

Fig. 5. Binance trading platform interface.

After activating and linking the trading robot application to the trading platform, one will move to the Binance trading platform in order to program the trading robot in order to initiate the trading process and choose the cryptocurrencies to be traded, as shown in Fig. 5. Programming the trading robot in order to determine entry points by buying, in addition to setting price targets and profit-taking points, as well as determining the loss margin and exit from trading by selling. After the trader applied the above points, the process of starting the trading process became ready, as the trading process was chosen at the one-hour level in addition to choosing the encrypted Bitcoin currency. Trading lasted for a period of six hours on 3/29/2024, and the results shown in the Table 1 were reached. Which was extracted through analysis of trading data from the trading platform and shown in the figures.

Table 1. Results of trading decisions made by the trading robot for six Japanese candles for a 6-h trading period.

Details	Candle (1)	Candle (2)	Candle (3)	Candle (4)	Candle (5)	Candle (6)
Trading candle type	Green	Green	Green	Red	Red	Green
Market trend	rising	rising	rising	Down	Down	rising
Trading time (hours)	Hour 10 29/3/2024	Hour 11 29/3/2024	Hour 12 29/3/2024	Hour 13 29/3/2024	Hour 14 29/3/2024	Hour 15 29/3/2024
Current price	69105.70	69324.11	69364.83	69665.15	69611.64	69421.63
Percentage increase in price (0.2%)	138.21	138.65	138.73	139.33	139.22	138.84
Entry price for initial trading (First Buy in Amount)	69243.91	69462.76	69503.56	69525.82	69750.86	69560.47
First reinforcement	0	0	0	0	69521.16	0
Second reinforcement	0	0	0	0	69389.96	0
Third reinforcement	0	0	0	0	69290.42	0
Average price	0	0	0	0	69488.10	0
Current price	69461.90	69597.89	69666.00	69525.82	69290.42	69822.02
Percentage decrease in price (0.1%)	69.4635	69.60	69.666	69.53	138.58	69.82
Exit price from the trade	69394.04	69528.29	69596.334	69595.35	69429.00	69752.2
The amount of profit achieved	150.13	65.53	92.774	69.53	-59.1	191.73
Figure number	10	11	12	13	14	15

Source: Table prepared by researchers based on the Figs. (12-13-14-15-16-17)

Determine the cryptocurrency that will be traded and click on it, which is represented by the Bitcoin cryptocurrency (BTC/USDT). Determine the specific amount of investment (Position Amount) in the cryptocurrency Bitcoin at $350,000. Determine the reinforcement operations (buying when the price decreases) (Numbers of Call Margin) at (4) times. Determine the entry amount for trading (First Buy in Amount) at $70,000. The opening price of the cryptocurrency Bitcoin (Current Price) reached $69,340.75. Profit targets (Take Profit Ratio) have been set at (0.5%) and the robot is ready to exit the current trading, but it will not exit trading until after the profits (Earnings Callback) decline by (0.1%). The return point to enter trading (Margin Call Drop) has been determined after the previous deal is closed and the robot is ready to buy again, but trading will not be entered until the price (Buy in Callback) rises by (0.2%). The Relative Strength Index (RSI) reached (22) points, which indicates that the price of Bitcoin is approaching the price bottom, which led to the trading robot preparing to enter buying trading after the price rebounded by (0.2%). The robot entered trading by buying (Call Margin Trigger Price) at a price of (69243.91) dollars, after rising by (0.2%) (Buy in Callback) from the opening price of (69105.70) dollars, with a difference of (138.21) dollars. The Relative Strength Index (RSI) reached (60.82) points, which indicates an increase in the price of Bitcoin above the price bottom, which led to the trading robot preparing to exit trading by selling after the price decreased by (0.1%). The robot exited trading by selling (Take Profit Ratio) at a selling price of (69,394.04) dollars after a decline in profits (Earnings Callback) by (0.1%) from the highest price of (69,461.90) dollars, with a difference of (69.4635) dollars. The trading robot made a profit for the trader of $150.13, after the price closed with a green candle, as shown Fig. 6.

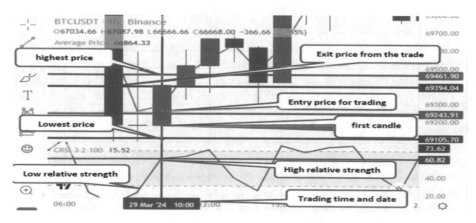

Fig. 6. The first trading candle.

The Relative Strength Index (RSI) reached (60.82) points, which indicates that the price of Bitcoin is approaching the price bottom, which led to the trading robot preparing to enter buying trading after the price rebounded by (0.2%). The robot entered trading by buying (Call Margin Trigger Price) at a price of (69324.11) dollars, after rising by (0.2%) dollars (Buy in Callback) from the opening price of (69462.76) dollars, with a difference of (138.65) dollars. The Relative Strength Index (RSI) reached (63.61) points, which indicates an increase in the price of Bitcoin from the price bottom, which led to the trading robot preparing to exit the sell trading after the price decreased by (0.1%).

The robot exited trading by selling (Take Profit Ratio) at a selling price of (69,528.29) dollars after a decline in profits (Earnings Callback) by (0.1%) from the highest price of (69,597.89) dollars, with a difference of (69.60) dollars. The trading robot made a profit of $65.53 for the trader, after the price closed with a green candle, as shown Fig. 7.

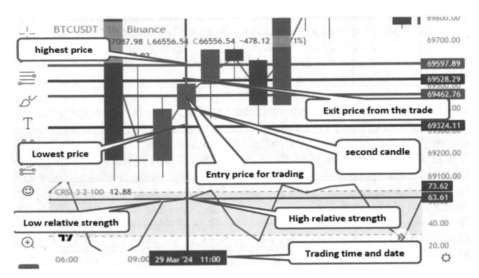

Fig. 7. The second trading candle.

The Relative Strength Index (RSI) reached (63.61) points, which indicates that the price of Bitcoin is approaching the price bottom, which led to the trading robot preparing to enter buying trading after the price rebounded by (0.2%). The robot entered trading by buying (Call Margin Trigger Price) at a price of (69503.56) dollars, after rising by (0.2%) dollars (Buy in Callback) from the opening price of (69364.83) dollars, with a difference of (138.73) dollars. The Relative Strength Index (RSI) reached (68.86) points, which indicates an increase in the price of Bitcoin from the price bottom, which led to the trading robot preparing to exit trading by selling after the price decreased by (0.1%). The robot exited trading by selling (Take Profit Ratio) at a selling price of (69,596.334) dollars after a decline in profits (Earnings Callback) by (0.1%) from the highest price of (69,666.00) dollars, with a difference of (69,666) dollars. The trading robot made a profit of $92,774 for the trader after the price closed with a green candle, as shown Fig. 8.

The Relative Strength Index (RSI) reached (68.86) points, which indicates that the price of Bitcoin is close to the price peak, which led to the trading robot preparing to exit trading by selling after the price rebounded by (0.2%), but the price did not reach It increased by (0.2%), but the price began to decline to price levels lower than the entry price for trading. The robot entered trading by buying (Call Margin Trigger Price) at a price of (69525.82) dollars, after declining by (0.2%) dollars (Buy in Callback) from the opening price of (69661.36) dollars, with a difference of (139.33) dollars. The Relative Strength Index (RSI) reached (38.52) points, which indicates a decline in the price of Bitcoin from the price peak, which led to the trading robot preparing to exit trading by

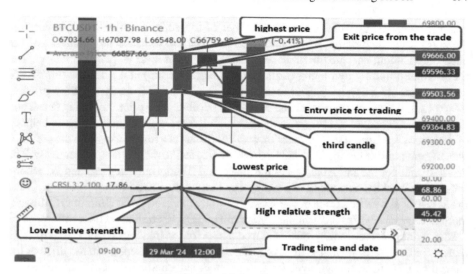

Fig. 8. The third trading candle.

selling after the price decreased by (0.1%). The robot exited trading by selling (Take Profit Ratio) at a selling price of (69,595.35) dollars after the profit (Earnings Callback) increased by (0.1%) from the highest price of (69,525.82) dollars, with a difference of (69,666) dollars, as shown Fig. 9.

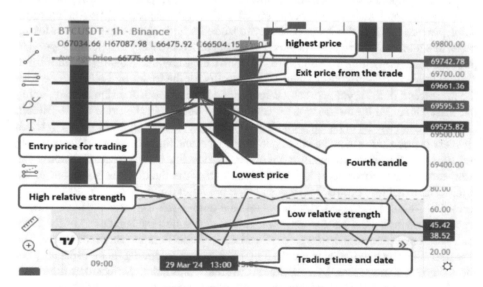

Fig. 9. The fourth trading candle.

The Relative Strength Index (RSI) reached (38.52) points, which indicates that the price of Bitcoin is close to the price bottom, which led the trading robot to prepare to exit trading by selling after the price rebounded by (0.2%), but the price did not reach It increased by (0.2%), but the price began to decline to price levels lower than the entry price for trading, which prompted the robot to enhance its purchases of the encrypted Bitcoin currency in preparation for reselling it after its price increased. The robot entered trading by buying (Call Margin Trigger Price) at a price of (69,750.86) dollars, after declining by (0.2%) dollars (Buy in Callback) from the opening price of (69,611.64) dollars, with a difference of (139.22) dollars. The price continued to fall and did not meet the condition for an increase, which led to the robot enhancing its balance of the encrypted currency by purchasing new quantities at a lower price than the previous one, in amounts amounting to (69521.16, 69389.96, 69290.42) dollars, respectively. The average price for the entire balance of the encrypted Bitcoin currency reached Which the robot purchased for an amount of $69,488.10. The robot exited trading by selling (Take Profit Ratio) at a selling price of (69,429.00) dollars after the profit (Earnings Callback) increased by (0.1%) from the highest price of (69,290.42) dollars, with a difference of (138.58) dollars. The trading robot achieved a loss for the trader of (−59.1) dollars. The reason for the loss is attributed to the significant decline that occurred in the price of the encrypted Bitcoin currency. However, the enhancements in the balance of the encrypted Bitcoin currency reduced the average purchase price, which reduced the size of the loss incurred by the robot. From the trading process, after it closed with a red candle.

5   Conclusions and Implications

Adopting artificial intelligence helps extract the psychological state and emotions that affect trading operations, which contributes to improving the quality of trading decisions made by traders, which reflects positively on their profits achieved. The trading robot helps provide traders with the opportunity to trade 24 h, 5 days a week without losing focus like humans. The trading robot is distinguished by its ability to quickly respond to price changes, in addition to its ability to deal with a large amount of data at the same time. The trading robot contributes to maximizing profits for the trader by keeping pace with the price trend of the cryptocurrency Bitcoin. The trading robot contributes to reducing the losses incurred by the trader as a result of the decline in the price of the Bitcoin encrypted currency by enhancing purchase operations at additional low prices to reach an average price lower than the initial purchase price, and then carrying out the reselling process when the price rises again and achieving profits that reduce losses. Previous. The short-term trading robot helps in taking advantage of the fluctuations that occur in the price trend to achieve profits from it despite the price trend continuing to rise in the long term. The Relative Strength Index shows that the trading robot entered the trade when it was high and exited the trade when it was low, which shows the agreement of the entry and exit signals for both Japanese candlesticks and the Relative Strength Index. The research suggests that traders adopt trading robots after practicing them, especially for those who are characterized by a psychological state that is quickly affected by the events that the cryptocurrency market is experiencing, which affects their trading decisions. The research suggests that traders practice how to set up a trading robot by understanding it and what powers should be given to it when carrying out the trading process. The research suggests trading in more than one currency to distribute and diversify the risks surrounding it, as one currency was adopted in the research due to the limited number of pages for the research.

References

Abbas, S., et al.: Antecedents of trustworthiness of social commerce platforms: a case of rural communities using multi group SEM & MCDM methods. Electron. Commer. Res. Appl. **62**, 101322 (2023)

Abednego, L., Nugraheni, C.E., Rinaldy, I.: Forex Trading Robot with Technical and Fundamental Analysis, Parahyangan Catholic University, Bandung, Indonesia (2018). https://doi.org/10.17706/jcp.13.9.1089-1097

Ahmed, A.D., Salih, M.M., Muhsen, Y.R.: Opinion weight criteria method (OWCM): a new method for weighting criteria with zero inconsistency. IEEE Access (2024)

Al-Enzi, S.H.Z., Abbas, S., Abbood, A.A., Muhsen, Y.R., Al-Hchaimi, A.A.J., Almosawi, Z.: Exploring research trends of metaverse: a bibliometric analysis. In: Al-Emran, M., Ali, J.H., Valeri, M., Alnoor, A., Hussien, Z.A. (eds.) IMDC-IST 2024, pp. 21–34. Springer, Cham (2023). https://doi.org/10.1007/978-3-031-51716-7_2

Ali, J., Hussain, K.N., Alnoor, A., Muhsen, Y.R., Atiyah, A.G.: Benchmarking methodology of banks based on financial sustainability using CRITIC and RAFSI techniques. Decis. Making Appl. Manag. Eng. **7**(1), 315–341 (2024)

Alnoor, A., Atiyah, A.G., Abbas, S.: Toward digitalization strategic perspective in the European food industry: non-linear nexuses analysis. Asia-Pac. J. Bus. Adm. (2023)

Alnoor, A., Atiyah, A.G., Abbas, S.: Unveiling the determinants of digital strategy from the perspective of entrepreneurial orientation theory: a two-stage SEM-ANN approach. Glob. J. Flex. Syst. Manag. 1–18 (2024a)

Alnoor, A., et al.: How positive and negative electronic word of mouth (eWOM) affects customers' intention to use social commerce? A dual-stage multi group-SEM and ANN analysis. Int. J. Hum.-Comput. Interact. **40**(3), 808–837 (2024b)

Atiyah, A.G.: Unveiling the quality perception of productivity from the senses of real-time multisensory social interactions strategies in metaverse. In: Al-Emran, M., Ali, J.H., Valeri, M., Alnoor, A., Hussien, Z.A. (eds.) IMDC-IST 2024, pp. 83–93. Springer, Cham (2023). https://doi.org/10.1007/978-3-031-51300-8_6

Atiyah, A.G., Zaidan, R.A.: Barriers to using social commerce. In: Alnoor, A., Wah, K.K., Hassan, A. (eds.) Artificial Neural Networks and Structural Equation Modeling, pp. 115–130. Springer, Singapore (2022). https://doi.org/10.1007/978-981-19-6509-8_7

Atiyah, A.G., Alhasnawi, M., Almasoodi, M.F.: Understanding metaverse adoption strategy from perspective of social presence and support theories: the moderating role of privacy risks. In: Al-Emran, M., Ali, J.H., Valeri, M., Alnoor, A., Hussien, Z.A. (eds.) IMDC-IST 2024, pp. 144–158. Springer, Cham (2023a). https://doi.org/10.1007/978-3-031-51300-8_10

Atiyah, A.G., All, N.D.A., Zaidan, A.S., Bayram, G.E.: Understating the social sustainability of metaverse by integrating adoption properties with users' satisfaction. In: Al-Emran, M., Ali, J.H., Valeri, M., Alnoor, A., Hussien, Z.A. (eds.) IMDC-IST 2024, pp. 95–107. Springer, Cham (2023b). https://doi.org/10.1007/978-3-031-51716-7_7

Atiyah, A.G., Faris, N.N., Rexhepi, G., Qasim, A.J.: Integrating ideal characteristics of chat-GPT mechanisms into the metaverse: knowledge, transparency, and ethics. In: Al-Emran, M., Ali, J.H., Valeri, M., Alnoor, A., Hussien, Z.A. (eds.) IMDC-IST 2024, pp. 131–141. Springer, Cham (2023c). https://doi.org/10.1007/978-3-031-51716-7_9

Chen, J.: What is a Trading Platform? Definition, Examples, and Features (2022). https://www.investopedia.com/terms/t/trading-platform.asp.

Choudhuri, S.: A research on trading of Sensex stocks by using RSI. Int. J. Innov. Technol. Exploring Eng. (IJITEE) **8**(9S2) (2019). ISSN 2278-3075

Husin, N.A., Abdulsaeed, A.A., Muhsen, Y.R., Zaidan, A.S., Alnoor, A., Al-mawla, Z.R.: Evaluation of metaverse tools based on privacy model using fuzzy MCDM approach. In: Al-Emran,

M., Ali, J.H., Valeri, M., Alnoor, A., Hussien, Z.A. (eds.) IMDC-IST 2024, pp. 1–20. Springer, Cham (2023). https://doi.org/10.1007/978-3-031-51716-7_1

Moroşan, A.Ţ.: The relative strength index revisited. Afr. J. Bus. Manag. **5**(14), 5855–5862 (2011). http://www.academicjournals.org/ajbm. ISSN 1993-8233 ©2011 Academic Journals

Muhsen, Y.R., Husin, N.A., Zolkepli, M.B., Manshor, N.: A systematic literature review of fuzzy-weighted zero-inconsistency and fuzzy-decision-by-opinion-score-methods: assessment of the past to inform the future. J. Intell. Fuzzy Syst. **45**(3), 4617–4638 (2023)

Wahid, M.S.B.: Automated Forex Trading Robot with FBH Robot on Metatrader 4 Platform (Expert Advisor), Universiti Teknologi Petronas, Bandar Seri Iskandar, 31750, Tronoh Perak Darul Ridzuan (2013)

Financing and Investing in Artificial Intelligence: The Lucrative Benefits in Terms of Sustainable Digitalization

Muntader Fadhil Saad and Nadwah Hilal Joudah[✉]

Department of Banking and Finance Sciences, College of Administration and Economics,
University of Basrah, Basrah, Iraq
muntader.saad@uobasrah.edu.iq, nadwah.hilal@uobaserah.edu.iq

Abstract. Artificial intelligence has many characteristics and advantages, as it can be utilized to solve certain problems in the absence of complete information. Artificial intelligence could use old experiences and employ them in new situations, process and perceive information, use trial and error to explore various matters, and learn and understand from previous experiences. It also can acquire and apply knowledge and react quickly to new situations and circumstances, i.e., it is distinguished by its ability to deal with difficult situations. Investment opportunities in artificial intelligence companies are limitless, especially for those with ambitious strategies and future plans. Investing in artificial intelligence will allow investors to benefit from modern technologies and achieve profits. Artificial intelligence in finance and investment is the most appropriate solution for dealing with financial data and ensuring the validity and accuracy of upcoming transactions in any company. Artificial intelligence can facilitate repetitive and traditional processes and save time, effort, and costs. As the report in 2022 indicated, which covered 158 countries around the world, including 16 Arab countries, some countries have expressed their readiness to adopt leading technology, including artificial intelligence, with the aim of clarifying their willingness to shift. The United Arab Emirates and the Kingdom of Saudi Arabia are considered among the leading countries in the field of artificial intelligence.

Keywords: Artificial intelligence · investment · finance · artificial intelligence stocks · trading

1 Introduction

Information technology has emerged as a tool for facilitating vital fields within active sectors in all modern societies. It has gradually developed to suit the given conditions of life and the various fields and sectors to facilitate the movement of sectors, activities, organizations, and countries alike. Accordingly, technologies have been developed to facilitate connections and relationships within the framework of joint business, which appear in a form called artificial intelligence. AI is the same as teaching a machine to be able to draw its own conclusions and understand what it should and shouldn't do. The encoding or encryption is not evident either, as it gives the machine the ability to move freely in a certain manner. Artificial intelligence includes several systems, including expert, neural network, algorithmic, fuzzy logic, and intelligent agent systems.

Investment and financing in artificial intelligence was chosen because of its vital role at present, especially with the emergence of essential stocks, including Nvidia, AMD, Micron, Amazon, Microsoft, Palantir, ADBEK, and IBM stocks. The UAE and Saudi Arabia had pioneering experiences in this field.

2   Study Methodology

2.1   Study Hypothesis

Artificial intelligence contributes to developing the finance sector, solving financial problems, and creating new ideas for firms and commercial institutions through financing and investment.

2.2   Study Importance

Intelligence has a role in developing innovative solutions and provides qualitative practices that help save costs and increase work efficiency.

2.3   Study Objectives

Investing in global financial trading companies is done through trading, estimating cash flows, and improving customer services through stocks.

2.4   Study Problem

The field of artificial intelligence is still evolving, and there is no guarantee that any particular technology or application will be successful. This may lead to significant risks for investors who put their money in this field or those who move to invest in artificial intelligence.

3   Theoretical Framework of Artificial Intelligence

3.1   The Concept of Artificial Intelligence

The Fourth Industrial Revolution is a technological shift that has an impact on cultures and economies all over the world. It reflects the creation and advancement of a broad range of modern technology that drives innovations and inventions across sectors while changing fundamental aspects of culture and society as we know it (Atiyah et al., 2023a, b, c; Alnoor et al., 2023). Elements of the Fourth Industrial Revolution include technologies, such as artificial intelligence, machine learning, automated control, Internet of Things, Big Data, Blockchain Data, government computing, and 3D printing (Atiyah and Zaidan, 2022).

Artificial intelligence is one of the sciences whose branching stems from computer science and is concerned with making computers perform tasks that are approximately similar to human intelligence processes, including learning, deduction, decision-making, and so on (Alnoor et al., 2024a, b, c). Other definitions are mentioned in books and references related to the science of artificial intelligence, the most prominent of which is the Arabic Encyclopedia Dictionary for Computers and the Internet (Atiyah, 2023). Artificial Intelligence, which is a term given to one of the latest computer sciences, belongs to the modern generation of computer generations and aims for the computer to simulate the intelligence processes that take place within the human mind so that the computer has the ability to solve problems and make decisions in a logical and organized manner in the same way as the human mind thinks (Zabout & Muhamad, 2022:14).

Artificial Intelligence is also known as the study of intellectual abilities through the use of computer models concerned with the method of simulating human thinking, and it is a technology used to build machines that can simulate humans in the processes of thinking, forming opinions, making judgments, and the ability to develop and learn (Al-Sharqawi, 2023:288). Artificial intelligence is explained as the ability of a machine to simulate the human mind and the way it works, such as its ability to think and explore (Alnoor et al., 2024a, b, c). With the tremendous developments of computers, it has become clear that they can carry out more complex tasks than we think, such as exploring and proving complex mathematical theories, playing chess with great skill, completing tasks with high accuracy, and having a large storage capacity (Ali et al., 2024). However, there has been no program that can keep up with the human mind, especially concerning the deductive and analytical tasks to which he/she is exposed (Institute of Banking Studies, 2021:3).

Different visions for defining the concept of artificial intelligence can be monitored by shedding light on four basic approaches (Hussein, 2023:119):

1- Thinking like a human: It is the science that makes computers think, i.e. a machine that has a mind.
2- Rational thinking: It is the science that carries out the tasks of the human mind through computing.
3- Acting like a human: It is the knowledge that enables a machine to carry out actions that, if carried out by humans, would require intelligence.
4- Rational acting: studying the design of intelligent functions through computing intelligence.

3.2 The Nature of Artificial Intelligence

Information technology has emerged as a tool that facilitates human life in all vital fields within the various sectors of society. This intelligence is represented by the human ability to make the machine, i.e., the computer, an intelligent machine. This intelligent machine came from the human being himself, and he granted it to the machine that takes the place of the human in thinking and processing information and data under the name artificial intelligence (Atiyah et al., 2023a, b, c). Therefore, artificial intelligence involves accumulated science and knowledge directed logically within the information technology system and according to specific algorithms that address issues requiring sharp intelligence.

In 1980 AD, an artificial intelligence system was developed to enable the user to interact with the machine, which in turn holds an integrated database through which commands can be received and stored and information can be retrieved and revealed to the user within the framework of the computer's information storage (Abbas et al., 2023). In the nineties of the last century, integrated artificial intelligence systems were developed into intelligence systems that enable one to interact artificially within a comprehensive information environment that is based on linking the machine to databases and global networks. That would enable dealing with the machine as if a human deals with another human in asking questions and obtaining more accurate, clearer, and richer information. In 2016, a conference was held at the White House in the United States of America on the future of artificial intelligence, its ethics, and the necessity of developing this event, which included the conduct of business and areas of individuals and groups, as well as countries alike. This was welcomed by countries all over the world in order to develop artificial intelligence systems in a more accurate and beneficial way for global organizations in particular and in accordance with the goals and demands of governments in a number of fields, the most important of which are the fields of international relations. Hence, serious work began to develop these systems to produce artificial intelligence (Al-Omari, 2021:311).

The automation system, however, is not the same as artificial intelligence. Automation is a machine-based system based on software programming that takes on well-defined tasks by using pre-defined logical sequences, so when symbol A logically follows symbol B, and so on. With regard to artificial intelligence, it is quite like teaching a machine to draw own conclusions and make out by itself what it should and what it should not do. The encryption or coding in the general principle purposely does not provide the machine with a complete view may, however, leave some margin for the machine to maneuver (Muhammad, 2021, 10).

3.3 Characteristics of Artificial Intelligence

Artificial intelligence has many characteristics and advantages (Borgdaou and Moussaoui, 2021:8), including:

- AI uses intelligence to solve present problems in the absence of complete information, and it also has the ability to use old experiences and employ them in new situations.
- AI can think and perceive, and it also uses trial and error to explore various matters
- Learning and understanding from previous experiences and expertise and having the ability to acquire and apply knowledge.
- AI is distinguished by its ability to respond quickly to new situations and circumstances, as it deals with difficult and complex situations.
- AI investigates the relative importance of the elements of the presented situation, and could visualize, understand, and perceive visual matters.
- AI provides information to support administrative decisions, in addition to dealing with ambiguous situations in the absence of information.

Artificial intelligence includes several systems, which could be mentioned as follows (Bosna & Hasnawi, 2021:10–14):

A- *Expert Systems*

They involve specialized information systems which are intended to reproduce the cognitive expertise of specialists from a certain field of knowledge. This definition consists of two important aspects: On one hand, the issue of information program verifiability which is a valid concern to computer programmers is important, and on the other hand, the field experience must be controlled (Muhsen et al., 2023; Ahmed et al., 2024). Here lies the application of knowledge-based approach that aims searching for efficiency. The expert system consists of several main components, which could be explained in the Fig. 1.

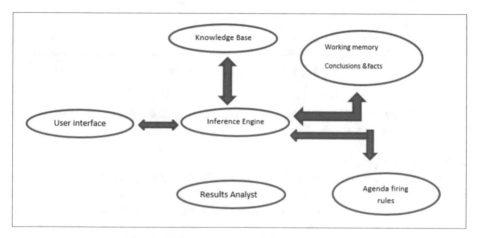

Fig. 1. Main components of the expert system.

B- *Neural Networks Systems*

Artificial neural networks are extremely complicated, learning aware structures which involve many sensors, devices, and other systems. These systems work in parallel due to their many processors that help in their collaborations (Alnoor et al., 2024a, b, c). Neural networks represent intelligence as it is based on knowledge bases and in alethic, ambiguous logic. The components of an artificial neuron can be summarized mathematically in the Fig. 2.

C- *Genetic Algorithm Systems*

Genetic algorithms are programs that mimic biological processes for the purpose of processing problems whose solutions are found in an evolutionary system. The genetic algorithm, which appeared in the current form in 1975 by John Holland from the University of Michigan and was developed very quickly in the early 1980s to become one of the most important and effective methods for complex trending and research. It was called the genetic process because of their heavy use of genetic solutions, which provided an almost perfect goal. The technology is an innovative one of a logic program running on a computer having as an idea a decision and possible answers for it. However, it is only those that stand up to the evolutionary struggle who come out on the top. It is especially

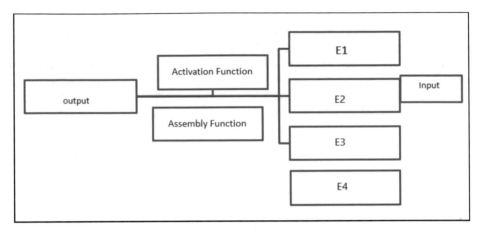

Fig. 2. Artificial Neuron Model.

used in the financial and banking sector, transportations and logistics, and the tracking of materials (Husin et al., 2023).

D- Fuzzy Logic Systems

It is also called fuzzy logic and refers to a method that relies on perception and simulates the way the human element perceives in terms of estimating values using non-fuzzy data (Data Fuzzy). It uses fuzzy logic techniques and systems with other integrated systems that work with artificial intelligence techniques, such as expert systems, that operate with fuzzy logic, neural networks with fuzzy logic, or fuzzy logic networks (Net Fuzz) in the most important areas. Those areas could include business, especially in banking applications, such as predicting the expected return on securities, risk management, liquidity cash planning, investment portfolio management, and other important applications.

E- Intelligent Agents Systems

An intelligent agent is considered as an entity that can have a perception of the environment which it exists through the sensors that this object carries out and then respond to it by means of the implementation's mechanisms. The smart agent is just one of the applications mining the data from the Internet or Internet databases. The classifying of smart agents is done through a program package that performs the repetitive or partially predictable tasks and duties of the beneficiary to support business activity and other software applications (Al-Enzi et al., 2023). Smart agents in the digital management system help reduce the electronic burden and guarantee the response speed to customer requests and the receipt of messages and comments concerning the quality of services and products that the organization provides. Sometimes, management decides to hire an experienced agent to read the e-mails and filter and sort the sales agents' reports or search for the cheapest flight at the best price, which are implemented by the branches last month, and do other tasks that are lacking intelligence and/or skill. At this time, intelligent agent programs are found in such multiple use areas as operating systems, application programs, network tools, internet business and e-commerce.

3.4 Motivation and Dimensions of AI Application

There are many motivations for interest in artificial intelligence, including the following (Borgdaou & Moussaoui, 2021:13):

- AI is the key to the health, education, and service industries due to the fact that they utterly depend on its technologies. Furthermore, it is the critical technology of other key industries, such as transport facilitated by unmanned aerial vehicles, autonomous vehicles, airborne taxis, underground systems, and all sorts of road and marine transport.
- Health as one of the beneficiaries of the rapid development of the Artificial Intelligence (AI) in recent years has been going through the various domains. Concerning the medical domain its capability to help medical workers in disease diagnosis and therapy, medication prescription, surgical procedures, and patient records handling using voice commands has promoted its growth in this field. Currently, patients have the choice of booking appointments via this app.
- AI has the potential to lead in setting new industries, providing affordable services at their best, enhancing security of systems, and assisting in developing policies and solutions for various problems such as cyberspace threats.
- AI reduces the suffering of human beings and removes some dangers like those that could be associated with ergonomic jobs and the determination of relief positions during natural disasters.
- AI is applied in the provision of legal consultations and also the comprehension of the interactive learning process. Meanwhile, the military and security fields are a promising sector for this technology.

Artificial intelligence has different dimensions that must be available in order to be ready for shifting, and among these dimensions are the following (Yaishi & Madari, 2022:10–11):

- *Infrastructure*
  The availability of cloud technology combined with high computing resources and infrastructure availability enables fast processing of Big Data at lower costs and efficient scalability. This means that organizations are ready to take advantage of artificial intelligence. Now, more than ever, the infrastructure represents various devices, hardware, and software, communication networks, databases, etc.
- *Social Intelligence*
  Individual customers can now access personal banking services 24/7, which is attributed to the extended hours that the banks offer. This includes providing facial recognition capabilities and voice commands for authentication in banking applications. Additionally, it can be used to examine costumers' behavior patterns and automatically perform customer segmentation, which would enable precise marketing strategies and enhancing customer experience and communications. The large increase in data increases the efficiency of models, leading to a gradual decrease in the level of required human intervention.
- *Social Responsibility*
  The emergence of online security threats in banking transactions has led to enhancing government regulations. Although these regulations are useful for monitoring

online financial transactions, they have limited the ability of banks to keep pace with digital transformation. Banks are no longer able to invest in technology, as they must maintain a capital adequacy ratio in accordance with international regulations. Artificial intelligence components are being added to existing systems to enable the identification of previously undetected transaction patterns, data anomalies, and suspicious relationships between individuals and entities. This allows for a more proactive approach, where AI is used to prevent fraud before it occurs, rather than the traditional reactive approach to fraud detection.

– *Data*

Artificial intelligence and other technological advances have made data the most valuable asset in financial services organizations. Recently, banks have begun to recognize the innovative and cost-effective solutions offered by AI. They also realize that having a large amount of assets is not enough in itself to create a thriving organization. The expanding big data market has had a significant impact on the banking sector as a result of evolving customer expectations. Customers now interact with their banks on a more digital level along with the traditional structured data, such as transaction data. Organizations, nowadays, collect large amounts of unstructured data, such as emails, text and voice messages, images and videos across customer service, social networks, global platforms, and other data collection media.

4 Artificial Intelligence Applications

Investment opportunities in artificial intelligence companies, especially companies with an ambitious strategy and future plan, will allow investors to benefit from modern technologies and achieve more profits. Artificial intelligence in finance and investment is the most appropriate solution for dealing with financial data and ensuring the validity and accuracy of transactions in any company. Artificial intelligence can facilitate repetitive and traditional processes and save a lot of time, effort and costs. This section demonstrates the most significant stocks invested in artificial intelligence (Riyad, 2024:3).

4.1 Nvidia Stocks - One of the Most Popular AI Stocks

Nvidia is often viewed as a leader in the AI field due to its production of exceptionally powerful graphics hardware with versatile applications. The company has played a leading role in driving the self-driving car revolution by developing its own artificial intelligence operating system, known as Driveworks. The company was founded in 1993 with the specific goal of developing computer solutions that utilized visual elements, primarily targeting the expanding video game industry. Over the years, the company has expanded the use of its graphics chips by utilizing technology to develop sophisticated machine learning APIs. As a result, Nvidia was a pioneer in emerging markets such as self-driving cars, electric vehicles, and cryptocurrency mining. Nvidia stock enjoys a strong upward trend and a significant amount of profit revenue.

4.2 AMD Stock - One of the Most Exciting AI Stocks

AMD, one of the most well-known semiconductor companies, is a pivotal cog in the machinery that runs artificial intelligence systems. AMD's tech products offering not only stimulates the growth of a strong friendship among this system of technological elite but also unites them together to further the cause of technology. Amazon, Microsoft, Google, and Oracle all utilize the AMD architecture's CPUs for their cloud computing activities. Since a growing number of companies are relocating their corporate organizations to the cloud, there has been a substantial increase in the number of AMD products used. AMD recently made custom processor chips for the current generation of gaming consoles, with Sony's PlayStation 5 and Microsoft's Xbox Series X S being two examples. Microsoft along with AMD is working to make this kind of processors for AI much better, which is the most interesting fact. Among the objectives of Microsoft is the plan to secure the required number of chips that will help assure the advancement of the growing AI workplace. On the other hand, AMD is devoted to creating a closer competitor to Nvidia, and this trend will be helpful to AMD. Like Nvidia stock, AMD stock is on the rise, so it's one of the most exciting AI stocks. However, both AMD and Nvidia stocks, as AI stocks, are currently trading at an impressive price, which indicates a short-term risk of overbought prices from a technical analysis standpoint.

4.3 Micron Stocks - One of the Best AI Stocks

Micron was established as a semiconductor design company during the 1970s. Over time, the company moved toward creating powerful memory chips capable of supporting artificial intelligence applications in smartphones and electric vehicles. In 2019, Micron purchased Fwdnxt, a company that specializes in developing artificial intelligence hardware and software. Micron was able to introduce the Micron Deep Learning Accelerator as a result of this acquisition, which integrates memory chip technology with Fwdnxt's AI technologies. This collaboration enables the company to create advanced machine learning solutions required for data analysis in advanced technologies, such as those used in autonomous vehicles. Micron stock has begun a new uptrend after finally breaking through the long-term resistance level at the $65 level.

4.4 Amazon Stock - One of the Most Popular AI Stocks

It can be said that no company uses artificial intelligence to a greater extent than Amazon. Jeff Bezos, founder and CEO, has been a prominent advocate of artificial intelligence and machine learning. Despite its origins as an e-commerce platform, Amazon has consistently prioritized technology as a core aspect of its operations. Nowadays, Amazon uses artificial intelligence across various aspects of its operations. AI is integral to multiple Amazon services, including Alexa, the popular voice-activated technology, and Amazon Go, a cashier-less grocery store. Furthermore, this company uses Amazon Web Services Sagemaker, a cloud infrastructure tool, and AI to facilitate the deployment of high-quality machine learning models for data scientists and developers. In this context, Amazon's e-commerce business also relies on artificial intelligence, with algorithms powering its e-commerce, video and music engines. Additionally, Amazon uses AI to determine product ratings, improve user experience, and improve its platform.

4.5 Microsoft Stocks - One of the Most Important and Promising Artificial Intelligence Stocks

Since 2019, Microsoft has been investing in OpenAI, the company behind the development of ChatGPT. The partnership started off with $1 billion investment and an exclusive Microsoft cloud computing services contract for the AI Lab. As early as January 2023, Microsoft manifests its plan to scale the partnership between them and OpenAI, which be a "multi-year, multi-billion-dollar investment." Microsoft also seek to enhance the flexibility of AI with its Azure global cloud computing system. Google and Microsoft also recently have released AI chatbots for Bing and the pervasive search engine, respectively. Alas, Bing's chatbot has experienced such errors as inaccuracies. According to Dmitri Brereton's report, the chatbot had wrongly stated financial information on Gap and Lululemon's quarterly reports. Microsoft stocks are one of the best stocks to buy for 2023, as its stock price has risen by almost 40%, and they are one of the future artificial intelligence stocks.

4.6 Palantir Stocks (PLTR) – One of the Biggest AI Stocks to Observe

Palantir specializes in operating AI data mining platforms specifically designed for government agencies and commercial companies. Its government platform - Gotham, leverages artificial intelligence to identify patterns in diverse data sets, enabling intelligence teams to effectively detect and respond to terrorist threats. Unconfirmed sources say Gotham played a role in the capture of Osama bin Laden in 2011. Palantir offers Foundry, a platform designed to store, transform, and process regulatory data, improve operations, and facilitate informed decision-making. In the first half of 2023, Palantir stocks saw a notable 60% increase, making it one of the biggest AI companies to consider.

4.7 Adobe Stocks (ADBE) – One of the Best AI Stocks Specializing in Marketing and Data Analysis

Adobe has become a company that produces all sorts of software used for all sorts of purposes, including content creation for marketing, data analysis, document management, publishing, and digital communication. It offers its main products, Creative Cloud library, the collection of the design software that is sold via subscription. In 2022, Adobe introduced a range of AI and ML tools in its Experience Cloud product, a collection of all marketing and analytical features under one roof. Among these developments is the predictive component which assists the sales and marketing operations in determining what manner of marketing campaign activities are the most influential in clients' purchase choices. These platforms are equipped with such tools and they can use them to evaluate campaigns and allocate the budget efficiently. Over the course of just two weeks, Adobe stocks have posted a 20% gain, and they appear poised to test their all-time highs over the summer. They could be considered one of the best AI stocks for investors seeking to explore the AI sector.

4.8   IBM Stocks - One of the Most Prestigious and Best Artificial Intelligence Stocks

IBM is a well-known technology company that offers a comprehensive range of hardware, software, and services to meet the needs of large enterprise customers. The company's mainframe systems are widely used in certain industries and are often involved in multi-year technology agreements worth hundreds of millions of dollars. IBM's approach to AI is to leverage technology to augment human intelligence, enhance operational efficiency, and reduce expenses. IBM's AI technology is used in healthcare to create individualized care strategies, accelerate drug development and distribution, and improve drug quality. IBM is a multi-faceted organization currently undergoing change, and artificial intelligence (AI) is just one avenue for its expansion. However, if an individual seeks to invest in AI companies that are strategically located to benefit from the gradual expansion of AI, IBM stock may be a suitable choice. Corporate investment in artificial intelligence, from mergers and acquisitions to public offerings, is a major contributor to artificial intelligence research and development and contributes to the impact of artificial intelligence on the economy.

Generally, AI is used in financing by examining cash accounts, credit accounts, and investment accounts to consider an individual's overall financial sound, keeping up with changes in real time, and creating personalized advice based on new incoming data. AI and ML benefit banks and fintech, as they can process vast amounts of information about customers. The data and information are then compared to get results about the right services/products that customers want, which essentially helped in developing relationships with the customer. We find that some countries have expressed their readiness to adopt leading technology, including artificial intelligence, as indicated in the report covering 158 countries around the world, including 16 Arab countries, with the aim of clarifying their readiness for shifting. Table 1 shows the ranking of Arab countries regarding obtaining funding.

The total global investment of companies in artificial intelligence was found to be growing between 2013–2021. In 2021, companies achieved the largest investments through private investments (a total of $93.3 billion), followed by mergers and acquisitions (about $72 billion), public offerings (about $9.5 billion), and minority shares (about $1.3 billion). In 2021, the investment value from mergers and acquisitions rose more than threefold against that of 2020, mainly driven by two companies in the health informatics and cybersecurity sectors. In 2021, the total global private investment in artificial intelligence climbed to $93.5 billion, which was more than twice as much as all the private investments in 2021 combined, with this year's total investment surpassing the total investment of 2014, which was the largest annual increase since 2014. Investment was more than two-fold greater than in the previous year, 2013. The total amount of financing and the number of Artificial Intelligence financing rounds had seen a substantial increase, especially for the companies with funding that ranged from $100 million to $500 million in 2021 in comparison to 2020, which were more than doubled, and financing rounds that ranged between $50 million and $100 million in 2020 more than doubled Investments in 2021, privity giant was 81.1% higher than in 2020, with the average size of a deal of private placing increasing from $60.6 million in 2020 to $110 million in 2021. Nevertheless, the number of AI companies that get a first investment

Table 1. The ranking of Arab countries regarding obtaining funding.

Country	Rank
Yemen	157
Iraq	153
Sudan	146
Libya	135
Algeria	119
Egypt	116
Saudi Arabia	69
Tunisia	50
Oman	47
Bahrain	46
Qatar	42
Jordan	41
UAE	38
Morocco	35
Kuwait	31
Lebanon	22

Source: Arab Forum for Smart Cities (2022). *Technology and Innovation Report 2021, Index of Countries' Readiness to Adopt Leading Technology, website* www.itcat.org.

declined from 762 in 2020 to 746 in 2021, the third year in a row that investments went down, with two of the new companies working in data management and two of the companies establishing robotics or driverless vehicles (Zaki, 2022, 20).

If we look at artificial intelligence from the other side, we find that most countries have begun developing their strategies to enter this new revolution. We find, for example, that China, Canada, the United States of America, Denmark, Finland, South Korea, Sweden, Norway, and Taiwan have developed their programs to clarify the mechanism for the population to have a greater understanding of artificial intelligence. On the other hand, France and Singapore have launched programs to develop the skills of the population by granting individuals sums of money but stipulating that they spend them on technological skills.

5 Arab Experiences in the Field of Artificial Intelligence

This section will focus on successful experiences in the field of artificial intelligence, as follows:

5.1 The UAE

The UAE has taken advanced steps in establishing the Ministry of Artificial Intelligence in a qualitative initiative that is the first of its kind in the world. This confirms its absolute commitment to adopting new-generation technologies. According to the results issued by the PwC Middle East study in the Emirates, the use of artificial intelligence applications saves about 40% of operational expenses in the banking sector. Sanjay Uppal, CEO of "Straits Bridge Advisors", confirmed that UAE banks are targeting artificial intelligence technology to help employ huge amounts of data in order to improve compliance, increase customer engagement, and enhance operational efficiency. Chatbots dominate the current stage of banks' move towards artificial intelligence, and it is expected that the next stage of adoption of artificial intelligence technologies by the UAE banking sector will be limited to the areas of business intelligence, cybersecurity, and financial management. The UAE experience points to the fact that only large banks have strong artificial intelligence applications for private development operations, and it also reflects the challenges facing the UAE banking sector, which also include the difficulty of attracting and retaining specialized competencies in the field of artificial intelligence and machine learning, for which the demand is growing enormously across various industries (Hisham & Al-Abdi, 2022:213).

A great achievement of the UAE is the remarkable ability to get the economy of the country grow by investing the money wisely in different fields. According to a global study, the national GDP will increase by 35 percent by 2031 and the governmental spending will be reduced to 50 percent on an annual basis by the virtue of AI technologies. This reduction, combined with saving waste from paper transactions and millions of hours lost during the completion of these forms, will be accomplished by minimizing waste in paper transactions. By 2030, the UAE is expected to become a global hub for artificial intelligence technologies widely adopted in different industries. The AI strategy of UAE that is expected to contribute to the economy of about 22.0 billion dirhams per year is through increased production efficiency across different sectors. The objectives will be achieved through the promotion of the overall productivity by 13% based on the fact that wastage of 396 million hours which was experienced in traditional transport methods and on roads will be negated and transport costs by 44% (equal to 900 million dirhams) will be reduced and emission of carbon and pollution will the reduced by 12% (equal to 1.5 billion dirhams), and reducing traffic accidents and resulting losses by 12%, achieving savings of two billion dirhams annually, reducing the need for parking by up to 20%, in addition to saving 18 billion dirhams by increasing the efficiency of the transportation sector in Dubai by 2030 (UAE Ministry of Economy, 2020:17).

In addition to achieving additional growth amounting to 335 billion UAE dirhams based on artificial intelligence technologies in services and data analysis at a rate of 100 percent by 2031, according to the artificial intelligence readiness index, the UAE ranks sixth in the world after Switzerland. Countries are making great efforts to improve public services by relying on artificial intelligence technologies. The "Panorama" Center for Artificial Intelligence and Big Data was opened at ADNOC's headquarters, in addition to establishing the "Salem Innovation Center", the first center of its kind to adopt self-operation for medical fitness examinations in the region.

5.2   The Kingdom of Saudi Arabia

In order to achieve the goals of "Kingdom's Vision 2030" and unleash the capabilities of the Kingdom of Saudi Arabia, the Authority looks forward to transforming the Kingdom's economy into a world-leading data-based economy by 2030. The Kingdom of Saudi Arabia witnessed the launch of the Artathon Artificial Intelligence Competition with the participation of a large number of artificial intelligence data experts from around the world to deliberate on creating the best works of art using artificial intelligence techniques. It is also the first bank to train all its employees in artificial intelligence. Alawwal Bank cooperated with Reactor in the field of technology, with the aim of providing all its employees with basic information about artificial intelligence, and becoming the first institution in the Middle East to train all its employees in artificial intelligence. By providing employees with basic information about artificial intelligence, Alawwal Bank seeks to lead the financial services sector in the region in using a technology that is expected to contribute to adding $320 billion (11%) to the gross domestic product in the Middle East by the year 2030 (Hisham & Al-Abdi, 2022:214).

In Saudi Arabia, the year 2019 witnessed the adoption of the strategy of the Saudi Data and Artificial Intelligence Authority (SDAIA) in order to support the achievement of the goals of the "Kingdom's Vision 2030" and unleash the Kingdom's capabilities, as the authority is looking forward to transforming the Kingdom's economy into a world-leading economy to discuss the innovation of the best businesses of technology using artificial intelligence techniques (Majdy, 2020:20).

So, it can be said that, at the level of Arab countries, the United Arab Emirates and the Kingdom of Saudi Arabia are considered among the leading countries in the field of artificial intelligence.

6   Opportunities and Gains of Applying Artificial Intelligence

Applying artificial intelligence will change the views of individuals. Investors might soon discover that medium-term returns will be much lower than expectations, especially after the current trend caused by quantitative easing ends. The following can also be observed:

1- Incorporating AI into stock trading offers investors a whole new range of tools and potential benefits that enhance performance and profits. Through data analytics, price prediction and risk management, investors will realize positive results and investment success in the stock market.
2- In the field of banking, it is possible to invest in some apps that can raise the alert of illicit banking operations, commercial masking or the software for the acceptance and renewal of credit cards.
3- It is also important to note that artificial intelligence is integrated into the investment strategies and data analysis that prompts economic decisions. Here, AI has the capability to analyze and comprehend market performances and forecast future trends as a significant tool.
4- AI-based algorithms which can analyze and find out different patterns from big data in the financial markets is the key component of financial market analysis in the present era.

5- Lastly, artificial intelligence is one of a kind technology that not only possesses the ability to solve any problem selected to tackle but also can learn from its mistakes and failures.
6- Over the years, artificial intelligence has evolved from mere machine learning algorithms to include advanced concepts, such as natural language processing and deep learning. This technological advancement helps AI solve issues like medical diagnosis, weather forecasts, and many more.

7 Conclusions

This study reaches the following conclusions:

1- The dangers of over-reliance on technology: It might happen that a society over-uses artificial intelligence technologies which may lead to the decline in human skills and capabilities in for example analysis and decision-making, this in turn might make it more difficult in the event of any malfunction in the artificial intelligence systems.
2- Uncertainty about the future: The field of artificial intelligence is still developing, and there is no guarantee of the success of any particular technology or application. This may lead to significant risks for investors investing their money in this area.
3- This science belongs to the modern generation of computers and aims to simulate the intelligence processes that occur within the human mind, so that the computer has the ability to solve problems and make decisions in a logical and organized manner.
4- At the level of Arab countries, the United Arab Emirates and the Kingdom of Saudi Arabia are considered among the leading countries in the field of artificial intelligence.
5- It is found that some countries have expressed their willingness or readiness to adopt leading technology, including artificial intelligence, as indicated in the report, which covered 158 countries around the world, including 16 Arab countries, with the aim of clarifying their readiness to shift to digitization and use artificial intelligence.
6- Artificial Intelligence is used in finance by examining cash accounts, credit accounts, and investment accounts to consider an individual's overall financial health, keeping up with changes in real time, and creating personalized advice based on new incoming data. AI and machine learning, generally, benefit banks and fintech.

8 Recommendations

This study suggests some recommendations, such as:

1- High costs: The development and application of artificial intelligence technologies require significant investments in areas such as data, computing, and human expertise. These costs may not be affordable for all investors and all countries.
2- Moral risks: Artificial intelligence technologies raise many moral risks, such as bias, discrimination, and violation of privacy. The use of these techniques may lead to undesirable or even harmful results.
3- The UAE and Saudi Arabia experience can be used as the basis for some countries to apply AI indicators, especially for some oil-producing countries.
4- Investing in human capabilities that support artificial intelligence and providing financial support to obtain innovation and creativity

5- Facilitate foreign investments to improve the development of artificial intelligence and cybersecurity systems in order for countries to be ready to access these technologies.
6- Disseminating awareness-raising programs to clarify what artificial intelligence is and understand its teaching and learning mechanisms.
7- Introducing most of the programs with modern technologies in banks, including automation and accounting programs, as well as learning about investment and trading.

References

Abbas, S., et al.: Antecedents of trustworthiness of social commerce platforms: a case of rural communities using multi group SEM & MCDM methods. Electron. Commer. Res. Appl. **62**, 101322 (2023)

Ahmed, A.D., Salih, M.M., Muhsen, Y.R.: Opinion weight criteria method (OWCM): a new method for weighting criteria with zero inconsistency. IEEE Access (2024)

Al-Enzi, S.H.Z., Abbas, S., Abbood, A.A., Muhsen, Y.R., Al-Hchaimi, A.A.J., Almosawi, Z.: Exploring research trends of metaverse: a bibliometric analysis. In: Al-Emran, M., Ali, J.H., Valeri, M., Alnoor, A., Hussien, Z.A. (eds.) IMDC-IST 2024, pp. 21–34. Springer, Cham (2023). https://doi.org/10.1007/978-3-031-51716-7_2

Ali, J., Hussain, K.N., Alnoor, A., Muhsen, Y.R., Atiyah, A.G.: Benchmarking methodology of banks based on financial sustainability using CRITIC and RAFSI techniques. Decis. Making Appl. Manag. Eng. **7**(1), 315–341 (2024)

Alnoor, A., Atiyah, A.G., Abbas, S.: Toward digitalization strategic perspective in the European food industry: non-linear nexuses analysis. Asia-Pac. J. Bus. Adm. (2023)

Alnoor, A., Atiyah, A.G., Abbas, S.: Unveiling the determinants of digital strategy from the perspective of entrepreneurial orientation theory: a two-stage SEM-ANN approach. Glob. J. Flex. Syst. Manag. 1–18 (2024a)

Alnoor, A., Chew, X., Khaw, K.W., Muhsen, Y.R., Sadaa, A.M.: Benchmarking of circular economy behaviors for Iraqi energy companies based on engagement modes with green technology and environmental, social, and governance rating. Environ. Sci. Pollut. Res. **31**(4), 5762–5783 (2024b)

Alnoor, A., et al.: How positive and negative electronic word of mouth (eWOM) affects customers' intention to use social commerce? A dual-stage multi group-SEM and ANN analysis. Int. J. Hum.-Comput. Interact. **40**(3), 808–837 (2024c)

Al-Omari, H.M.: Artificial intelligence and its role in international relations. Arab J. Sci. Publishing (AJSP) (29) (2021)

Al-Sharqawi, M.: Economic dimensions of artificial intelligence: assessing the readiness of the Egyptian economy. J. Legal Econ. Stud. **9**(1) (2023)

Arab Forum for Smart Cities. Technology and Innovation Report 2021, Index of Countries' Readiness to Adopt Leading Technology (2022). www.itcat.org

Atiyah, A.G.: Unveiling the quality perception of productivity from the senses of real-time multisensory social interactions strategies in metaverse. In: Al-Emran, M., Ali, J.H., Valeri, M., Alnoor, A., Hussien, Z.A. (eds.) IMDC-IST 2024, pp. 83–93. Springer, Cham (2023). https://doi.org/10.1007/978-3-031-51300-8_6

Atiyah, A.G., Zaidan, R.A.: Barriers to using social commerce. In: Alnoor, A., Wah, K.K., Hassan, A. (eds.) Artificial Neural Networks and Structural Equation Modeling: Marketing and Consumer Research Applications, pp. 115–130. Springer, Singapore (2022). https://doi.org/10.1007/978-981-19-6509-8_7

Atiyah, A.G., Alhasnawi, M., Almasoodi, M.F.: Understanding metaverse adoption strategy from perspective of social presence and support theories: the moderating role of privacy risks. In: Al-Emran, M., Ali, J.H., Valeri, M., Alnoor, A., Hussien, Z.A. (eds.) IMDC-IST 2024, pp. 144–158. Springer, Cham (2023a). https://doi.org/10.1007/978-3-031-51300-8_10

Atiyah, A.G., All, N.D.A., Zaidan, A.S., Bayram, G.E.: Understating the social sustainability of metaverse by integrating adoption properties with users' satisfaction. In: Al-Emran, M., Ali, J.H., Valeri, M., Alnoor, A., Hussien, Z.A. (eds.) IMDC-IST 2024, pp. 95–107. Springer, Cham (2023b). https://doi.org/10.1007/978-3-031-51716-7_7

Atiyah, A.G., Faris, N.N., Rexhepi, G., Qasim, A.J.: Integrating ideal characteristics of chat-GPT mechanisms into the metaverse: knowledge, transparency, and ethics. In: Al-Emran, M., Ali, J.H., Valeri, M., Alnoor, A., Hussien, Z.A. (eds.) IMDC-IST 2024, pp. 131–141. Springer, Cham (2023c). https://doi.org/10.1007/978-3-031-51716-7_9

Borgdaou, A., Moussaoui, H.: The importance of artificial intelligence in bank financing for international trade: an OCR case study. (MA thesis), Mohamed Al-Bishri Al-Ibrahimi University (2021)

Bosna, Kh., Hasnawi, S.: Uses of Artificial Intelligence in Banking Operations: A Case Study of the Foreign Bank of Algeria (BEA), Bordj Bou Arreridj Agency, (MA thesis), Mohamed Bishri Brahimi University (2021)

Hisham, E., Al-Abdi, D.: Applications of artificial intelligence in financial institutions as an entry point for activating digital financial inclusion: an analytical study of international experiences in the field of bank digitization. Namaa J. Econ. Trade **6**(2) (2022)

Husin, N.A., Abdulsaeed, A.A., Muhsen, Y.R., Zaidan, A.S., Alnoor, A., Al-mawla, Z.R.: Evaluation of metaverse tools based on privacy model using fuzzy MCDM approach. In: Al-Emran, M., Ali, J.H., Valeri, M., Alnoor, A., Hussien, Z.A. (eds.) IMDC-IST 2024, pp. 1–20. Springer, Cham (2023). https://doi.org/10.1007/978-3-031-51716-7_1

Institute of Banking Studies. Highlights on Artificial Intelligence. Kuwait, Series 13, Issue 4 (2021)

Islamic Development Bank. The Future of Finance, Redefining the Role of Finance in the World of the Fourth Industrial Revolution, UAE (2022)

Majdy, N.: Artificial Intelligence and Machine Learning, Introductory Booklets Series 3, Arab Monetary Fund (2020)

Muhammad, H.: Artificial intelligence systems and the future of education. J. Stud. Univ. Educ. **52** (2021)

Muhsen, Y.R., Husin, N.A., Zolkepli, M.B., Manshor, N.: A systematic literature review of fuzzy-weighted zero-inconsistency and fuzzy-decision-by-opinion-score-methods: assessment of the past to inform the future. J. Intell. Fuzzy Syst. **45**(3), 4617–4638 (2023)

Riyad, F.: Trading and investing in artificial intelligence (2021). https://capex.com/ar/academy/alestesmar-fi-alzkah-alestenae

UAE Ministry of Economy. Artificial Intelligence in the United Arab Emirates (2020)

Yaishi, S., Madari, M.: The role of artificial intelligence in developing financial technology in financial institutions. (MA thesis), Ahmed Draya University - Adrar: Algeria (2022)

Zabout, S.A., Muhamad, A.: Computing Systems and Artificial Intelligence Systems: Advanced Artificial Intelligence (2022). https://shorturl.at/FQVZ0

Zaki, N.: Artificial Intelligence Report. Stanford University (2022). https://aiindex.stanford.edu/wp-content/uploads/2022/03/2022-AI-Index

Sustaining an Agile Supply Chain by Adopting Industry Technologies (4.0)

Ali Hussein Ali, Fatima Saddam Merhej, and Abbas Gatea Atiyah(✉) ⓘD

College of Administration and Economic, University of Thi-Qar, Nasiriyah, Iraq
madaraking2000@gmail.com, abbas-al-khalidi@utq.edu.iq

Abstract. Industry 4.0 technologies are a technical method that creates the lean and flexible world. It supports companies cooperate and determine how to sustain his agile supply chain. So, Industry 4.0 technologies greatly expands the process of interaction among companies and among companies with customers. It is central to response how reaching the sustainability for agile supply chain. This study designed to shape a theoretical and practical model to attain the sustainability of agile supply chain. Depending on a set of vital elements that affect the companies. Three hypotheses have been assumed between the factors cited. The consequences presented that big data analytics, Internet of Things, and Cloud Computing, are a key element in achieving the sustainability. The study decided that the sustainability of agile supply chain lies in Industry 4.0 technologies.

Keywords: Agile Supply Chains · Industry Technologies (4.0) · Big Data · Internet of Things · Cloud Computing · Sustainability

1 Introduction

Many companies today are suffering the change the path of supply chains (Chua et al., 2024). Which requires companies to adopt a lean approach to supply chains (Dubey et al., 2022; Abbas et al., 2023; Ali et al., 2024; Atiyah et al., 2022). Supply chain agility means the ability to respond quickly and efficiently to changes in the path of goods and their arrival to the final consumer (Zhao et al., 2023; Atiyah, 2023a, b). The producing company is a key part of the supply chain. Its ability to adapt to differences demand on speed in processing, appropriate timing of delivery, and effectively deal with geopolitical crises, labor shortages, shortages of raw materials, as well as natural events, are all necessary elements for an agile supply chain (Tian et al., 2024). Therefore, companies adopted this type of supply chain. So, it has become necessary to think about how to sustain this type of supply chain. Since supply chain agility is directly linked to sudden events, it is necessary for companies to possess modern technologies in the industry (Ramirez-Peña et al., 2020; Alnoor et al., 2023). Industry 4.0 technologies are a very important means of sustaining the agile supply chain. Because it enables the organization to draw possible future scenarios, and helps in building a huge database that the company's management can refer to in building its scenarios. It also reduces risks by diversifying them between different possibilities and thus choosing the least risky

A. Alnoor et al. (Eds.): AIRDS 2024, LNNS 1033, pp. 218–228, 2024.
https://doi.org/10.1007/978-3-031-63717-9_14

possibilities (Duan et al., 2024). All of this can be achieved by successfully dealing with big data and the Internet of Things, as well as cloud computing, which together constitute Industry 4.0 technologies (Lambourdière et al., 2022). Although previous studies showed how to create agile supply chains, they did not answer the question about the extent of their sustainability (Annosi et al., 2021). The originality of this study is that it examines how to answer the question posed. This study provided some theoretical and applied contributions that can be benefited from in the future.

2 Hypothesis Development

2.1 Big Data Analytics and Sustaining Agile Supply Chains

Big data analytics means collecting and extracting information from a very large set of diverse data of different periods (Xu et al., 2023). The ability to analyze this type of data requires very high speed in the analysis process (Shafique et al., 2024; Atiyah, 2020; Alnoor et al., 2024a, b). Because the nature of the data is very complex and its sources are also diverse, it either comes from various web sources, from social networks, or from stored databases, etc. (Jaouadi, 2022). Analyzing this complex data helps make more rational decisions (Patrucco et al., 2023). Because these data have varying durations and carry various variables and changing situations, they are considered reliable resources for making sound and sober decisions (Zheng et al., 2023). Therefore, these decisions will certainly mitigate future risks resulting from geopolitical changes or changes resulting from other factors (Gupta et al., 2022; Alnoor et al., 2024a, b). This helps in the sustainability of the supply chain (Bag et al., 2023). Therefore, we can assume the following:

H1: Big data analytics has a positive impact on sustaining supply chain agility.

2.2 Internet of Things and Sustaining Agile Supply Chains

The term Internet of Things means collective and shared communication between several parties using smart devices (Oliveira-Dias et al., 2023). One of the elements that give great importance to the Internet of Things is sudden situations that require immediate treatment, especially in the field of goods transportation (Pal, 2023; Atiyah, 2023a, b). Companies rely heavily on the Internet of Things for various tasks, such as tracking the movement of goods, in which they rely on tracking devices, following up on inventory movement, in which they rely on tablets connected via the Internet between the company's various facilities, tracking the order and providing it to the customer, relying on smart watches (Sadeghi et al., 2024). These devices and others greatly facilitate the possibility of maintaining a highly resilient supply chain (Ali et al., 2023; Atiyah, 2022). It facilitates avoiding errors resulting from crowding at work (Khan et al., 2023), in addition to sudden accidents that require tools that facilitate the process of implementing quick decisions and immediate solutions (Prajapati et al., 2024). Therefore, we assume the following:

H2: Internet of Things has a positive impact on sustaining supply chain agility.

2.3   Cloud Computing and Sustaining Agile Supply Chains

The comparison of a computer server to the cloud came from the enormous ability to save large data and transfer it at a very high speed (Shu et al., 2021). This type of computing makes it possible to save and transfer data with great ease and speed. So, it is called cloud computing (Prajapati et al., 2024). Cloud computing means providing a single data storage server in different places to perform different tasks simultaneously (Alamsjah and Yunus, 2022). Cloud computing provides a great benefit to companies in all their activities, as it helps provide economies of scale in the devices and data they contain (Abdulrahman and Yuvaraj, 2023). Cloud computing facilitates the process of implementing activities with high flexibility, as it provides data regardless of the user's location (Liu et al., 2018). Cloud computing helps reduce the need to communicate with multiple data providers, as through the push of a button it is possible to obtain different data at the same time and for multiple conditions (Hasani, 2021). Such kind of services provided by cloud computing help companies to build future scenarios, reduce risks and diversify opportunities (Shu et al., 2021; Atiyah et al., 2023a, b, c). By adopting cloud computing, companies can continuously communicate the change occurring in the course of previous and current events and compare cases to obtain the best decision (Prajapati et al., 2024). Therefore, we can assume the following:

H3: Cloud Computing has a positive impact on sustaining supply chain agility.

3   Research Methodology

In this study, the researchers adopted the survey method. Primary data was collected using a questionnaire (Ahmed et al., 2024). The study sample consisted of 167 individuals who were managers in the food industry company, namely (Grain Processing Corporation (GPC)), where the researchers received 143 responses. Table 1 shows the demographics of participants. 0.15% of the taster managers were top, followed by 0.29% of middle-level managers, and 0.56% of low-level managers. 0.27% of participants earned a master's degree or higher, and 0.73% earned an under-graduate degree. Most of the participants were between 40 and 49 years old.

Table 1. Demographics of the participations

Demographics	Frequency (n = 143)	Percent %
Gender		
Male	98	.69
Female	45	.31
position		
High level	22	.15

(*continued*)

Table 1. (*continued*)

Demographics	Frequency (n = 143)	Percent %
Mid-level	41	.29
Low-level	80	.56
Age		
30–39	24	.17
40–49	58	.40
50–59	33	.23
≥60	28	.20
Academic qualifications		
Undergraduate	105	.73
Post-graduate	38	.27
Organizational tenure		
3–6	33	.23
7–10	46	.32
≥11	64	.45

The three basic elements (big data analytics, Internet of Things, and cloud computing) represent the independent variables that can affect the dependent variable (agile supply chain). In this study, well-established standards in solid literature were adopted to measure the research model. (5) items were used to measure the big data analytics process, taken from (Dubey et al., 2019). As for the Internet of Things, it was measured by (Umair et al., 2021), which adopted (8) items. Cloud computing was measured by a scale (Sachdeva et al., 2024), which consists of (6) items. Regarding the dependent variable, supply chain agility sustainability, the scale (Oliveira-Dias et al., 2023) was adopted, which consists of (5) items. A 7-point scale (1 = strongly disagree, 7 = strongly agree) was used to operationalize all concepts.

4 Data Analysis

In order to ensure the validity and reliability of the measurement model used in this research, a set of important statistical indicators were adopted based on (Hair et al., 2021; Muhsen et al., 2023). The average variance extracted (AVE) and the composite reliability (CR) must be greater than (0.5, 0.7) consecutively. Factor loading (FL) values must be higher than (0.70). As shown in the results of Fig. 1, the values of (Big DA1, Big DA2), (IoT1, IoT2, IoT8) and (CC1) were less than required. So, we re-moved them and recalculated another time and our results as in Fig. 2 and Table 2, as all values for the required indicators were acceptable.

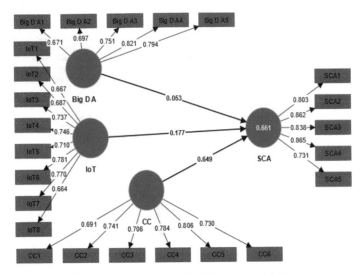

Fig. 1. Measurement of validity and reliability

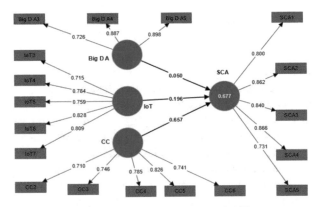

Fig. 2. Corrected validity and reliability

Moreover, Fornell and Larcker (1981) criterion was adopted to degree discriminant validity. The square root of (AVE), as is shown in Table 3, all variables are clearly different from each other (Hair et al., 2010; Al-Enzi et al., 2023; Husin et al., 2023).

Table 2. Result of measurement model

Variables	Items	FL	CR	AVE
Big data analytics (Dubey et al., 2019)	Big D A3	0.726	0.878	0.707
	Big D A4	0.887		
	Big D A5	0.898		
Cloud computing (Sachdeva et al., 2024)	CC2	0.710	0.874	0.582
	CC3	0.746		
	CC4	0.785		
	CC5	0.826		
	CC6	0.741		
Internet of Things (Umair et al., 2021)	IoT3	0.715	0.886	0.609
	IoT4	0.784		
	IoT5	0.759		
	IoT6	0.828		
	IoT7	0.809		
Supply chain agility (Oliveira-Dias et al., 2023)	SCA1	0.800	0.912	0.675
	SCA2	0.862		
	SCA3	0.840		
	SCA4	0.866		
	SCA5	0.731		

Table 3. Discriminant validity

	Big D A	CC	IoT	SCA
Big D A	**0.841**			
CC	0.334	**0.763**		
IoT	0.414	0.680	**0.780**	
SCA	0.350	0.807	0.663	**0.821**

Table 4 displays the results of the structural evaluation of the model. The research model is based on three hypotheses. The results shown in Table 4 and Fig. 3 indicate that there is a direct impact of big data analytics, the Internet of Things, and cloud computing on the sustainability of the agile supply chain.

Table 4. Hypotheses test.

	(O)	(M)	(STDEV)	(|O/STDEV|)	P values	Results
Big D A -> SCA	0.562	0.056	0.848	0.663	0.002	yes
CC -> SCA	0.657	0.659	0.067	9.754	0.000	yes
IoT -> SCA	0.196	0.193	0.082	2.397	0.017	yes

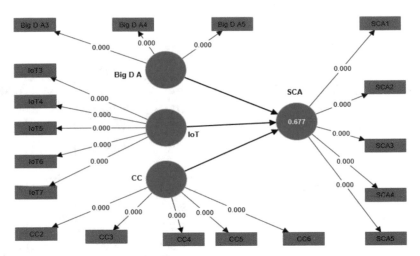

Fig. 3. Path coefficient of the model.

5   Discussion

The results of the study indicate that there is a pivotal role of modern Industry 4.0 applications in the sustainability of the agile supply chain (Shu et al., 2021; Atiyah et al., 2023a, b, c). Regarding big data analysis, we found that the results of the study show that there is an influential role for it (H1), which reinforces what the literature of this study has shown about the influential role of big data analysis (Xu et al., 2023). On the other hand, the results showed that the Internet of Things contributes significantly to the sustainability of the agile supply chain (H2). Oliveira-Dias et al. (2023) argued that the Internet of Things represents the most important and greatest means of sustainability for companies' work and communication with customers. The results of the study also showed that cloud computing has a role in maintaining supply chain agility (H3), and this enhances the originality of the study's literature (Abdul Rahman and Yuvaraj, 2023). The possibility of maintaining a strong economic structure by adopting modern technologies in industry. This means generating employment and increasing investments (Oliveira-Dias et al., 2023). Therefore, it can be said that modern Industry 4.0 applications contribute to improving product quality, reducing costs, and rationalizing supply and distribution processes, which enhances companies' competitive potential and expands growth opportunities (Shu et al., 2021).

6 Conclusion

This study contributes to uncovering digital tools that facilitate agile supply chain sustainability. Therefore, applying the results of this study in the reality of competition facilitates the possibility of survival and sustainability of supply chains, especially in light of rapid changes and events. When a company possesses huge historical data on various facts and events, it can build future scenarios without much effort and based on realistic facts. Also, the company's ability to preserve and retrieve its data quickly makes it unlikely that any error will occur. All of this constitutes a solid foundation for maintaining successful and advanced performance in light of the presence of innovative and flexible supply chains. Which enhances management effectiveness and drives innovation forward (Xu et al., 2023). Therefore, companies must pay attention to this type of creativity and not limit it to other activities and neglect the supply chain because the flexible supply chain expresses the organization's ability to fulfill its obligations to customers and thus achieve sustainable growth (Ramirez Peña et al., 2020; Attia et al., 2023; Zheng et al., 2023).

References

Abbas, S., et al.: Antecedents of trustworthiness of social commerce platforms: a case of rural communities using multi group SEM & MCDM methods. Electron. Commer. Res. Appl. **62**, 101322 (2023)

Abdulrahman, M.D., Yuvaraj, M.: Improving agility and resilience of automotive spares supply chain: the additive manufacturing enabled truck model. Socioecon. Plann. Sci. **85**, 101401 (2023)

Ahmed, A.D., Salih, M.M., Muhsen, Y.R.: Opinion weight criteria method (OWCM): a new method for weighting criteria with zero inconsistency. IEEE Access (2024)

Alamsjah, F., Yunus, E.N.: Achieving supply chain 4.0 and the importance of agility, ambidexterity, and organizational culture: a Case of Indonesia. J. Open Innov. Technol. Market Complex. **8**(2), 83 (2022)

Al-Enzi, S.H.Z., Abbas, S., Abbood, A.A., Muhsen, Y.R., Al-Hchaimi, A.A.J., Almosawi, Z.: Exploring research trends of metaverse: a bibliometric analysis. In: Al-Emran, M., Ali, J.H., Valeri, M., Alnoor, A., Hussien, Z.A. (eds.) IMDC-IST 2024, pp. 21–34. Springer, Cham (2023). https://doi.org/10.1007/978-3-031-51716-7_2

Ali, J., Hussain, K.N., Alnoor, A., Muhsen, Y.R., Atiyah, A.G.: Benchmarking methodology of banks based on financial sustainability using CRITIC and RAFSI techniques. Decis. Making Appl. Manag. Eng. **7**(1), 315–341 (2024)

Ali, S.M., et al.: Drivers for Internet of Things (IoT) adoption in supply chains: implications for sustainability in the post-pandemic era. Comput. Ind. Eng. **183**, 109515 (2023)

Alnoor, A., Atiyah, A.G., Abbas, S.: Toward digitalization strategic perspective in the European food industry: non-linear nexuses analysis. Asia-Pac. J. Bus. Adm. (2023)

Alnoor, A., Atiyah, A.G., Abbas, S.: Unveiling the determinants of digital strategy from the perspective of entrepreneurial orientation theory: a two-stage SEM-ANN approach. Glob. J. Flex. Syst. Manag. 1–18 (2024a)

Alnoor, A., et al.: How positive and negative electronic word of mouth (eWOM) affects customers' intention to use social commerce? A dual-stage multi group-SEM and ANN analysis. Int. J. Hum.-Comput. Interact. **40**(3), 808–837 (2024b)

Annosi, M.C., Brunetta, F., Bimbo, F., Kostoula, M.: Digitalization within food supply chains to prevent food waste. Drivers, barriers and collaboration practices. Ind. Mark. Manag. **93**, 208–220 (2021)

Atiyah, A.G.: Impact of knowledge workers characteristics in promoting organizational creativity: an applied study in a sample of Smart organizations. PalArch's J. Archaeol. Egypt/Egyptol. **17**(6), 16626–16637 (2020)

Atiyah, A.G.: Effect of temporal and spatial myopia on managerial performance. J. La Bisecoman **3**(4), 140–150 (2022)

Atiyah, A.G.: Strategic network and psychological contract breach: the mediating effect of role ambiguity. Int. J. Res. Manag. Stud. (IJRMS) **13**(1) (2023a)

Atiyah, A.G.: Unveiling the quality perception of productivity from the senses of real-time multisensory social interactions strategies in metaverse. In: Al-Emran, M., Ali, J.H., Valeri, M., Alnoor, A., Hussien, Z.A. (eds.) IMDC-IST 2024, pp. 83–93. Springer, Cham (2023b). https://doi.org/10.1007/978-3-031-51300-8_6

Atiyah, A.G., Zaidan, R.A.: Barriers to using social commerce. In: Alnoor, A., Wah, K.K., Hassan, A. (eds.) Artificial Neural Networks and Structural Equation Modeling: Marketing and Consumer Research Applications, pp. 115–130. Springer, Singapore (2022). https://doi.org/10.1007/978-981-19-6509-8_7

Atiyah, A.G., Alhasnawi, M., Almasoodi, M.F.: Understanding metaverse adoption strategy from perspective of social presence and support theories: the moderating role of privacy risks. In: Al-Emran, M., Ali, J.H., Valeri, M., Alnoor, A., Hussien, Z.A. (eds.) IMDC-IST 2024, pp. 144–158. Springer, Cham (2023a). https://doi.org/10.1007/978-3-031-51300-8_10

Atiyah, A.G., All, N.D.A., Zaidan, A.S., Bayram, G.E.: Understating the social sustainability of metaverse by integrating adoption properties with users' satisfaction. In: Al-Emran, M., Ali, J.H., Valeri, M., Alnoor, A., Hussien, Z.A. (eds.) IMDC-IST 2024, pp. 95–107. Springer, Cham (2023b). https://doi.org/10.1007/978-3-031-51716-7_7

Atiyah, A.G., Faris, N.N., Rexhepi, G., Qasim, A.J.: Integrating ideal characteristics of chat-GPT mechanisms into the metaverse: knowledge, transparency, and ethics. In: Al-Emran, M., Ali, J.H., Valeri, M., Alnoor, A., Hussien, Z.A. (eds.) IMDC-IST 2024, pp. 131–141. Springer, Cham (2023c). https://doi.org/10.1007/978-3-031-51716-7_9

Bag, S., Dhamija, P., Singh, R.K., Rahman, M.S., Sreedharan, V.R.: Big data analytics and artificial intelligence technologies based collaborative platform empowering absorptive capacity in health care supply chain: an empirical study. J. Bus. Res. **154**, 113315 (2023)

Chua, G.A., Davis, E.D., Viswanathan, S.: Business models and supply chain strategy for digital manufacturing. In: Digital Manufacturing, pp. 377–408. Elsevier (2024)

de Oliveira-Dias, D., Maqueira-Marin, J.M., Moyano-Fuentes, J., Carvalho, H.: Implications of using Industry 4.0 base technologies for lean and agile supply chains and performance. Int. J. Prod. Econ. **262**, 108916 (2023)

Duan, K., Pang, G., Lin, Y.: Exploring the current status and future opportunities of blockchain technology adoption and application in supply chain management. J. Digit. Econ. (2024)

Dubey, R., Bryde, D.J., Dwivedi, Y.K., Graham, G., Foropon, C.: Impact of artificial intelligence-driven big data analytics culture on agility and resilience in humanitarian supply chain: a practice-based view. Int. J. Prod. Econ. **250**, 108618 (2022)

Dubey, R., Gunasekaran, A., Childe, S.J.: Big data analytics capability in supply chain agility: the moderating effect of organizational flexibility. Manag. Decis. **57**(8), 2092–2112 (2019)

Fornell, C., Larcker, D.F.: Evaluating structural equation models with unobservable variables and measurement error. J. Mark. Res. **18**(1), 39–50 (1981)

Gupta, S., Bag, S., Modgil, S., de Sousa Jabbour, A.B.L., Kumar, A.: Examining the influence of big data analytics and additive manufacturing on supply chain risk control and resilience: an empirical study. Comput. Ind. Eng. **172**, 108629 (2022)

Hasani, A.: Resilience cloud-based global supply chain network design under uncertainty: Resource-based approach. Comput. Ind. Eng. **158**, 107382 (2021)

Husin, N.A., Abdulsaeed, A.A., Muhsen, Y.R., Zaidan, A.S., Alnoor, A., Al-mawla, Z.R.: Evaluation of metaverse tools based on privacy model using fuzzy MCDM approach. In: Al-Emran, M., Ali, J.H., Valeri, M., Alnoor, A., Hussien, Z.A. (eds.) IMDC-IST 2024, pp. 1–20. Springer, Cham (2023). https://doi.org/10.1007/978-3-031-51716-7_1

Jaouadi, M.H.O.: Investigating the influence of big data analytics capabilities and human resource factors in achieving supply chain innovativeness. Comput. Ind. Eng. **168**, 108055 (2022)

Khan, S., Singh, R., Khan, S., Ngah, A.H.: Unearthing the barriers of internet of things adoption in food supply chain: a developing country perspective. Green Technol. Sustain. **1**(2), 100023 (2023)

Lambourdière, E., Corbin, E., Verny, J.: Reconceptualizing supply chain strategy for the digital era: achieving digital ambidexterity through dynamic capabilities. In: The Digital Supply Chain, pp. 419–434. Elsevier (2022)

Liu, S., Chan, F.T., Yang, J., Niu, B.: Understanding the effect of cloud computing on organizational agility: an empirical examination. Int. J. Inf. Manag. **43**, 98–111 (2018)

Muhsen, Y.R., Husin, N.A., Zolkepli, M.B., Manshor, N.: A systematic literature review of fuzzy-weighted zero-inconsistency and fuzzy-decision-by-opinion-score-methods: assessment of the past to inform the future. J. Intell. Fuzzy Syst. **45**(3), 4617–4638 (2023)

Pal, K.: Internet of Things impact on supply chain management. Procedia Comput. Sci. **220**, 478–485 (2023)

Patrucco, A.S., Marzi, G., Trabucchi, D.: The role of absorptive capacity and big data analytics in strategic purchasing and supply chain management decisions. Technovation **126**, 102814 (2023)

Prajapati, D.K., Mathiyazhagan, K., Agarwal, V., Khorana, S., Gunasekaran, A.: Enabling industry 4.0: assessing technologies and prioritization framework for agile manufacturing in India. J. Cleaner Prod. **447**, 141488 (2024)

Ramirez-Peña, M., Sotano, A.J.S., Pérez-Fernandez, V., Abad, F.J., Batista, M.: Achieving a sustainable shipbuilding supply chain under I4. 0 perspective. J. Cleaner Prod. **244**, 118789 (2020)

Sachdeva, S., et al.: Unraveling the role of cloud computing in health care system and biomedical sciences. Heliyon (2024)

Sadeghi, K., Ojha, D., Kaur, P., Mahto, R.V., Dhir, A.: Explainable artificial intelligence and agile decision-making in supply chain cyber resilience. Decis. Support. Syst. **180**, 114194 (2024)

Shafique, M.N., Yeo, S.F., Tan, C.L.: Roles of top management support and compatibility in big data predictive analytics for supply chain collaboration and supply chain performance. Technol. Forecast. Soc. Chang. **199**, 123074 (2024)

Shu, W., Cai, K., Xiong, N.N.: Research on strong agile response task scheduling optimization enhancement with optimal resource usage in green cloud computing. Futur. Gener. Comput. Syst. **124**, 12–20 (2021)

Tian, S., Wu, L., Ciano, M.P., Ardolino, M., Pawar, K.S.: Enhancing innovativeness and performance of the manufacturing supply chain through datafication: the role of resilience. Comput. Ind. Eng. **188**, 109841 (2024)

Umair, M., Cheema, M.A., Cheema, O., Li, H., Lu, H.: Impact of COVID-19 on IoT adoption in healthcare, smart homes, smart buildings, smart cities, transportation and industrial IoT. Sensors **21**(11), 3838 (2021)

Xu, J., Pero, M., Fabbri, M.: Unfolding the link between big data analytics and supply chain planning. Technol. Forecast. Soc. Chang. **196**, 122805 (2023)

Zhao, N., Hong, J., Lau, K.H.: Impact of supply chain digitalization on supply chain resilience and performance: a multi-mediation model. Int. J. Prod. Econ. **259**, 108817 (2023)

Zheng, J., Alzaman, C., Diabat, A.: Big data analytics in flexible supply chain networks. Comput. Ind. Eng. **178**, 109098 (2023)

The Investment of Human Capital in Digital Green Economy Transition to Support Artificial Intelligence Technologies at Private Hospitals

Ali Razzaq Chyad Al-Abedi[1]([⊠]) and Qais Ibrahim Hussein[2]

[1] Department of Business Administration, College of Administration and Economics,
University of Kufa, Kufa, Iraq
alir.alabed@uokufa.edu.iq

[2] Department of Business Administration, College of Administration and Economic, Al Iraqia
University, Baghdad, Iraq

Abstract. The purpose of this research is to examine the effects of digital human capital on artificial intelligence technologies at private hospitals. Data were collected using questionnaires distributed to 118 doctors working in hospitals for the Al-Najaf Governorate. The research's framework model studies the direct relationship between digital human capital and artificial intelligence. This study controlled the effect level and tested all the hypotheses using AMOS V.25 - SPSS V.26 statistical programs. The results indicate the acceptance of all the four hypotheses in the research model which demonstrate a positive and significant relationship between digital human capital and artificial intelligence. Furthermore, knowledge-based digital leisure was shown to strengthen the direct impact on artificial intelligence whilst perceived digital autonomy weakens the direct impact on artificial intelligence.

This research contributes to the knowledge of doctors in hospitals by exploring the impact of digital human capital on artificial intelligence technologies. Secondly, this research develops a new framework to explain the relationship between digital human capital behaviors and artificial intelligence application in the health sector.

Keywords: Digital human capital · Artificial intelligence · Private hospitals

1 Introduction

Deep learning refers to machine learning technology enhanced by artificial intelligence that goes beyond its ability to know clips, images, speech recognition, computer vision, natural language processing, and various industrial products (Al-Abedi and Al-Orbawi, 2024). Artificial networks are considered a part of the modern developments that use artificial intelligence algorithms to understand and study digital human capital ways by stimulating a human brain structure known as the neural network (Atiyah et al., 2023). Digital human capital and artificial intelligence work seamlessly through neural networks, mainly driven by the accessibility of big amounts of data in cloud (Ali

© The Author(s), under exclusive license to Springer Nature Switzerland AG 2024
A. Alnoor et al. (Eds.): AIRDS 2024, LNNS 1033, pp. 229–243, 2024.
https://doi.org/10.1007/978-3-031-63717-9_15

et al., 2024). By benefiting from big data, it analyzes and interprets the data so that they can be used meaningfully (Alnoor et al., 2024). Deep learning can increase the productivity of human resource, hence increasing organizational productivity and reducing wasted time in collecting, sorting, and assembling demands of applicants and determination of each employee's potential to ensure professional and personal growth (Al-Abedi and Al-Shimmery, 2021).

Digital human capital is considered a crucial knowledge source for organizations. It can improve employees' skills in programming, data analysis, digital marketing, and machine learning, and also lead to an increase in efficiency, innovation, and economic growth (Atiyah and Zaidan, 2022). In digital societies, golden-collar individuals can become digital human capital, playing a crucial role in promoting innovation and providing solutions for economic and social problems and challenges (Atiyah et al., 2023). These individuals take part in digital opportunities and are highly beneficial as a result of their knowledge, skills, and digital competences that support high-tech applications of rapid development, including artificial intelligence applications (Bach et al., 2013).

Artificial intelligence helps organizations make better and faster decisions by applying artificial intelligence programs in order to benefit from accelerating digital work and to create greater efficiency in organizational functions, especially in human resource management and digital tasks (Al-Abedi and Al-Orbawi, 2024). Digital human resource applications that rely on artificial intelligence have the ability to enhance employees' productivity as well as help human resource to become learning advisors. They also enhance performance as well as the ability to evaluate, predict, diagnose, and locate the most competent employees who are able to perform various tasks which take up most of the managers' time. Therefore, organizations can save money and time by using artificial intelligence technologies (Biliavska et al., 2022).

Iraqi researchers can use results of this research to make different choices about the value of digital human capital. The results have written down this useful research for private health hospitals that contribute to develop of the health sector. Artificial intelligence is important for the readiness or competence of human capital to adapt with artificial intelligence, because it will accelerate digital innovation in hospitals.

This research will provide context on almost original ideas about the future and the necessary approaches and strategies that can help in apply the artificial intelligence technologies at private hospitals.

2 Literature Review

2.1 Digital Human Capital

The digital human capital concept is still very limited with very little discussion between academics and researchers. Hence, it is very important to add contributions and scientific literature in order to develop the concept. Digital human capital refers to individual characteristics that are considered valuable in the digital production process, including but not limited to employees' knowledge, skills, and technological competence (Ter Beek et al., 2022).

Digital human capital refers to the digital transformation of human resource services and processes through the use of social, mobile, analytical, and cloud technologies, and represents a radical change albeit occurring continuously from a rapid change in the progress of institutional work (Daniel, 2023; Atiyah et al., 2023).

Digital human capital is the use of computer systems, communications networks, and interactive electronic media to perform human resource management functions through the digital transformation of human resource management based on data and digital technologies (Zhang & Chen, 2023).

Digital human capital also refers to the knowledge, skills, and abilities of individuals with regards to digital technologies such as a software coding, artificial intelligence, or machine learning (Grimpe et al., 2023).

It is basically the knowledge, skills, and abilities that enable individuals and organizations to use digital technologies effectively and innovatively, and is a form of human capital which is often defined as the stock of competencies, knowledge, social, and personal attributes as well as creativity embodied in the ability to perform work that will produce economic value (Von Krogh, 2018). Digital human capital is important for the following reasons:

First, reduces recruitment and training costs by attracting and retaining the best talents. Human resource management also increases employees' satisfaction and engagement, which leads to increased productivity and better business results (Grimpe et al., 2023).

Second, makes it easier to identify and develop high-potential employees, especially those with knowledge and skills relevant to digital transformation, which is becoming increasingly important to the competitiveness of multinational companies. However, digital human capital shortage in many host countries puts pressure on multinational companies' subsidiaries to prevent these employees from leaving their job (Colbert et al., 2016).

Third, facilitates access to information and skills development, which can enhance wealth of individuals and societies. Broadband use can also affect results for the wider society, with ripple and spillover effects on local labor markets and institutions, and richer information networks to encourage innovation (Al-Abedi and Al-Shimmery, 2021).

2.2 Dimensions of Digital Human Capital

When we talk about digital human capital, we are referring to the value of knowledge and skills possessed by individuals in the digitalization age because digital human capital is considered one of the components in modern knowledge economy. There are four dimensions of digital human capital (Price, 2022):

Perceived Digital Autonomy: The ability to access and use digital technology independently and effectively. Digital human capital is considered an important source of digital autonomy, where individuals with strong technological skills can interact and excel in diverse digital environments. Perceived digital autonomy enhances digital human capital value and the ability to improve skills and knowledge of digital technology, as well as develop abilities to deal with autonomy on the Internet and modern technology. Perceived digital autonomy and digital human capital are closely linked in a digitally connected world where investment in developing individuals' technological and cognitive capabilities are essential aspects of promoting autonomy and mastery in digital environment.

Perceived Digital Competence: Perceived digital competence is closely linked to digital human capital. When individuals and organizations have advanced and diverse digital skills, they are more likely to use and benefit from digital technology effectively. Hence, they can enhance productivity and adapt to technological changes. For example, individuals who possess perceived digital competence (such as an understanding of data analytics, the programming, and digital marketing) are more equipped to develop and manage digital solutions and deliver services effectively. In general, realized digital competence and digital human capital contribute to build a strong and advanced digital economy, as individuals and institutions are able to innovate, adapt to technological changes, and benefit greatly.

Informal Digital Browsing: Digital human capital can have a positive impact through informal browsing, as individuals can gain knowledge and skills personally and directly, as well as connect with others and share ideas and experiences. Informal browsing contributes to developing digital human capital through access to online learning resources such as educational courses, lectures, and study materials. It may also have an impact on the development of technical skills and the understanding of developments and innovations in the digital world. However, we must note that informal browsing may not necessarily replace formal education and expertise. Therefore, individuals must strike a balance between the benefits of informal browsing and the attainment of formal education and training which enable them to develop skills and increase digital competence.

Knowledge-Based Digital Leisure: This refers to making use of information and digital content to provide entertainment experiences of cognitive value to beneficiaries. This includes virtual reality simulations, digital publishing sites for news and general information, and development of digital human capital by providing learning and personal development opportunities as well as improving digital work skills and innovation. Advanced digital technology that uses knowledge and entertainment content can be transformed into successful business products and services. A knowledge-based digital vision allows individuals to develop skills and create new job opportunities in digital media and information technology industries.

2.3 Artificial Intelligence (AI)

The term artificial intelligence (AI) was first used in 1956 by John McCarthy. Currently, artificial intelligence is employed in all human resource departments institutions and business companies according to digital transformation experience with Industry 4.0. This shift has implications for human resource management practices and behaviors (Biliavska et al., 2022).

Artificial intelligence can be defined as "the science and engineering of making intelligent machines". Artificial intelligence has been described with certain approaches in relation to human intelligence or intelligence in general (Mukherjee, 2022), which reflects the simulation of human intelligence in programmed machines. This term applies to those machines that exhibit human mental traits such as learning and problem solving (Choubey & Zohuri, 2021).

Artificial intelligence refers to the idea of building machines that are capable of performing tasks which are normally performed by humans. It is a subfield of artificial intelligence. Algorithms are applied to learn patterns and intrinsic statistical structures in data, thus allowing predictions of unseen data. Artificial intelligence is a branch of computational science that is concerned with developing systems and programs which are capable of carrying out tasks that require human intelligence.

Artificial intelligence aims to create systems that are capable of learning, thinking, and making decisions independently in order to solve problems and achieve goals. Artificial intelligence techniques rely on a set of methods and tools such as machine learning, artificial neural networks, statistical analysis, and others (Schwendicke et al., 2020).

Artificial intelligence plays an important role in social development by predicting errors and facilitating effective planning for social development goals. It has revolutionized various aspects of life and work environments. Unsurprisingly, it has also had an impact on my conversion to human resource management through our use of multi-directional systems, sustainability use of resource, and its processes, as well as making decisions about technology informed by broader sustainability objectives (Sova et al., 2023).

This technology can radically change data, and generate valuable visions that enable us to make deductive decisions that are independent of the rest of the social and social challenges on top of making strategic goals based on the analysis of human resource information.

2.4 Dimensions of Artificial Intelligence (AI)

There are three dimensions of artificial intelligence in human resource (Purwaamijaya & Prasetyo, 2022):

Machine Learning Algorithm: Machine learning is the systematic study of algorithms and models that use a computer to remember tasks without explicitly programming them. For example, studied web engines like Google work well and know algorithms that know how to rank websites. These algorithms can be used in a number of ways, including digital analytics, image processing, and data mining. Once an algorithm learns how to access data, it can perform its main tasks, which is a key feature of learning. It also allows diverse managers with real information to support management in making the appropriate decision with digital human capital.

Deep Learning: Deep learning refers to machine learning technology enhanced by artificial intelligence that goes beyond its ability to understand clips, images, speech recognition, computer vision, natural language processing, and various industrial products. Artificial neural networks are considered a part of the latest developments that use artificial intelligence algorithms to understand and study human capital habits by stimulating a human brain structure known as the neural network. Human capital and artificial intelligence work seamlessly through neural networks, mainly driven by the availability of large amounts of data in cloud (storage). By benefiting from big data, it analyzes and interprets the data so that they can be used meaningfully. Deep learning can increase the productivity of human resource, hence increasing organizational productivity and reducing wasted time in collecting, sorting, and assembling demands of applicants and determination of each employee's potential to ensure professional and personal growth.

Big Data: The large and voluminous data can be managed by digital human capital; human resource can improve the actual performance of company, whereby the method in which data sources evaluate digital human resource in real time helps in identifying and developing knowledge that contributes to organizational performance and enhances company competence.

3   Hypothesis and Research Framework

The hypothesizes of the research are provided below (see Fig. 1):

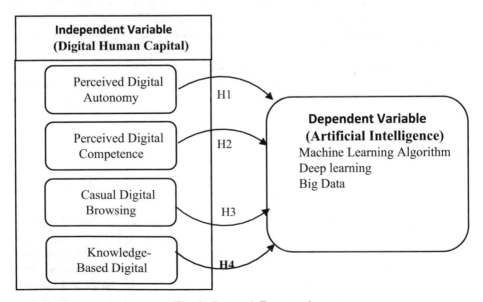

Fig. 1. Research Framework

H1: Adoption of perceived digital autonomy positively impacts artificial intelligence.

H2: Adoption of perceived digital competence positively impacts artificial intelligence.

H3: Adoption of informal digital browsing positively impacts artificial intelligence.

H4: Adoption of knowledge-based digital leisure positively impacts artificial intelligence.

4 Method

The research population is represented by private hospitals operating in Najaf city (Al-Hayat Hospital, Amal Al-Hayat Hospital, Al-Batoul Hospital, Al-Amir Hospital, Al-Ghadeer Hospital, Imam Ali Hospital) because they have specialized consultant doctors who continue to develop self-skills in a way that is compatible with technology in the field of medicine. The participants in this research are anyone working in the private hospitals. The samples are expected to represent the wider population so as to ease data collection. The sample of 142 of doctors to ensure representativeness of the total population, arriving at more validity data, and to guarantee data consistency. Ultimately, a total of 118 responses were derived.

The measurement items were derived from the questionnaire. The scale for digital human capital which is the independent variable was adopted from Price (2022). The items for measuring digital human capital were adopted from the digital capital scale developed by DiMaggio et al. (2004). The items for artificial intelligence were adapted from Purwaamijaya and Prasetyo (2022). All the items were measured using a Likert scale from 1 = Strongly Disagree to 5 = Strongly Agree.

To test the hypotheses and analyze the respondents' data, we relied on AMOS V. 25 and SPSS V.27. The reliability of the measures was confirmed, and descriptive statistics were presented.

5 Data Analysis and Results

"Reliability" refers to the stability of a study's measure and the ability to obtain consistent results across different time periods. The structural reliability of the measurement tool was evaluated using Cronbach's alpha test. Based on Table 1, the Cronbach's alpha coefficient values range between (0.931–8690). All the constructs exhibit values exceeding 0.7, indicating strong reliability (Hair et al., 2014; Al-Abedie and Al-Sultani, 2021).

Table 1. Reliability of variables

Variables	Cronbach's alpha
Perceived Digital Autonomy	0.914
Perceived Digital Competence	0.886
Casual Digital Browsing	0.894
Knowledge-Based Digital Leisure	0.8910
Digital human capital	0.920
Machine Learning Algorithm	0.918
Deep learning	0.913
Data big	0.8690
Artificial intelligence	0.931

Confirmatory factor analysis (CFA) was performed to assess the factor structure of the main variables' items that measured the two central constructs namely digital human capital and artificial intelligence, using estimates of critical ratio (CR) (see Tables 2 and 3; Figs. 2 and 3).

Table 2. Estimates of Digital Human Capital

Items	Path	Sub-Variables	Estimate	C.R.	P
PDA1	<---	Perceived Digital Autonomy	.909	11.837	***
PDA2	<---		.830		
PDA3	<---		.845	11.103	***
PDA4	<---		.894	12.533	***
PDA5	<---		.884	11.988	***
PDC1	<---	Perceived Digital Competence	.743		
PDC2	<---		.825	9.144	***
PDC3	<---		.833	9.227	***
PDC4	<---		.845	9.370	***
PDC5	<---		.823	9.111	***
CDB1	<---	Casual Digital Browsing	.813		
CDB2	<---		.827	13.892	***
CDB3	<---		.901	11.216	***
CDB4	<---		.935	12.624	***
CDB5	<---		.898	11.548	***

(*continued*)

Table 2. (*continued*)

Items	Path	Sub-Variables	Estimate	C.R.	P
KBDV1	<---	Knowledge-Based Digital Leisure	.903		
KBDV2	<---		.848	12.991	***
KBDV3	<---		.804	10.805	***
KBDV4	<---		.860	13.433	***
KBDV5	<---		.806	13.175	***

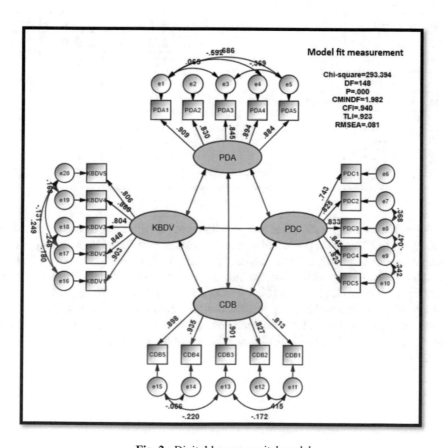

Fig. 2. Digital human capital model

Descriptive statistics are reported in Table 4 below.

SPSS was used to test the hypotheses in order to determine the effects of digital human capital on artificial intelligence. Table 5 shows a direct effect. The results indicate the acceptance of all the four hypotheses in the research framework, which were found to have a positive and significant relationship with perceived digital autonomy (H1) (b =

Table 3. Estimates of Artificial Intelligence

Items	Path	Sub-Variables	Estimate	C.R.	P
MLA1	--->	Machine Learning Algorithm	.941	18.225	***
MLA2	--->		.954		
MLA3	--->		.925	19.422	***
MLA4	--->		.885	16.513	***
MLA5	--->		.884	14.735	***
DL1	--->	Deep learning	.875		
DL2	--->		.875	13.164	***
DL3	--->		.876	12.203	***
DL4	--->		.863	12.112	***
DL5	--->		.867	12.343	***
BD1	--->	Big data	.813		
BD2	--->		.807	12.343	***
BD3	--->		.843	10.823	***
BD4	--->		.852	11.335	***
BD5	--->		.764	9.315	***

0.682, t = 10.061 and p < 0.000) as well as with perceived digital competence (H2) (b = 0.795, t = 14.093 and p < 0.000). Moreover, a hospital was found to have a significant positive relationship with informal digital browsing (H3) (b = 0.657, t = 16.763 and p < 0.000). Finally, a positive and significant relationship was found with knowledge-based digital leisure (H4) (b = 0.455, t = 20.48 and p < 0.000). The regression model explains 46.6% of the variance for perceived digital autonomy ($R^2 = 0.466$), 63.1% of the variance for perceived digital competence ($R^2 = 0.631$), 70.8% of the variance for informal digital browsing ($R^2 = 0.708$), and 78.3% of the variance for knowledge-based digital leisure ($R^2 = 0.783$).

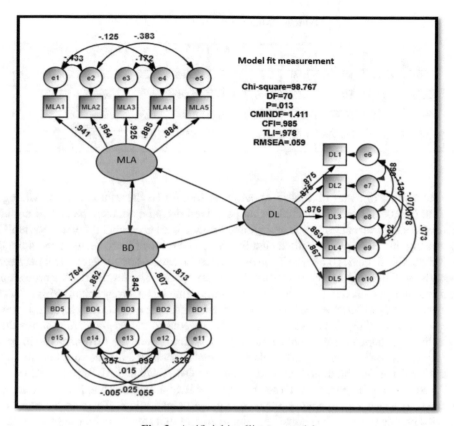

Fig. 3. Artificial intelligence model

Table 4. Descriptive Statistics N = (118)

CV	SD	Mean	Variables
23.93	0.864	3.610	Perceived Digital Autonomy
23.50	0.863	3.673	Perceived Digital Competence
23.79	0.920	3.868	Casual Digital Browsing
23.48	0.874	3.722	Knowledge-Based Digital Leisure
21.28	0.791	3.718	**Digital human capital**
28.56	1.038	3.636	Machine Learning Algorithm
22.31	0.848	3.798	Deep earning
23.43	0.866	3.695	Big Data
23.27	0.863	3.710	**Artificial intelligence**

Table 5. Multiple Regression Results

sig	t	F	Adj R²	R²	R	B	α	Digital Human Capital	
0.000	10.061	101.227	0.461	0.466	0.683	0.682	1.247	Perceived digital autonomy	Artificial Intelligence
0.000	14.093	198.601	0.628	0.631	0.795	0.795	0.791	Perceived digital competence	
0.000	16.763	281.003	0.705	0.708	0.841	0.789	0.657	Informal digital browsing	
0.000	20.481	419.469	0.782	0.783	0.885	0.874	0.455	Knowledge-based digital Leisure	

According to result of Table 5 above, it is found that the value of Sig F. Change is .000 (<0.05), it can be concluded that perceived digital autonomy, perceived digital competence, informal digital browsing, and knowledge-based digital Leisure variables have a significant impact to artificial intelligence. The value of R (correlation coefficient) is (0.683, 0.795, 0.841, 0.885) so it can be concluded that the level of agreement between perceived digital autonomy, perceived digital competence, informal digital browsing, and knowledge-based digital Leisure at the same time has a strong relationship. There are interesting differences between these studies and previous research, as previous studies were conducted by researchers in different countries, and gave different results. Specifically, adaptation to digital human capital, which is outcome-driven, has an impact or impact on artificial intelligence. This means that artificial intelligence implementation is highly dependent on digital HR in a given area (Al-Abedi and Al-Orbawi, 2024).

Artificial Intelligence is still a seemingly new field in the health industry in Iraq, so it needs to attract a lot of attention from many health institutions. There are no research studies on this topic, so it is hoped that the results of this study will be used by farmers or organizations in Iraq but the cost of human resources is aimed at health institutions (Husin et al., 2023). This could be the beginning of useful research, especially regarding the specific health benefits that serve as a valuable contribution to increasing treatment learning for clinicians in hospitals. Because artificial intelligence is one of the fastest developing human resources, and because training in new skills is required to adapt to general work, it is important to learn that human resource is ready or qualified to adapt psychologically to the experiences and skills. The main goal of this research will be to provide additional explanations on the impact of digital capital on artificial intelligence and how it can have a positive impact on the ability through better reliance on human scrutiny in masking the effects on industry (Al-Enzi et al., 2023).

Accordingly, we started writing this research, the use of artificial intelligence is a good concept for Iraq industrial as a form of systemic cooperation and skills to help, especially in the special health sector. Adaptation of human capital through perceived digital independence, perceived digital competence, casual digital browsing, and digital entertainment variants can outperform artificial intelligence technologies. Refer to the search results obtained and certainly, the term digital human capital writer can yet have compensation and does not exist in the field of artificial intelligence. Many industries in Iraq, with real talents are already evident. There are several uses of artificial intelligence that the human resource can change at this time and that hospitals can provide by the health hospitals, and other strategies will be discussed in results with similar studies such as (Purwaamijaya, & Prasetyo, 2022 and Price, 2022).

6 Implications and Future Studies

The current research answers questions about the effect of digital human capital on artificial intelligence technologies at private hospitals. This research contributes to hospitals by exploring the impact of digital human capital on artificial intelligence technologies in the health sector. The results indicate a positive and significant relationship between digital human capital and artificial intelligence.

Digital models and platforms for digital human capital based on digital technology can be adopted to keep pace with rapid developments in the health sector and to benefit from experiences and learning curves towards developing personal talents to achieve customized educational paths through online training, knowledge sharing, performance evaluation, and personalized feedback. This provides flexible training and development.

7 Conclusion

The private hospitals have adopted digital human capital components, which will be reflected positively in supporting artificial intelligence technologies. The adoption of digital human capital has a significant positive impact on artificial intelligence. Private hospitals can direct their medical staff to adopt various artificial intelligence technologies by employing digital human capital that is necessary to maintain their human resource talents.

References

Abbas, S., et al.: Antecedents of trustworthiness of social commerce platforms: a case of rural communities using multi group SEM & MCDM methods. Electron. Commer. Res. Appl. **62**, 101322 (2023)

Aiyub, A., Adnan, A.: Digital human: human capital development formulation to achieve excellence industrial competitiveness. Sci. Res. J. Econ. Bus. Manage. **2**(1) (2022)

Al-Abedie, A.R., Al-Sultani, A.H.R.: The influence of the organizational physics on organizations vitality. J. Manag. Inf. Decis. Sci. **24**(Special Issue 1), 1–10 (2021)

Al-Abedi, A.R.C., Al-Shimmery, L.L.: Strategic anxiety: the influence of organizational conspiracy theory beliefs and employee's withdrawal. J. Leg. Ethical Regul. Issues **24**(Special Issue 1), 1–11 (2021)

Al-Abedi, A.R.C., Al-Orbawi, M.J.: Investment elite human capital and its impact on building postmodern organizations: reconnaissance study in a sample of private schools/Najaf. In: 2nd International Conference on Engineering and Science to Achieve the Sustainable Development Goals. AIP Conference Proceedings, vol. 3092, pp. 080013-1–080013-14 (2024). https://doi.org/10.1063/5.0199641

Al-Enzi, S.H.Z., Abbas, S., Abbood, A.A., Muhsen, Y.R., Al-Hchaimi, A.A.J., Almosawi, Z.: Exploring research trends of metaverse: a bibliometric analysis. In: Al-Emran, M., Ali, J.H., Valeri, M., Alnoor, A., Hussien, Z.A. (eds.) Beyond Reality: Navigating the Power of Metaverse and Its Applications (IMDC-IST 2024). LNNS, vol. 895, pp. 21–34. Springer, Cham (2023). https://doi.org/10.1007/978-3-031-51716-7_2

Ali, J., Hussain, K.N., Alnoor, A., Muhsen, Y.R., Atiyah, A.G.: Benchmarking methodology of banks based on financial sustainability using CRITIC and RAFSI techniques. Decis. Mak. Appl. Manage. Eng. **7**(1), 315–341 (2024)

Alnoor, A., Atiyah, A.G., Abbas, S.: Unveiling the determinants of digital strategy from the perspective of entrepreneurial orientation theory: a two-stage SEM-ANN approach. Glob. J. Flex. Syst. Manage. 1–18 (2024)

Alnoor, A., et al.: How positive and negative electronic word of mouth (eWOM) affects customers' intention to use social commerce? A dual-stage multi group-SEM and ANN analysis. Int. J. Hum. Comput. Interact. **40**(3), 808–837 (2024)

Atiyah, A.G.: Unveiling the quality perception of productivity from the senses of real-time multisensory social interactions strategies in metaverse. In: Al-Emran, M., Ali, J.H., Valeri, M., Alnoor, A., Hussien, Z.A. (eds.) Beyond Reality: Navigating the Power of Metaverse and Its Applications (IMDC-IST 2024). LNNS, vol. 876, pp. 83–93. Springer, Cham (2023). https://doi.org/10.1007/978-3-031-51300-8_6

Atiyah, A.G., Zaidan, R.A.: Barriers to using social commerce. In: Alnoor, A., Wah, K.K., Hassan, A. (eds.) Artificial Neural Networks and Structural Equation Modeling, pp. 115–130. Springer, Singapore (2022). https://doi.org/10.1007/978-981-19-6509-8_7

Atiyah, A.G., Alhasnawi, M., Almasoodi, M.F.: Understanding metaverse adoption strategy from perspective of social presence and support theories: the moderating role of privacy risks. In: Al-Emran, M., Ali, J.H., Valeri, M., Alnoor, A., Hussien, Z.A. (eds.) Beyond Reality: Navigating the Power of Metaverse and Its Applications (IMDC-IST 2024). LNNS, vol. 876, pp. 144–158. Springer, Cham (2023). https://doi.org/10.1007/978-3-031-51300-8_10

Atiyah, A.G., All, N.D.A., Zaidan, A.S., Bayram, G.E.: Understating the social sustainability of metaverse by integrating adoption properties with users' satisfaction. In: Al-Emran, M., Ali, J.H., Valeri, M., Alnoor, A., Hussien, Z.A. (eds.) Beyond Reality: Navigating the Power of Metaverse and Its Applications (IMDC-IST 2024). LNNS, vol. 895, pp. 95–107. Springer, Cham (2023). https://doi.org/10.1007/978-3-031-51716-7_7

Atiyah, A.G., Faris, N.N., Rexhepi, G., Qasim, A.J.: Integrating ideal characteristics of chat-GPT mechanisms into the metaverse: knowledge, transparency, and ethics. In: Al-Emran, M., Ali, J.H., Valeri, M., Alnoor, A., Hussien, Z.A. (eds.) Beyond Reality: Navigating the Power of Metaverse and Its Applications (IMDC-IST 2024). LNNS, vol. 895, pp. 131–141. Springer, Cham (2023). https://doi.org/10.1007/978-3-031-51716-7_9

Bach, A., Shaffer, G., Wolfson, T.: Digital human capital: developing a framework for understanding the economic impact of digital exclusion in low-income communities. J. Inf. Policy **3**, 247–266 (2013)

Eubanksm, B.: Artificial Intelligence for HR Use AI to Support and Develop a Successful Workplace. 1st Edition, Kogan pa Publish (2020)

Choubey, S., Zohuri, B.: Merits and demerits of AI in HR. Management **9**(5) (2021)

Colbert, A., Yee, N., George, G.: The digital workforce and the workplace of the future. Acad. Manage. J. **59**(3) (2016)

Daniel, D.: Digital HR (TechTarget Editorial) (2023). https://www.techtarget.com/searchhrsoftware/definition/digital-HR

DiMaggio, P., Hargittai, E., Celeste, C., Shafer, S.: Digital inequality: from unequal access to differentiated use. In: Social Inequality. Neckerman, K. (ed.). Russell Sage Foundation, New York, pp. 355–400 (2004)

Grimpe, C., Sofka, W., Kaiser, U.: Competing for digital human capital: the retention effect of digital expertise in MNC subsidiaries. J. Int. Bus. Stud. **54**(4) (2023)

Hair, J.F., Jr., Sarstedt, M., Hopkins, L., Kuppelwieser, V.G.: Partial least squares structural equation modeling (PLS-SEM): an emerging tool in business research. Eur. Bus. Rev. **26**(2), 106–121 (2014). https://doi.org/10.1108/EBR-10-2013-0128

Husin, N.A., Abdulsaeed, A.A., Muhsen, Y.R., Zaidan, A.S., Alnoor, A., Al-mawla, Z.R.: Evaluation of metaverse tools based on privacy model using fuzzy MCDM approach. In: Al-Emran, M., Ali, J.H., Valeri, M., Alnoor, A., Hussien, Z.A. (eds.) Beyond Reality: Navigating the Power of Metaverse and Its Applications (IMDC-IST 2024). LNNS, vol. 895, pp. 1–20. Springer, Cham (2023). https://doi.org/10.1007/978-3-031-51716-7_1

Muhsen, Y.R., Husin, N.A., Zolkepli, M.B., Manshor, N.: A systematic literature review of fuzzy-weighted zero-inconsistency and fuzzy-decision-by-opinion-score-methods: assessment of the past to inform the future. J. Intell. Fuzzy Syst. **45**(3), 4617–4638 (2023)

Mukherjee, A.N.: Application of artificial intelligence: benefits and limitations for human potential and labor-intensive economy–an empirical investigation into pandemic ridden Indian industry. Manage. Matters **19**(2) (2022)

Price, R.: Beyond a divide: reconceptualizing digital capital and links to academic proficiency. Electronic Theses and Dissertations, University of Louisville (2022)

Purwaamijaya, B.M., Prasetyo, Y.: The effect of artificial intelligence (AI) on human capital management in Indonesia. Jurnal Manajemen dan Kewirausahaan **10**(2) (2022)

Schwendicke, F.A., Samek, W., Krois, J.: Artificial intelligence in dentistry: chances and challenges. J. Dent. Res. **99**(7) (2020)

Sova, O., Bieliaieva, N., Antypenko, N., Drozd, N.: Impact of artificial intelligence and digital HRM on the resource consumption within sustainable development perspective. In: E3S Web of Conferences 408, p. 01006. EDP Sciences (2023)

Tan, G.W.H., et al.: Metaverse in marketing and logistics: the state of the art and the path forward. Asia Pac. J. Mark. Logist. **35**(12), 2932–2946 (2023)

Ter Beek, M., Wopereis, I., Schildkamp, K.: Don't wait, innovate! Preparing students and lecturers in higher education for the future labor market. Educ. Sci. **12**(9) (2022)

Tlili, A., et al.: Is metaverse in education a blessing or a curse: a combined content and bibliometric analysis. Smart Learn. Environ. **9**(1), 1–31 (2022)

Biliavska, V., Castanho, R.A., Vulevic, A.: Analysis of the impact of artificial intelligence in enhancing the human resource practices. J. Intell. Manage. Decis. **1**, 128–136 (2022). https://doi.org/10.56578/jimd010206

Von Krogh, G.: Artificial intelligence in organizations: new opportunities for phenomenon-based theorizing. Acad. Manage. Discov. **4**(4) (2018)

Vrontis, D., Christofi, M., Pereira, V., Tarba, S., Makrides, A., Trichina, E.: Artificial intelligence, robotics, advanced technologies and human resource management: a systematic review. Int. J. Hum. Resour. Manage. **33**(6) (2022)

Wang, F.Y., Qin, R., Wang, X., Hu, B.: Metasocieties in metaverse: metaeconomics and meta management for meta enterprises and megacities. IEEE Trans. Comput. Soc. Syst. **9**(1), 2–7 (2022)

Wang, Y., et al.: A survey on metaverse: Fundamentals, security, and privacy. IEEE Commun. Surv. Tutor. (2022). https://doi.org/10.36227/techrxiv.19255058.v1

Yadav, M., Rahman, Z.: Measuring consumer perception of social media marketing activities in e-commerce industry: scale development & validation. Telemat. Inform. **34**(7), 1294–1307 (2017)

Yim, M.Y.C., Chu, S.C., Sauer, P.L.: Is augmented reality technology an effective tool for e-commerce? An interactivity and vividness perspective. J. Interact. Mark. **39**(1), 89–103 (2017)

Zhang, D., Chadwick, S., Liu, L.: The metaverse: opportunities and challenges for marketing in Web3 (2022). Available at SSRN 4278498

Zhang, J., Chen, Z.: Exploring human resource management digital transformation in the digital age. J. Knowl. Econ. (2023). https://doi.org/10.1007/s13132-023-01214-y

Zhao, Y., et al.: Metaverse: perspectives from graphics, interactions, and visualization. Vis. Inform. **6**(1), 56–67 (2022)

Interpretive Structural Modeling (ISM) as an Artificial Intelligence System for Improving Sustainable Performance

Mohammed N. Al-Salim and Hadi Al-Abrrow[(✉)]

College of Administration and Economics, University of Basrah, Basrah, Iraq
{pgs.muhammed.nabeel,Hadi.abdulimmam}@uobasrah.edu.iq

Abstract. Given the importance of the health sector and its impact on society in general, this study is made to improve performance in organizations working in this sector by using artificial intelligence, which has occupied a large part of our world today, as medical robots and medical and administrative systems would develop many medical procedures and work. And private administration in the medical field. In this study, the Interpretative Structural Modeling (ISM) system is used to clarify the impact of the most important standards for improving performance and determine the levels of those standards in addition to the direct impact of each standard on the other standards and on the overall performance of the health organization. The study is concluded that patient satisfaction is the basic factor that the health organization looks for in the process of improving its performance, that is, it is the most important criterion for improving performance. The standards come in the form of a conical matrix that shows the impact of the standards among them on overall performance. The process is carried out by simulating the system for the data obtained by experts working in the health sector and providing the system with the results that help it to improve its performance by relying on the most important standards.

Keywords: Artificial Intelligence · Improving Performance · Expert Systems · Healthcare

1 Introduction

Throughout history, the pursuit of good health has remained a central concern for humanity. When illness strikes, it not only impacts an individual's physical well-being but also reshapes their interactions within society. Various factors, including the type of illness, socio-economic status, treatment methods, quality of care, and the state of the healthcare system, all play pivotal roles in the healing journey. Hence, it is essential to delve into the multifaceted nature of this issue. An essential aspect of patient care revolves around the challenge of evaluating the quality of healthcare services. Patients often place their trust in healthcare professionals, relying on their expertise for diagnoses and treatment recommendations. Complex medical interventions are typically administered within specialized institutions, with hospitals standing out as prominent examples. Within such

A. Alnoor et al. (Eds.): AIRDS 2024, LNNS 1033, pp. 244–255, 2024.
https://doi.org/10.1007/978-3-031-63717-9_16

environments, the primary goal is to deliver comprehensive care tailored to the needs of diverse patient populations (Jashnabadi et al., 2023; Al-Enzi et al., 2023). Irrespective of country's circumstances, maintaining healthcare services at a certain standard of quality is imperative due to their pivotal role in human life (Behdioglu et al., 2017: 21; Husin et al., 2024). The patient assumes a central role within the hospital environment, and the provision of hospital services revolves around their presence (Ghoushchi et al., 2021; Alnoor et al., 2024). Consequently, to a significant extent, patient satisfaction serves as an indicator of the effective delivery of services, a feat not solely achievable through technology but largely reliant on the behavior and performance of individuals (Cathron et al., 2021). The complex phenomenon of satisfaction manifests in the alignment of patient needs, expectations, and experiences in receiving healthcare services (Ghuoshchi et al., 2021; Ahmed et al., 2024).

Artificial Intelligence (AI) and data analytics are experiencing a rising presence within the healthcare sector, demonstrating their efficacy in benefiting patients and assisting healthcare professionals. These technologies are instrumental in diminishing patient wait times and improving the affordability of healthcare services. Additionally, they empower physicians to provide more precise diagnoses while alleviating burnout resulting from administrative duties and paperwork. Key clinical applications of health AI hold the promise of generating annual savings of $150 billion for the global healthcare economy by 2026 (Pham et al., 2024). Simulation emerges as a pivotal tool and methodology for analyzing intricate structures. It represents a scientific method for comprehending the behavior of actual systems without perturbing their natural settings. With applications spanning diverse domains such as transportation and ports (Dieckmann et al., 2020: 2), simulation is pivotal. Expert systems are subsequently devised through the integration of artificial intelligence techniques and simulation tools. These systems harness intelligent evolutionary algorithms to formulate schedules within a simulation model (Nagdee et al., 2022; Muhsen et al., 2023).

Expert systems find utility across diverse applications and are utilized by decision-makers at all levels within organizations, including managers, accountants, financial analysts, consultants, and strategic planners (Shariat et al., 2013). The Interpretive Structural Modeling (ISM) system can help in policy selection and identifying influential points that affect the model more than others (Abbas et al., 2023; Atiyah, 2024). ISM is a method that determines levels of influence and criterion impact, useful for complementing the systemic approach, The question we want to answer in this study is to what extent can an interpretive structural modeling system improve the performance of organizations in the health sector? In addition to supplement the existing literature, this study provides practical guidance for professionals in the field (Alnoor et al., 2024). It delves into the impact of AI adoption in large hospitals from various perspectives, including its effects on operations, the workload of healthcare personnel, and both operating and financial performance (Atiyah et al., 2023). For large hospitals, embracing breakthrough AI technology is nearly indispensable for continuous performance improvement, while smaller hospitals may find benefit in implementing traditional, non-AI projects to enhance performance (Atiyah et al., 2023). The study has found that the most successful and profitable hospitals in the world use artificial intelligence more to achieve patient satisfaction by improving their performance and supporting their decisions with artificial intelligence (Alnoor et al., 2024).

2 Literature Review

In the middle of the twentieth century, artificial intelligence was recognized as a tool similar to the human mind in terms of perception and reasoning, and at the Dartmouth Conference specifically in 1956, artificial intelligence was recognized as part of the fields of scientific research. (McCarthy et al., 1956). Artificial intelligence began to develop rapidly, and algorithms began to develop gradually and rapidly after the development in the process of dealing with the huge amount of data and designing applications that rely on artificial intelligence, which are used in many fields, including the health sector (Silver et al., 2016; Atiyah et al., 2023). In health sector, artificial intelligence was relied upon in the late twentieth century, as it was used in diagnosis, medical decision-making, and patient care. This is considered a radical change for work in the health sector (Shortliffe and Buchanan, 1975). In the twenty-first century, artificial intelligence was used in the health field to find solutions to health problems by digitizing patients' health data and relying on machine learning techniques, which led to improving the accuracy of detecting diseases, including cardiovascular diseases and cancer (Esteva et al., 2019).

In recent years, the scope of artificial intelligence work in the field of health care has expanded by increasing the systems that are used in predictive analytics and patient monitoring (Topol, 2019; Atiyah and Zaidan, 2022). Artificial intelligence technologies have great potential to enhance patient satisfaction by improving healthcare efficiency and reducing expenses (Obermeyer and Emanuel, 2016). The use of artificial intelligence in the health field may develop further in the future by providing more accurate solutions in health goals and disease management (Beam and Kohane, 2018; Ali et al., 2024). Due to the current great development of artificial intelligence, this could revolutionize the field of health care to improve all aspects of the health sector and achieve patient satisfaction (Zaidan et al., 2024). Artificial intelligence systems help workers in the health sector easily access, interpret, evaluate, and benefit from data and information and lead to the possibility of making high-quality decisions and finding solutions to health problems. Perhaps hospitals information system is one of the prominent systems for hospitals management in general by supervising administrative and clinical functions and organizational (Moghaddam et al., 2019; Alnoor et al., 2023).

Perhaps the Corona Covid-19 pandemic is the most prominent challenge facing the health sector at the present time, due to the small number of health workers in addition to their lack of the necessary training to deal with crises. Simulation is used to improve the efficiency of the emergency department's performance by studying internal operations and staffing levels which is helped to reduce patient waiting times and improve the department's efficiency (Taleb et al., 2023; Rouhonen et al., 2019). Simulation was also used to improve the structure of the emergency department and helped reduce waiting time by predicting the arrival time of patients (Chang et al., 2010; Al-Refaie et al., 2014). In conclusion, artificial intelligence shows us a distinguished history through innovation and continuous development, and it is likely to go deeper into all areas of industry and service. This leads to business development through interaction between humans and machines and encourages researchers to research and develop in the field of artificial intelligence.

2.1 Research Background

To enhance the current situation of the health sector, structural solutions are necessary in the sector's work, starting from the individuals covered by hospital care, all the way to facilitating business operations in the various health care sectors. Leveraging Information and Communication Technology (ICT) can facilitate cost rationalization and boost hospital revenues (Božić, 2022). Another study delves into the impact of the Hospital Information System (HIS) on the operational efficiency of management units in public hospitals in Ahvaz, Iran. The HIS facilitates hospital operations at practical, tactical, and strategic levels, bridging patient care with administrative information across all hospital functions. The study findings underscore that an automated HIS can serve as a potent tool, aiding managers in hospital management processes and decision-making, ultimately leading to notable enhancements in hospital performance (Wang et al., 2023). The urgency to enhance health systems' capacity to swiftly assimilate and apply evidence has transitioned from an opportunity to a necessity. Health systems globally face strain, with high-quality primary care becoming increasingly elusive. Adopting a Learning Health System approach enables health systems to identify requisite capabilities for implementing learning elements, including pertinent inquiries and methodologies, ensuring a systematic approach to learning and achieving equity-centered quadruple aim metrics (Reid et al., 2024).

In the medical realm, the integration of artificial intelligence (AI) and expert systems holds immense potential in realizing the healthcare sector's five-year objectives (Cresswell et al., 2023). AI-powered computer systems can undertake tasks typically reliant on human cognitive abilities, encompassing visual perception, pattern recognition, speech processing, rapid data analysis, multilingual translation, and optimal decision-making among alternative choices. The landscape of research and development in AI has witnessed remarkable growth over the past decade, with profound impacts across various domains, notably the medical sector. In conclusion, physicians possess substantial knowledge concerning artificial intelligence and display openness to enhancing their medical competencies in alignment with AI advancements (Berrami et al., 2023).

2.2 Research Gap

Numerous studies have explored patients' experiences using dynamic systems, yet these investigations often lack comprehensiveness, tending to prioritize certain aspects over others. While some have delved into issues within the medical system using system dynamics, medical entrepreneurship, nursing staff ratios, and healthcare shortages, a gap emerged in designing a holistic system model centered around hospital performance improvement, considering patients' mental understanding of care quality and medical activity criteria. This study is aimed to elevate patient satisfaction by enhancing the performance of medical organizations and cultivating their workforce across various specialties, employing dynamic expert systems to inform senior management decision-making and medical field operations.

Other standards can also be used to improve organizational performance and achieve more comprehensive results in the health care sector, such as drug safety, accuracy in diagnosis, and clinical quality. In the context of the study, patient satisfaction is addressed while neglecting environmental changes. However, it can be relied on the dynamic systems approach because it provides an accurate understanding of patient satisfaction, in addition to revealing dynamic behaviors as time changes (Maidin et al., 2019). Davaheli et al. (2020) presented a study illustrating the process of exploring a practical application that simulates a dynamic health care system, and reached a conclusion that focused on patient access, obesity, and required employment in addition to AIDS, with a focus on patient satisfaction and concerns. At the same time, (Lui et al., 2021) presented a study that directly focused on the loyalty of healthcare customers in public hospitals in China, they focused on patient satisfaction, loyalty, and trust, and did not delve deeply into the results of improving performance, staff experience, quality, or communication methods.

3 Methodology

The current study employs a mixed methods approach, integrating both quantitative and qualitative research methodologies. This approach offers several advantages, including the ability to align with diverse research requirements and yield valuable insights, particularly when informed by the researcher's expertise in the subject matter. By leveraging the strengths of each method, the mixed methods approach facilitates broader learning, deeper understanding, and the development of a comprehensive knowledge base (Schulze, 2003). Furthermore, it establishes important foundations and presents a significant opportunity for innovation and development, driven by systematic knowledge acquisition and core technical proficiency (Bazeley, 2003), as shown in Fig. 1.

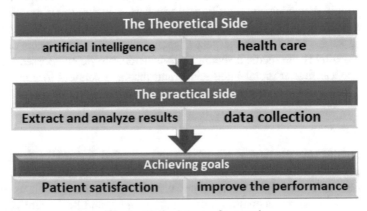

Fig. 1. Methodology of research.

The system's algorithm depends on the following equations to find the best criteria results:

a Identification of the criteria based on the literature review and experts' opinions.

b Establishment of the relative relationship between the pair-wise criteria based on the opinions of experts.

c Establishment of the structural self-interaction matrix. The structural self-interaction matrix can be considered as follows.

In this step, the initial reachability matrix is established by substituting the elements of the structural self-interaction matrix based on the binary rules shown in Table 2. The vectors A12, A21, and A32 represent the influence of criteria within matrix A, with A being the super matrix and I representing the identity matrix. When simplifying the super matrix A, it's crucial to evaluate the principal eigenvalue, qi, to determine the priority limits of a simplified stochastic matrix. Equation (3) serves as a straightforward illustration of a simplified stochastic matrix within a three-level hierarchy. Determining the structural levels of the criteria using the final reachability matrix, reachability set, and antecedent set. Design a digraph determined from the final reachability matrix.

The data collection process and identification of key criteria are conducted by drawing from existing literature, where a comprehensive set of criteria for evaluating performance in healthcare organizations was extracted. Subsequently, direct interviews are conducted with administrative, technical, and medical experts actively engaged in the healthcare sector across more than five hospitals in Basra Governorate. Through this collaborative effort, the most significant criteria are discerned and integrated into the system.

Self-interaction matrices, reflecting the interactions among the identified criteria, and accessibility matrices are established. These matrices enabled mathematical operations to be performed on the experts' responses. Following the conversion of symbols into numerical values, the results are extracted and meticulously analyzed to discern the most crucial criteria and ascertain their respective impacts.

Based on the final findings, recommendations are formulated and presented to the relevant authorities. These recommendations aim to guide efforts aimed at enhancing the performance of healthcare organizations operating within Basra Governorate.

The current study rely on the Interpretative Structure Model (ISM) system to determine the relationship between the criteria, and the results will be found after performing the following steps:

1. Determine criteria by conducting personal interviews with experts in the health field and relying on scientific literature.
2. Conducting comparisons between criteria by experts as well and finding a relative relationship between those criteria
3. Creating an interaction matrix between the criteria, where the criteria are interconnected and the preference is chosen by the expert's choices of a symbol that has quantitative connotations, as shown in Table 1.
4. Convert the data in the interaction matrix between the criteria into numbers to extract the results as shown in Table 2.
5. Then the results are extracted using the (ISM) system after making comparisons between all criteria and knowing the results of the impact between them and how each standard affects or does not affect the other criteria, as shown in Table 1.

Table 1. The Symbols list for expert judgment.

Symbol	Description	No.
O	Criterion i & j are unrelated	1
X	Criterion i & j are will lead to achieving each other	2
A	Criterion j will help to attain criterion	3
V	Criterion i will help to attain criterion j	4

Table 2. The binary rules for the establishment of the initial reachability matrix.

(j, i)	(i, j)	Symbol	No.
0	1	O	1
1	0	X	2
1	1	A	3
0	0	V	4

At this stage, using research background and interviews with experts, key variables and effective factors of improving performance are identified and extracted, then causal and circular diagrams are designed. To better understand the causal and circular model, the accepted assumptions have been drawn separately in Table 3.

Table 3. Criteria for developing performance in the hospital.

No.	Criteria
1	Patient satisfaction
2	Efficiency
3	Financial performance
4	Staff efficiency
5	Technology integration

4   Results

By using the Smart Interpretive Structural Modeling system to analyze the answers of the expert in the medical field, the following results in Table 4 were reached.

Table 4. Structural Self-Interaction Matrix (SSIM).

Variables	1	2	3	4	5
Patient Satisfaction		A	A	A	A
Efficiency			V	X	A
Financial performance				O	A
Staff efficiency					A
Technology integration					

The self-interaction matrix, through which the experts determined what are the levels of impact of the standards on each other, as the study finds that all standards can be affected by patient satisfaction, so it is prepared as a main objective of the study. It finds that financial performance can affect efficiency, in addition to that efficiency can affect and be affected by efficiency. Staff and technological integration can affect the efficiency of the health service in general. The matrix is also made clear that there is no direct impact relationship between financial performance and staff efficiency and that financial performance can be affected by technological integration, as is the case with staff efficiency which is also affected by technological integration, as sown in Table 5.

Table 5. Reachability Matrix (RM).

Variables	1	2	3	4	5	Driving Power
Patient Satisfaction	1	0	0	0	0	1
Efficiency	1	1	1	1	0	4
Financial performance	1	0	1	0	0	2
Staff efficiency	1	1	0	1	0	3
Technology integration	1	1	1	1	1	5
Dependence Power	5	3	3	3	1	

Reachability matrix, through which symbols are converted into numbers, as shown in Table 5, to be simulated through the system's algorithms and extract results that determine the most important criteria, while calculating the frequency of choices. The final form of the Reachability matrix through which the driving power values and the dependence power values are determined. The conical matrix specifies the locations of the criteria and the preference values for each criterion in Table 6.

Table 6. Reduced Conical Matrix (RCM).

Variables	1	3	2	4	5	Driving Power	Level
Patient Satisfaction	1	0	0	0	0	1	1
Efficiency	1	1	0	0	0	2	2
Financial performance	0	1	1	1	0	4	3
Staff efficiency	0	1*	1	1	0	4	3
Technology integration	0	0	1	1	1	5	4
Dependence Power	5	4	3	3	1		
Level	1	2	3	3	4		

Reduced conical matrix according to the importance of the criteria and in sequence from most important to least important. The first step is writing the expert's answers about the comparison matrix between the criteria and according to the choices in Table 6, the self-interaction matrix is created and then the accessibility matrix is formed, which leads us to the final accessibility matrix. Based on the frequencies, a table is formed to divide the levels and according to the frequencies of the answers to determine the level of each criterion. From the criteria, and from the table, the conical matrix and the reduced conical matrix are designed to illustrate the criteria levels in the form of a hypothetical diagram, after which the final resulting form of the matrix is drawn, which shows the following:

- All criteria affect the final result on patient satisfaction.
- Technological integration may affect the efficiency of the staff and the general efficiency of the service provided by the hospital.
- Staff efficiency and public service efficiency affect each other.
- The efficiency of the staff and the efficiency of public services affect the financial performance of the hospital.

5   Conclusion

The Interpretative Structural Modeling (ISM) system has an important role in developing the overall performance of medical services by identifying the standards that have the greatest impact on performance and thus focusing on these standards more than others and following the theory of (the most important then the most important). Through the data that has been entered, relying on experts working in the medical field, the Interpretative Structural Modeling (ISM) system has been able to determine what are the most important standards, the importance rates, the comparison between these standards, and their final impact on patient satisfaction, which has become an utmost necessity for organizations working in the health field because achieving patient satisfaction. It will help the organization achieve a complete recovery for the patient, and thus the work will continue to continue and develop and achieve the goals desired by the health organization's senior management. This is done by paying attention to technological integration and keeping pace with technological developments in the world, in addition to paying attention to the efficiency of the health organization's cadres, medical, administrative and technical, because this will lead to the efficiency of the health service provided to the patient and thus achieve the desired financial return in addition to achieving patient satisfaction.

References

Al-Refaie, A., Fouad, R.H., Li, M.H., Shurrab, M.: Applying simulation and DEA to improve performance of emergency department in a Jordanian hospital. Simul. Model. Pract. Theory **41**, 59–72 (2014)

Abbas, S., et al.: Antecedents of trustworthiness of social commerce platforms: a case of rural communities using multi group SEM & MCDM methods. Electron. Commer. Res. Appl. **62**, 101322 (2023)

Ahmed, A.D., Salih, M.M., Muhsen, Y.R.: Opinion weight criteria method (OWCM): a new method for weighting criteria with zero inconsistency. IEEE Access (2024)

Al-Enzi, S.H.Z., Abbas, S., Abbood, A.A., Muhsen, Y.R., Al-Hchaimi, A.A.J., Almosawi, Z.: Exploring research trends of metaverse: a bibliometric analysis. In: Al-Emran, M., Ali, J.H., Valeri, M., Alnoor, A., Hussien, Z.A. (eds.) Beyond Reality: Navigating the Power of Metaverse and Its Applications (IMDC-IST 2024). LNNS, vol. 895, pp. 21–34. Springer, Cham (2023). https://doi.org/10.1007/978-3-031-51716-7_2

Ali, J., Hussain, K.N., Alnoor, A., Muhsen, Y.R., Atiyah, A.G.: Benchmarking methodology of banks based on financial sustainability using CRITIC and RAFSI techniques. Decis. Mak. Appl. Manage. Eng. **7**(1), 315–341 (2024)

Alnoor, A., Atiyah, A.G., Abbas, S.: Toward digitalization strategic perspective in the European food industry: non-linear nexuses analysis. Asia Pac. J. Bus. Adm. (2023)

Alnoor, A., Atiyah, A.G., Abbas, S.: Unveiling the determinants of digital strategy from the perspective of entrepreneurial orientation theory: a two-stage SEM-ANN approach. Glob. J. Flex. Syst. Manage. 1–18 (2024)

Alnoor, A., Chew, X., Khaw, K.W., Muhsen, Y.R., Sadaa, A.M.: Benchmarking of circular economy behaviors for Iraqi energy companies based on engagement modes with green technology and environmental, social, and governance rating. Environ. Sci. Pollut. Res. **31**(4), 5762–5783 (2024)

Alnoor, A., et al.: How positive and negative electronic word of mouth (eWOM) affects customers' intention to use social commerce? A dual-stage multi group-SEM and ANN analysis. Int. J. Hum. Comput. Interact. **40**(3), 808–837 (2024)

Atiyah, A.G.: Unveiling the quality perception of productivity from the senses of real-time multisensory social interactions strategies in metaverse. In: Al-Emran, M., Ali, J.H., Valeri, M., Alnoor, A., Hussien, Z.A. (eds.) Beyond Reality: Navigating the Power of Metaverse and Its Applications (IMDC-IST 2024). LNNS, vol. 876, pp. 83–93. Springer, Cham (2023). https://doi.org/10.1007/978-3-031-51300-8_6

Atiyah, A.G., Zaidan, R.A.: Barriers to using social commerce. In: Artificial Neural Networks and Structural Equation Modeling: Marketing and Consumer Research Applications, pp. 115–130. Springer, Singapore (2022). https://doi.org/10.1007/978-981-19-6509-8

Atiyah, A.G., Alhasnawi, M., Almasoodi, M.F.: Understanding metaverse adoption strategy from perspective of social presence and support theories: the moderating role of privacy risks. In: Al-Emran, M., Ali, J.H., Valeri, M., Alnoor, A., Hussien, Z.A. (eds.) Beyond Reality: Navigating the Power of Metaverse and Its Applications (IMDC-IST 2024). LNNS, vol. 876, pp. 144–158. Springer, Cham (2023). https://doi.org/10.1007/978-3-031-51300-8_10

Atiyah, A.G., All, N.D.A., Zaidan, A.S., Bayram, G.E.: Understating the social sustainability of metaverse by integrating adoption properties with users' satisfaction. In: Al-Emran, M., Ali,

J.H., Valeri, M., Alnoor, A., Hussien, Z.A. (eds.) Beyond Reality: Navigating the Power of Metaverse and Its Applications (IMDC-IST 2024). LNNS, vol. 895, pp. 95–107. Springer, Cham (2023). https://doi.org/10.1007/978-3-031-51716-7_7

Atiyah, A.G., Faris, N.N., Rexhepi, G., Qasim, A.J.: Integrating ideal characteristics of Chat-GPT mechanisms into the metaverse: knowledge, transparency, and ethics. In: Al-Emran, M., Ali, J.H., Valeri, M., Alnoor, A., Hussien, Z.A. (eds.) Beyond Reality: Navigating the Power of Metaverse and Its Applications (IMDC-IST 2024). LNNS, vol. 895, pp. 131–141. Springer, Cham (2023). https://doi.org/10.1007/978-3-031-51716-7_9

Baldi, P., Brunak, S.: Bioinformatics: The Machine Learning Approach. MIT Press (2001)

Beam, A.L., Kohane, I.S.: Big Data and Machine Learning in Health Care (2018)

Cheraghalipour, A., Paydar, M.M., Hajiaghaei-Keshteli, M.: Applying a hybrid BWM-VIKOR approach to supplier selection: a case study in the Iranian agricultural implements industry. Int. J. Appl. Decis. Sci. (2018)

Clancey, W.J.: The epistemology of a rule-based expert system: a framework for explanation. Artificial Intelligence (1983)

Esteva, A., Kuprel, B., Novoa, R.A., et al.: Dermatologist-level classification of skin cancer with deep neural networks. Nature (2019)

Ghoushchi, S.J., Bonab, S.R., Ghiaci, A.M., Haseli, G., Tomaskova, H., Hajiaghaei-Keshteli, M.: Landfill site selection for medical waste using an integrated, SWARA-WASPAS framework based on spherical fuzzy set. Sustainability (2021)

Husin, N.A., Abdulsaeed, A.A., Muhsen, Y.R., Zaidan, A.S., Alnoor, A., Al-mawla, Z.R.: Evaluation of metaverse tools based on privacy model using fuzzy MCDM approach. In: Al-Emran, M., Ali, J.H., Valeri, M., Alnoor, A., Hussien, Z.A. (eds.) Beyond Reality: Navigating the Power of Metaverse and Its Applications (IMDC-IST 2024). LNNS, vol. 895, pp. 1–20. Springer, Cham (2023). https://doi.org/10.1007/978-3-031-51716-7_1

Jafarzadeh Ghoushchi, S., Memarpour Ghiaci, A., Rahnamay Bonab, S., Ranjbarzadeh, R.: Barriers to circular economy implementation in designing (2022)

Nazarian-Jashnabadi, J., Bonab, S.R., Haseli, G., Tomaskova, H., Hajiaghaei-Keshteli, M.: A dynamic expert system to increase patient satisfaction with an integrated approach of system dynamics, ISM, and ANP methods. Expert Syst. Appl. **234**, 121010 (2023)

Maidin, A., Sidin, I., Rivai, F., Palutturi, S.: Patient satisfaction based on Bugis philosophy at the Siwa hospital in Wajo district, South Sulawesi. Enfermería Clínica (2020)

McCarthy, J., Minsky, M.L., Rochester, N., Shannon, C.E.: A Proposal for the Dartmouth Summer Research Project on Artificial Intelligence. Dartmouth College (1956)

Muhsen, Y.R., Husin, N.A., Zolkepli, M.B., Manshor, N.: A systematic literature review of fuzzy-weighted zero-inconsistency and fuzzy-decision-by-opinion-score-methods: assessment of the past to inform the future. J. Intell. Fuzzy Syst. **45**(3), 4617–4638 (2023)

Nagdee, N., Sebothoma, B., Madahana, M., Khoza-Shangase, K., Moroe, N.: Simulations as a mode of clinical training in healthcare professions: a scoping review to guide planning in speech-language pathology and audiology during the COVID-19 pandemic and beyond. S. Afr. J. Commun. Disord. **69**(2), 905 (2022)

Markazi-Moghaddam, N., Kazemi, A., Alimoradnori, M.: Using the importance-performance analysis to improve hospital information system attributes based on nurses' perceptions. Inform. Med. Unlocked **17**, 100251 (2019)

Obermeyer, Z., Emanuel, E.J.: Predicting the future — big data, machine learning, and clinical medicine. N. Engl. J. Med. (2016)

Dieckmann, P., Torgeirsen, K., Qvindesland, S.A., Thomas, L., Bushell, V., Langli Ersdal, H.: The use of simulation to prepare and improve responses to infectious disease outbreaks like COVID-19: practical tips and resources from Norway, Denmark, and the UK. Adv. Simul. **5**, 1–10 (2020)

Phuoc Pham, P., Zhang, H., Gao, W., Zhu, X.: Determinants and performance outcomes of artificial intelligence adoption: evidence from US hospitals. J. Bus. Res. **172**, 114402 (2024)

Ranjbarzadeh, R., Caputo, A., Tirkolaee, E.B., Ghoushchi, S.J., Bendechache, M.: Brain tumor segmentation of MRI images: a comprehensive review on the application of artificial intelligence tools. Comput. Biol. Med. **152**, 106405 (2023)

Saaty, T.L.: Theory and applications of the analytic network process: Decision making with benefits, opportunities, costs, and risks. RWS. Publications (2005)

Behdioğlu, S., Duran, C., Akti, Ü., Boz, D.: Conscientious intelligence: DPÜ faculty of economics and administrative sciences business administration students and students of Islamic science faculty example. In: International Applied Social Sciences Congress (IASOS)

Shariat, S., Mazloumi, S., Khodabakhshi, A., Vahdani, M., Roudposhti, R.: Expert system and its application in management. Singaporean J. Bus. Econ. Manage. Stud. (2013)

Wang, S., Li, L., Liu, C., Huang, L., Chuang, Y.C., Jin, Y.: Applying a multi-criteria decision-making approach to identify key satisfaction gaps in hospital nurses' work environment. Heliyon **9**(3) (2023)

Shortliffe, E.H., Buchanan, B.G.: A model of inexact reasoning in medicine. Mathematical Biosciences (1975)

Silver, D., Huang, A., Maddison, C.J., et al.: Mastering the game of go with deep neural networks and tree search (2016)

Itchhaporia, D.: The evolution of the quintuple aim: health equity, health outcomes, and the economy. J. Am. Coll. Cardiol. **78**(22), 2262–2264 (2021)

Topol, E.J.: High-performance medicine: the convergence of human and artificial intelligence. Nature Medicine (2019)

Zaidan, A.S., Alshammary, K.M., Khaw, K.W., Yousif, M., Chew, X.: Investigating behavior of using metaverse by integrating UTAUT2 and self-efficacy. In: Al-Emran, M., Ali, J.H., Valeri, M., Alnoor, A., Hussien, Z.A. (eds.) Beyond Reality: Navigating the Power of Metaverse and Its Applications (IMDC-IST 2024). LNNS, vol. 895, pp. 81–94. Springer, Cham (2023). https://doi.org/10.1007/978-3-031-51716-7_6

The Role of Artificial Intelligence in Promoting the Environmental, Social and Governance (ESG) Practices

Mushtaq Taleb and Hussein Jawad Kadhum[✉]

Department of Banking and Finance Sciences, College of Administration and Economics,
University of Basrah, Basrah, Iraq
`hussein.kadum@uobasrah.edu.iq`

Abstract. Environmental, social, and governance (ESG) issues have had a prominent position in recent years, especially with the increasing use of artificial intelligence (AI) in various aspects of life. It is crucial to comprehend how AI can help build a sustainable future by enhancing ESG trends. Through reviewing the mechanism of connection and impact of AI on ESG practices in the current study, it is evident that this integration works to enhance sustainable growth and increases social well-being while, at the same time, resulting in negative environmental and social impacts besides the presence of financial and institutional challenges that accompany the merging process. However, data and reports on the positive impact of applying AI systems in enhancing ESG practices outperform its negative effects. To ensure that the sustainable impact of AI continues to be enhanced, policymakers will need to balance new liability frameworks that stabilize the benefits of encouraging innovation with the moral imperative to protect society from technical risks to ensure that the risks of excessive liability do not hinder innovation.

Keywords: Environmental · social and governance practices (ESG) · artificial intelligence (AI) · environmental impact · social impact · governance

1 Introduction

Environmental, Social, and Governance (ESG) practices have received increasing attention as a result of the environmental, social, and institutional side effects of technological development, modernity, and resource exploitation in response to increasing demand as well as the pressures placed by environmental activists on companies to integrate ESG criteria into their daily operations. The latter is an integral part of these processes and a basic condition for integration into the global economy and competition. Here, research and commercial developments began to show the necessity of achieving this sustainable change. However, this integration of ESG standards has become a major challenge, as ESG information often includes huge and complex data that traditional analysis methods are not able to analyze efficiently. Besides, the data provided by rating agencies and based on human analysis regarding ESG standards has led to these ratings being loosely linked to each other as a result of the methodological choices adopted by each agency, which are not reviewed frequently and are based on linking them with the company's stronger financial performance. This leads to large contradictions between service providers, and these contradictions increase with the rise in information available to the public, causing a widening of the discrepancy gap in the classifications of these companies in terms of ESG indicators. It now requires more than just focusing on technology, but rather utilizing digital analytics through the development and implementation of artificial intelligence (AI) and machine learning models to solve ESG problems in the work of companies. Hence, Environmental, Social, and Governance Intelligence (ESGAI) emerged, which revolves around the specialized application of AI in processing and analyzing ESG data.

AI includes advanced algorithms capable of not only processing but also generating predictive insights from ESG data, which is crucial in making informed and ethical decisions in sustainable finance. Developments in the field of AI and machine learning technologies have led to the creation of a new type of ESG data providers who collect, analyze and extract large amounts of unstructured data from various Internet sources without necessarily relying on information provided by companies. This is done by using ESG technologies such as Natural Language Processing (NLP), which allows machines to understand human language and analyze unstructured data (such as articles published on the Internet, social media, and financial reports). By using NLP algorithms, these companies can identify ESG risks and evaluate the sustainability performance of potential investments, in addition to developing a large number of text analysis programs, including RepRisk and Tru Value Labs, which make it possible to accurately measure disagreements related to companies on various issues of environmental policies, working conditions, child labor, corruption, etc. These technologies have the advantage of being refined frequently and incorporating real-time company information. Although integrating AI into ESG data analysis not only promotes economic growth and social well-being and thus helps achieve global sustainability goals, training and deploying AI systems can require vast amounts of computational resources, each with their own environmental impacts, adding to that the enormous institutional and financial challenges. The main challenge facing most AI applications is attributed to their lack to generalize different contexts, as they may encounter situations referred to as corner cases where the system is not trained to deal with (Taeihagh, 2021; Alnoor et al., 2023). This requires that software designers are subjected to responsibility for designing and manufacturing defects. It is widely recognized that the risks of excessive liability can hinder innovation and that there must, therefore, be a balancing act between new liability frameworks that balance the benefits of encouraging innovation with the moral imperative to protect society from technical risks (Atiyah et al., 2023; Abbas et al., 2023).

Therefore, this study is significant because it reviews and analyzes the impact of AI technologies on ESG, the underlying mechanism of this impact, its positive and negative consequences, and the main challenges associated with this integration.

2   Artificial Intelligence (AI)

AI is a branch of computer science that involves dealing with all aspects of simulating cognitive functions to solve real-world problems and build systems that have the ability to learn and think like humans (Holzinger et al., 2019). Therefore, it is often called machine intelligence because it includes the intersection of cognitive science and computer science and, at the same time, aims to develop software that can automatically learn from previous data to gain knowledge from experience and gradually improve learning behavior to make predictions about new data (Atiyah, 2023; Alnoor et al., 2023). It is defined as "a field of science and engineering concerned with the computational understanding of what is commonly called intelligent behavior. Intelligent behavior in computer science is the ability to achieve human-level performance on cognitive tasks (Holmes et al., 2004; Ali et al., 2024). AI aims to expand and increase humanity's ability and efficiency in reshaping nature and governing society through intelligent machines, with the ultimate goal of achieving a society in which people and machines coexist harmoniously together (Liu et al., 2018; Atiyah and Zaidan, 2022).

Historically, Ertel refers to the stages of development of AI. In the 1950s, Newell and Simon presented the first automated theorem, showing that by using computers, which actually only worked with numbers, one could also manipulate symbols. At the same time, McCarthy introduced a programming language, using LISP, specifically created to process symbolic structures. Both systems were presented in 1956 at the historic Dartmouth Conference, which marked the birth of the AI RoboCup (Ertel, 2018). Accordingly, the historical development of AI can be shown in Table 1.

Table 1. The historical development of AI.

Year	Development
1956	"AI" as a term was coined at the Dartmouth Workshop
1966	The ELIZA chatbot was created by Joseph Weizenbaum
1986	The self-driving robotic truck was created by Ernst Dickmanns
1994	Chinook checkers win championship against human
1995	ALICE chatbot was developed by Richard Wallace
1997	IBM's Deep Blue program beats world chess champion Garry Kasparov
2000	The US Food and Drug Administration has approved the first robotic surgeon to perform operations laparoscopically
2002	The automatic vacuum cleaner was introduced by the Roomba iRobot Foundation
2011	IBM's Watson computer beats humans at Jeopardy game
2011	Apple launches SIRI (Speech Interpretation and Recognition Interface) on iPhone
2014	Amazon is launching its own speech recognition assistant, Alexa
2018	Facebook uses AI to help filter explicit content
2019	AI improves lung cancer detection, outperforms radiologists
2020	Waymo driverless taxis start operating in southwest Phoenix suburbs

Source: Prepared by the researcher

AI technologies are increasingly being adopted in finance, such as asset management, algorithmic trading, underwriting credit, or blockchain-based finance. This is made possible by the abundance of available data and affordable computing power. Machine learning models use data to learn and improve predictability and performance automatically through experience and data without being programmed to do so by humans (Belhaj & Hachaichi, 2002; Alnoor et al., 2024). The following are the most significant areas of finance supported by AI.

1. *Algorithmic trading*: Algorithmic trading uses a computer program that follows a specific set of instructions (algorithms) to conduct a trading process. These instructions depend on timing, price, quantity, or any mathematical model, regardless of the profit opportunities for the trader. Automated trading makes markets more liquid and trading more regular by eliminating the influence of human emotions on trading activities (Napate & Thakur, 2020; Muhsen et al., 2023). One way to facilitate the division of large orders in the market and distribute their execution over time is to reduce their implicit transaction costs (Ahmed et al., 2024). The decision-making process takes place without any human intervention, as the order trading parameters are determined with regard to timing, pricing, and quantity determination, in addition to managing the orders after they are submitted (Mandes, 2016; Husin et al., 2023).

2. *Risk management*: Through data processing, efficiency is improved by reducing costs via automating daily assistance and guidance in risk management and forecasting processes. This gives people more awareness of new exposures, increasing preventive risk advice, and faster response time in critical situations, in addition to business decisions related to making better decisions through greater predictive and risk visibility and also to senior management (Naim, 2022; Alnoor et al., 2024). AI and machine learning will provide flexibility and efficiency that organizations may not have in the field of risk assessment, as shown in Fig. 1 (Žigienė et al., 2019).

3. *Fraud detection*: AI-based technologies have emerged as a crucial mechanism for measuring and monitoring the increasing cases of economic crimes and financial fraud. AI can be used to analyze large numbers of transactions in order to detect anomalies, financial crime trends, unexpected developments, etc. It can play an effective role in fraud detection and prevention in a timely manner (Mohanty & Mishra, 2023).

In today's competitive financial market, detecting fraudulent financial activities is one of the key success requirements for any financial institution. Financial crimes include different types of fraud, including fraud in checks, credit cards, healthcare cards, point of sale, tax evasion, identity theft, and cyber-attacks, the most important of which is money laundering. The smart methods used in the field of money laundering can be explained in Fig. 2 (Rouhollahi, 2021).

4. *Customer service:* It is essential to focus on customers and come up with better service techniques. AI and big data analytics have enabled data-driven systems to serve customers in a better way. This made it easy to get a range of features, thus enhancing customer engagement. According to McKinsey's Global AI Survey, the implementation of virtual assistants and conversational interfaces in front offices

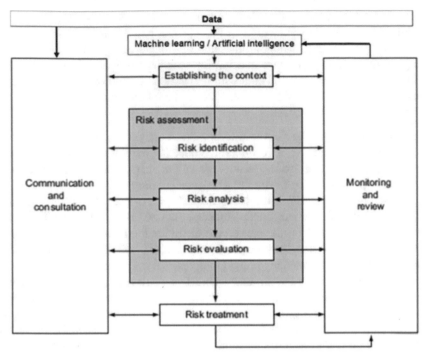

Fig. 1. Machine learning or AI-based content for risk management and risk assessment. Source: Žigienė, G., Rybakovas, E., & Alzbutas, R. (2019). Artificial intelligence based commercial risk management framework for SMEs. Sustainability, 11(16), 4501.

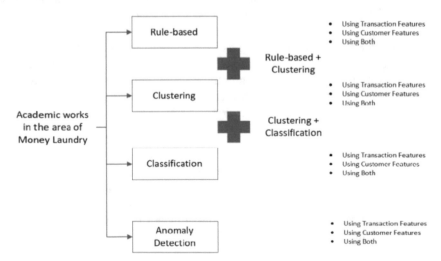

Fig. 2. Smart methods used in combating money laundering. Source: Rouhollahi, Z. (2021). Towards artificial intelligence enabled financial crime detection. arXiv preprint arXiv:2105.10866.

accounts for about 32% of all AI technologies (Bhattacharya & Sinha, 2022; Atiyah et al., 2023).

5. *Credit evaluation*: Recent studies have revealed that emerging AI techniques, such as decision tree (DT), support vector machine (SVM), genetic algorithm (GA) and artificial neural networks (ANN), are useful for statistical models and techniques for improving credit risk evaluation, as these methods automatically extract knowledge from training samples. AI methods are superior to statistical methods in dealing with the problems of assessing corporate credit risks, especially with regard to classifying non-linear patterns. The application of the above techniques has been investigated in several works (Ghodselahi & Amirmadhi, 2011).

On the other hand, there is ambiguity surrounding some AI methods. Some algorithms are viewed as "black boxes" that link predictions about the target variable to a set of predictions, without revealing the origin and lineage of these predictions, which raises serious ethical and legal concerns. This is a matter of concern from the perspective of financial regulation, given that these models are used to guide decisions that affect the lives of individuals or companies, most notably the granting of credit (Sadok, 2022).

6. *Portfolio Management*: AI techniques, especially algorithms determining the optimal allocation of diverse assets, are a major area of research. Among these algorithms are the most famous artificial neural networks (ANN). Because the investor's decision-making process is unstructured and the market environment is uncertain, the financial market is a rather complex nonlinear dynamic system, and ANN can correctly quantify the nonlinear property, which is definitely a suitable way to manage portfolios. Artificial neural networks have been successfully applied in different markets, such as the mortgage market, the stock market, the real estate market, and accounting, which have been proven to have higher performance than traditional models (Zhang & Chen, 2017). Creating a model for portfolio management using artificial neural networks (ANN) depends on three main components: the algorithm for selecting the assets (stocks) that will form the investment portfolio, the efficient Markowitz limit to find the optimal allocation for each asset on the basis of risk adjustment, and the NN model for predicting asset returns. To provide a better estimate of expected returns versus the historical arithmetic average, see Fig. 3 (Abdelazim & Wahba, 2006).

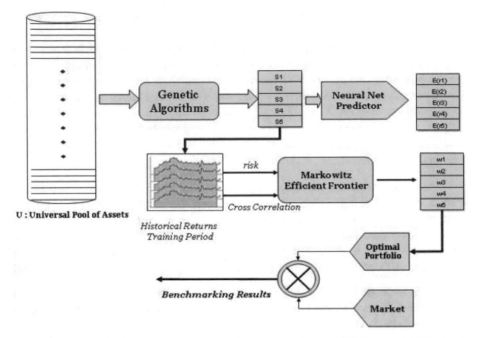

Fig. 3. Proposed portfolio management model using neural networks. Source: Abdelazim, H. Y., & Wahba, K. (2006). An artificial intelligence approach to portfolio selection and management. International Journal of Financial Services Management, 1(2–3), 243–254.

3   Environmental, Social and Governance (ESG) Practices

The concept of ESG is seen as the process of taking into account ESG factors when making investment and financing decisions, which leads to increased long-term investments in sustainable economic activities and projects (Boffo & Patalano, 2020). It is also defined as the integration of ESG principles into business decisions, economic development, and investment strategies (IMF, 2019). The previous definitions of sustainable financing include implications, the most important of which is achieving long-term value, which constitutes an important motivation for exporters and investors alike. Many investors are looking for better financial value in the long term rather than maximizing shareholder profits, as was the prevailing thought in the past (Powaski et al., 2021; Atiyah et al., 2023).

Nowadays, in addition to long-term financial returns, investors demand trust, transparency, ethics, respect for the environment, and greater social responsibility. Hence, ESG is adopted in evaluating the company's performance, profitability, and returns when making financial decisions. The concept of ESG also includes environmental, social, and administrative issues that affect the behavior of companies in their investment decisions. Environmental issues refer to awareness of climate change, population growth, and its harmful impact on the natural environment. Social issues include corporate social responsibility (CSR), which represents a response to the impact of its activities on the communities in which it operates. While governance issues include executive benefits, compensation, bribery, corruption, shareholder rights, work ethics, board diversity, board structure, independent directors, risk management, whistleblowing schemes, dialogue with stakeholders, and disclosure (Armstrong, 2022).

The term ESG is not new, as it has witnessed remarkable historical development that reflects the increasing interest in ESG standards. It was reflected in the shift in public awareness and the changes that occurred in policies and legislation in this regard. In the 1970s, following a marked increase in concern about the social and environmental impacts of business activities, the beginning of corporate social responsibility reporting initiatives was seen. These initiatives focused on assessing the ethical, environmental, philanthropic, and economic impact of companies. Initially, these reports focused on immediate and attractive actions, such as participating in charitable work, volunteering, and contributing to strengthening society. These reports were initially separate from companies' financial reports, which highlights the lack of understanding at that time of the interrelation between economic performance and external factors related to environmental, social, and management standards (Kostić & Hujdur, 2023).

The idea is gradually evolving with the UN Charter based on the call of the former Secretary-General of the United Nations, Kofi Annan, which he launched at the World Economic Forum in 1999, when business leaders were invited to cooperate with the organization to create a "global agreement" based on common values and principles. With the aim of giving the global market a humane dimension. Today, this agreement plays a crucial role in strengthening trade cooperation with the United Nations and constitutes a call for companies to align their strategies and operations with ten global principles centered around human rights, labor, environment, and anti-corruption (Kingo, 2019).

In 2004, in implementation of the provisions of the Global Compact, a joint initiative was launched by financial institutions and was invited by the Secretary-General of the United Nations, Kofi Annan, to develop guidelines and provide recommendations on how to better integrate environmental, social, and corporate governance issues into asset management and brokerage services in Securities and Related Research (UN, 2004). In 2006, another initiative was presented under the title "Principles of Responsible Investment" by a number of investors in partnership with the United Nations Global Compact and the Funding Initiative of the United Nations Environment Program, which produced six principles for socially responsible investment (www.unpri.org). In 2007, the Impact Investment Initiative was launched when the Rockefeller Foundation invited finance, philanthropy, and development leaders to its Bellagio Center in Italy to discuss the need for ways to build a global industry that strives for investments with a positive social and environmental impact (Jackson, 2012). In 2015, two initiatives were put forward: the Paris Climate Conference 2015, which emphasized reducing carbon emissions to levels less than 2 degrees Celsius above pre-industrial levels (UN, 2015). The second initiative was represented by the historic summit of the United Nations General Assembly in September 2015, when the 2030 Sustainable Development Plan was adopted by the United Nations, which included 17 goals with the participation of UN member states. These goals varied between environmental, social, and economic issues.

The environmental aspect of these goals is in line with the adoptions of the Paris Climate Conference in terms of reducing carbon dioxide emissions, waste management, energy consumption, resource efficiency, water management, and pollution emissions (Radu et al., 2023). The 1999 United Nations Charter, the 2004 United Nations initiative to develop a framework for ESG, the 2006 Principles for Responsible Investment, the provisions of the Human Rights and Business Council Agreements of 2008 and 2011, the International Finance Corporation standards of 2008 and 2012, and the Paris Climate Conference of 2015, are considered connected paths that have contributed effectively in forming the foundations of ESG as a path to achieving sustainable development goals. Li et al. (2021) and Guevara & Dib (2022) agree that ESG principles are generally divided into three areas as shown in Table 2.

Table 2. ESG Principles

Environmental Area	Social Area	Governance
Reducing and offsetting greenhouse gas emissions	Gender equality	Combating tax evasion
Illegal deforestation	Respect human rights	Clear criteria on wages and professional plans
Responsible water consumption	compliance with labor laws	Anti-corruption policy development
waste management	Cooperation for the development of the communities in which they are set	Enhancing transparency and ethics
Energy Efficiency	Combating child labor and slavery-like labor	Independence and diversity in the selection of board members
Promote and conserve biodiversity	Managing Privacy and Data Protection	Maintain reliable audits

Source: Prepared by the researchers

As for ESG goals, many researchers agree that they should be viewed through sustainable development goals. Sadiq et al. (2023), Ziolo et al. (2020), Khaled et al. (2021), and Chien (2023) believe that ESG practices by companies ensure their contribution to achieving sustainable development goals as ultimate goals.

While ESG emerges from the concept of sustainability, the focus has shifted from the impact of activities and businesses on society and the environment to the impact on returns and risks for investors so that the goal of integrating ESG with financial management decisions is to align environmental and social benefits and impacts with financial returns (Delgado-Ceballos et al., 2023).

4 ESG Strategies

There are a number of strategies that organizations adopt in ESG practices, which can be summarized as follows (Boffo & Patalano, 2020).

1. Negative Screening
2. Positive Screening (best-in-class)
3. ESG integration

1. Negative Screening: This strategy is based on excluding some industries from the field of investment. These industries could be, for example, alcohol, tobacco, cluster munitions, atomic weapons, fossil fuels, etc. (Ögren & Forslund, 2017). This strategy is described as first-generation, as it was adopted in the socially responsible investment policy as a form of sustainable financing, which depends on excluding assets that are not in line with the SRI policy (Torricelli & Bertelli, 2022).
2. Positive Screening: This strategy Identifies companies that make positive contributions to society and the environment, and may include screening companies based on specific ESG criteria. For example, investing in companies that have strong performance in terms of ESG aspects compared to their peers (Del Vitto et al., 2023). In contrast to negative screening, which is primarily driven by ESG risk mitigation, positive screening is viewed as a value creation opportunity. It is most prominently followed among venture capital fund managers who consider ESG criteria due to their importance to investment performance or as part of their company's investment policy (Botsari & Lang, 2020).
3. ESG Integration: This strategy refers to the explicit inclusion by asset managers of ESG risks and opportunities into traditional financial analysis and investment decisions based on a systematic process and appropriate research sources (Folqué et al., 2021) and (Van Duuren et al., 2016). In the broad sense, this type of integration strategy is seen as a strategy that covers environmental, social, and administrative factors in addition to financial factors in the prevailing analysis of investments. The integration process focuses on the potential impact of ESG issues on a company's financial statements (positive and negative), which in turn may influence the investment decision (Schramade, 2016).

5 Literature Review: The Relationship Between AI and ESG Practices

Studies that have investigated the nature of the relationship between ESG and AI are very limited. Brusseau's study dealt with designing an ESG investment model for companies that rely on AI. The researcher believes that traditional ESG indicators are insufficient to evaluate the performance of this type of company. The model included three elements: personal freedom, social well-being, and the artistic merit of the scale. The goal was creating a model for human investment in AI-intensive companies that is intellectually robust, manageable by analysts, useful to portfolio managers, and credible to investors (Brusseau, 2023). Guo et al. (2020) proposed a deep learning framework, which is one of the branches of AI and includes natural language processing, known as risk2esg, to predict stock volatility using ESG news. de la Barcena Grau (2021) discussed the application of machine learning to generate social impact returns by building a proposed model, and the model proved its ability to drive social impact returns at the level of small and medium enterprises, large companies, or the country level. Maree & Omlin proposed an enhanced learning model to manage and improve the performance of the ESG portfolio (Maree & Omlin, 2022). Meier and Danzinger used a genetic algorithm to choose a portfolio that complies with ESG standards based on sustainable development goals (Meier & Danzinger, 2022). In contrast, other studies have dealt with mapping ESG trends historically using neural language models, such as the study of Raman et al. (2020).

Other studies, like Huang et al. (2023), have dealt with developing a linguistic model to extract information from ESG discussions of analyst reports to identify positive and negative sentiments toward ESG issues. In another study, a methodology was proposed to evaluate the performance of companies in the field of ESG that combines AHP and TOPSIS techniques in a neutrosophic environment, and the proposal was tested through a real-life case study of leading companies in the oil and gas industry (Reig-Mullor et al., 2022). Bussmann et al.'s study examined a proposed interpretable machine learning model to evaluate the impact of ESG factors on a company's ex-ante cost of capital, a measure that reflects investors' perceptions of a company's risk, by adopting the XGBoost machine learning model (Bussmann et al., 2023).

In their study, De Lucia et al. (2020) investigated whether good ESG performance is associated with better financial performance using a set of deep learning techniques. Khan and Ahmad proposed integrating ESG, and machine learning considerations into decentralized finance through a new application program, DCarbonX, to address a number of unresolved issues in the ESG sector, such as accountability, greenwashing, traceability, impact assessment, and carbon credit trading (Khan & Ahmad, 2022). In a recent study conducted by Adeoye et al. (2024), the integration of AI technologies is addressed in ESG investment strategies and demonstrates its profound impact on decision-making processes and financial results by taking advantage of advanced data analytics and machine learning algorithms. It concluded that AI works to enable investors into analyzing comprehensive data sets related to ESG, extracting actionable insights, and identifying investment opportunities aligned with sustainability goals.

In their study, Minkkinen et al. (2024) addressed ESG analyzes as tools for ethics-based AI auditing within two research questions: (1) How is corporate responsible use of AI included in ESG investment analyses, and (2) What connections can be found between responsible AI principles and ESG rating criteria. The results indicate that AI is still a relatively unknown topic for investors, and that taking responsible use of AI into account in ESG analyses is not a well-established practice.

As mentioned above, regarding the relationship between ESG and AI, it is revealed that there is a research gap on the nature of the relationship between ESG and AI and the extent of the degree of connection between them. Most of these studies consisted of building proposed models, developing a proposed framework, or a proposed methodology based on one of the branches of AI, such as machine learning, deep learning, natural language, and genetic algorithms, which work to analyze ESG data, predict its paths, or evaluate the performance of companies that adhere to ESG standards and how to benefit from AI to accelerate the pace of compliance with these standards and consolidate them among economic units, individuals and institutions in line with the goals of sustainable development.

6 Stages of Using AI in ESG

AI-powered ESG tools typically follow three steps to produce the desired and useful outputs: harvesting, organizing, and analysis (TSE et al., 2023).

1. Harvesting: It starts with using AI to search and extract company data from a range of sources, including news coverage, social media messages and posts, expert analyses, third-party assessments, and ESG reports.
2. Organizing: It includes examining and entering data. AI can collect, organize, and process data that follows environmental ESG frameworks established by users and investors, which can vary significantly since different investors have different goals, investment philosophies, rating standards, and appetites for risk. Accordingly, investors put themselves in a position to continuously produce and update their own ratings.
3. Analysis: The final stage is about discovering and gathering valuable insights from the structured data set. This includes the development of various natural language processing (NLP) techniques, such as those related to taxonomies. It also uses analytics that capture emotional, contextual, and semantic factors embedded in the collected data.

7 AI to Enhance ESG

As ESG issues have gained prominence over recent years with the increasing use of AI in various aspects of society, it is crucial to understand how AI can help build a sustainable future by enhancing ESG practices and how these practices relate to AI systems. Add to this the challenges that accompany the process of integrating AI technologies with ESG according to the following aspects (Pelesaraei, et.al, (2018), https://assets.ey.com, Lim (2024), Chen, et.al (2023), OECD (2022).

7.1 Assessing Environmental Impact

Assessing environmental pollutants is a productive process in terms of providing solutions to reduce energy consumption, pollutant emissions, and other areas of the environment. Many researchers focus on this aspect because it has a direct impact on human health (Kaab et al., 2019). An example of this is the adoption of data from the agricultural products life cycle approach (raw materials, energy, pesticides, land area, etc.) as input to artificial neural networks (one of the Al branches) for the purpose of assessing and predicting environmental impact, as shown in Fig. 4 (Nabavi-Pelesaraei et al., 2018).

Despite the positive environmental impact that AI comes from the integration with environmental practices, policy makers must make ensure that AI is part of the solution to achieving global sustainability goals. As datasets and models become more complex, the energy required to train and run AI models becomes enormous. This increase in energy use directly affects greenhouse gas emissions, exacerbating climate change. According to OpenAI researchers, since 2012, the amount of computing power required to train advanced AI models has doubled every 3.4 months. By 2040, emissions from the ICT industry as a whole are expected to reach 14% of global emissions. The majority of these emissions come from ICT infrastructure, especially data centers and telecommunications networks.

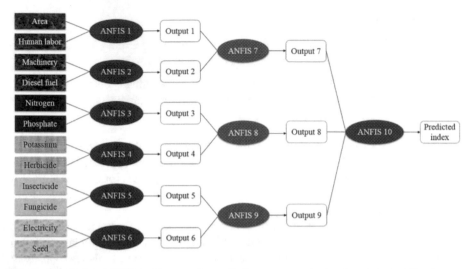

Fig. 4. A three-level ANN architecture for predicting energy output and environment for rice crop production. Source: Nabavi-Pelesaraei, A., Rafiee, S., Mohtasebi, S. S., Hosseinzadeh-Bandbafha, H., & Chau, K. W. (2018). Integration of artificial intelligence methods and life cycle assessment to predict energy output and environmental impacts of paddy production. Science of the total environment, 631, 1279–1294.

7.2   Improving Resource Utilization

AI has the ability to play an appropriate role in transforming the industry into sustainable production through data analysis, leading to optimal production planning, thus reducing depletion, waste management, and rationalizing water and energy consumption (Waltersmann et al., 2021). Efficient use of resources is based on preserving biological diversity, which is the diversity of all life forms on Earth; its depletion, though, means a threat to the basics of human life in the long term. In this sense, AI models have been built to improve the protection of biodiversity in a complex and rapidly developing world by developing tools for systematic conservation planning based on monitoring biological species within a limited financial budget, experimental analyses, and a simulation process of reality based on the reinforcement learning (RL) mechanism based on the Conservation Area Prioritization Through Artificial INtelligence model, which is symbolized by (CAPTAIN), as shown in Fig. 5 (Silvestro et al., 2022).

Through AI algorithms, Biodiversity management helps preserve and manage ecosystems. For example, machine learning algorithms can help identify and track endangered species by analyzing camera trap images or audio recordings. AI can also help predict the impact of human activities on biodiversity and support decision-making processes. For example, robots with AI capabilities can be used to sort recyclable materials from waste. Rainforest Connection, a Bay Area nonprofit, uses AI tools like Google's TensorFlow in conservation efforts around the world. Its platform can detect illegal logging in vulnerable forest areas by analyzing acoustic sensor data. Other applications include using satellite images to predict the paths and behavior of illegal fishing vessels.

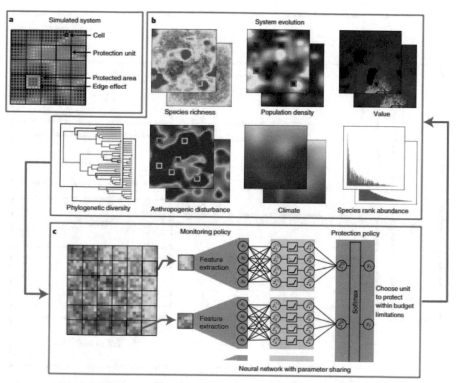

Fig. 5. CAPTAIN reinforcement learning framework. 2022 Source: Silvestro, D., Goria, S., Sterner, T., & Antonelli, A. (). Improving biodiversity protection through artificial intelligence. Nature sustainability, 5(5), 415–424.

7.3 Managing Supply Chain

AI has shown significant potential for improving human decision-making processes and subsequent productivity in various business endeavors due to its ability to recognize business patterns, learn business phenomena, search for information, and analyze data intelligently (Min, 2010). AI is one of the vital motivations for improving the efficiency of supply chain management at all stages through its ability to store data and analyze operations (Sharma et al., 2022). Aspects of the impact of AI within this field can be demonstrated through a dual point of view that includes environmental scanning and the comprehensive supply chain, as shown in Fig. 6 (Modgil et al., 2022).

7.4 Improving the Efficiency of Resource Management

AI algorithms can optimize the use of resources, such as energy, water, and materials, to reduce waste and environmental impact. For example, AI can optimize energy consumption in buildings by analyzing data on occupancy patterns, weather conditions, and energy use to automatically adjust heating, cooling, and lighting systems. AI-powered models can also optimize supply chain logistics to reduce transportation emissions and

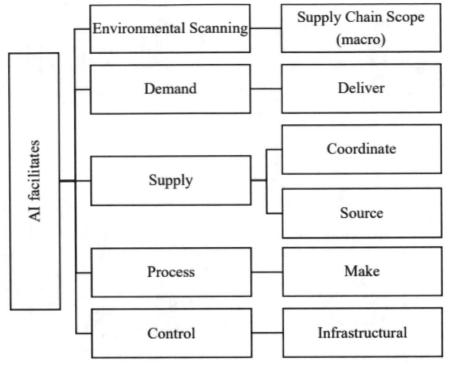

Fig. 6. The impact of AI on various dimensions of supply chains. Source: Modgil, S., Singh, R. K., & Hannibal, C. (2022). Artificial intelligence for supply chain resilience: learning from Covid-19. The International Journal of Logistics Management, 33(4), 1246–1268.

reduce inefficiencies. Some companies, such as Google and Huawei, have already implemented AI solutions to control energy consumption in their data centers. Google reduced its energy consumption by 40% by using AI to analyze the times of day when people perform energy-consuming searches and improved the cooling of its data centers. Huawei improved the energy efficiency of its data centers by using AI to identify and process factors that contribute to increased energy consumption, in addition to predicting the future energy efficiency of its data centers. Microsoft has partnered with Vattenfall to develop a smart grid management solution that optimizes renewable energy production based on demand.

7.5 Climate Change Modeling and Forecasting

AI can help predict extreme weather events, assess the effects of climate change on different regions, and develop strategies for climate adaptation and mitigation. This is done by analyzing large climate data sets and simulating complex climate models to improve our understanding of climate change dynamics. AI-powered models can also help improve renewable energy systems, such as solar and wind farms, to enhance their efficiency and integrate with existing energy grids.

7.6 Assessing the Social Impact of AI

The capabilities of AI in social impact are particularly relevant to four large areas (Perrault et al., 2020), such as health, hunger, education, security, justice, equality and inclusion, where the potential frequency of usage is high, and a large number of the target population is typically affected. In health, for example, if they become sufficiently accessible, AI-enabled wearables could help more than 400 million people with the disease around the world, which can actually detect potential early signs of diabetes through the heart rate sensor data with an accuracy of up to 85%. In the same context, researchers at the Universities of Heidelberg and Stanford created an AI system to detect diseases, using visual diagnosis of natural images, such as images of skin lesions, to determine whether they are cancerous, the system has outperformed professional dermatologists. Other use cases include combining different types of alternative data sources such as geospatial data, social media data, communications data, online search data, and vaccination data to help predict virus and disease transmission patterns (Chui et al., 2020).

In education, more than 1.5 billion students can benefit from the application of adaptive learning technology, which tailors' content to students based on their abilities. In the field of security and justice, AI technologies include preventing harm - whether from crime or other physical dangers - as well as tracking criminals and reducing bias in police forces. This field focuses on security, policing and criminal justice issues as a unique category adjacent to public sector management. An example of this is using AI to create solutions that help firefighters determine safe paths in case of burnings using data from IoT devices. On the equality and inclusion side, AI technologies include addressing the challenges of equality, inclusion, and self-determination, such as reducing or eliminating bias based on race, sexual orientation, religion, citizenship, and disability. Based on the work of Affectiva, which spun out of the Media Lab at MIT, and Autism Glass, a research project at Stanford University, one usage case involves using AI to automate emotion recognition and provide social cues to help individuals along the autism spectrum interact in social environments. Another example is creating an alternative identity verification system for individuals who do not have traditional forms of identification, such as driver's licenses. Other elements included in AI can be added related to economic empowerment, infrastructure management, public and social sector management, and crisis response. However, expanding the use of AI for social good will require many challenges in terms of access to data and talent capable of working in this field. In many cases, sensitive or monetizable data with societal applications is owned by private sector or is only available in commercial contexts where it has commercial value and must be purchased, i.e., cannot be easily accessed by social or non-governmental organizations.

In other cases, bureaucratic inertia keeps data that could be used to enable solutions locked. Therefore, social impact work requires extra effort and additional considerations when evaluating its contributions. This is reflected in the Association for the Advancement of AI's 2019 conference and the 2020 AI Social Impact Track call for papers, which states three main aspects: First: AI for social impact requires greater effort than AI that is entirely focused on improving algorithms. Collecting data can be costly and time consuming. Second, problem modeling may require significant collaboration with domain experts. Third, assessing social impact may require time-consuming and complex field studies. AI for social impact researchers invests their resources differently to make contributions to problems of significant social importance (Chui et al., 2020).

Another set of concerns stems from the effects of AI on the social side of the increasingly sophisticated technology behind the so-called "Big Data" (Perrault et al., 2020), which is defined as "extremely large data sets that can be computationally analyzed to reveal patterns, trends and correlations, especially concerning human behavior and interactions". These data sets enable companies to identify consumer purchasing patterns and target them with their products. This big data works in close alliance with consumer capitalism and is often experienced as "someone is watching me"; however, this is only true metaphorically, as this information is generated by an automated set of 0 and 1 algorithms. In the context of the social impact, a set of questions are raised that are necessary to answer in order to avoid the side effects of AI on the social context, such as: Is meaning constructed in the digital worlds that we deal with daily? How do we balance our privacy needs with our desire for the convenience and creative opportunities that digital technology enables and enhances? What strategies should governments, media, educational institutions, churches and other agencies develop to maintain values and behaviors that contribute to building healthy, cohesive and compassionate societies? Or should these agencies remain outside the process, leaving such choices entirely to the individual?

7.7 Evaluating the Impact of AI on Governance

AI includes various tools and techniques that provide data analysis, predictions, data consistency, quantify uncertainty, anticipate users' data needs and suggest courses of action, all the way to supporting the decision-making process and thus supporting companies by assuming responsibility at the highest and lowest levels (Kaya, 2022). Regarding decision-making by a company's board of directors, there are three ways to introduce AI into the boardroom: assistance, augmentation, and automation (Baburaj, 2021). In general, AI in corporate governance systems is seen as a concept that describes how information is planned, collected, created, organized, used, controlled, disseminated and removed, as well as managed through analysis, modeling, ensuring quality characteristics and protecting the information in a way that ensures its use by interested parties, as shown in the Fig. 7 (Turluev & Hadjieva, 2021).

In the same context, the OECD expert team on AI has developed a framework for AI called the "Artificial Intelligence Lifecycle Framework" within the field of corporate governance, which includes four stages: The first stage relates to research and design of the AI system. The second stage is data collection and documentation, while the third stage establishes the system in the production stage. The fourth stage works to operate and monitor the AI system during deployment and evaluate its outputs and effects based on the designers' initial intentions and metrics. This can be explained in Fig. 8 (Cihon et al., 2021).

In general, AI can play different vital roles in corporate governance, which includes the systems and processes that govern and supervise the company's operations. Here are some of the key roles that AI can play in corporate governance:

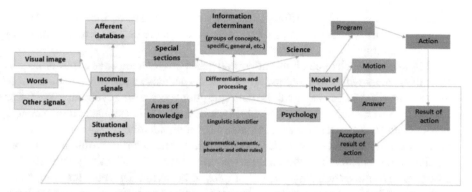

Fig. 7. AI system model in corporate governance *Source: Turluev, R., & Hadjieva, L. (2021). Artificial Intelligence in Corporate Governance Systems. In SHS Web of Conferences (Vol. 93, p. 03015). EDP Sciences.*

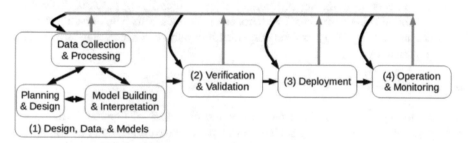

Fig. 8. Artificial Intelligence Lifecycle Framework in corporate. Source: Cihon, P., Schuett, J., & Baum, S. D. (2021). Corporate governance of artificial intelligence in the public interest. Information, 12(7), 275.

1- AI can help identify and manage risks associated with corporate governance. By analyzing vast amounts of data, including regulatory requirements, legal documents, financial reports, and industry standards, AI can help detect potential compliance violations, irregularities, or deviations from governance frameworks. This supports proactive risk management and enables companies to adhere to legal and ethical obligations.

2- AI can provide decision support to corporate boards and executives by analyzing complex data sets and generating insights. AI algorithms can process information from various sources, including market trends, financial indicators and stakeholder feedback, to help make strategic decisions. This can help boards evaluate potential investments, evaluate merger and acquisition opportunities, and make informed decisions about corporate governance structures.

3- AI can help enhance the effectiveness and diversity of corporate boards. By analyzing data related to board composition, skills, and performance, AI algorithms can identify potential gaps in experience and diversity. This information can assist in board selection processes, support efforts to build boards that are more inclusive and comprehensive, take into account a broader range of perspectives, and promote effective decision-making.

4- AI technologies, such as chatbots and virtual assistants, can improve communication and engagement with stakeholders. AI-powered systems can provide personal information, answer queries, and facilitate dialogue about corporate governance practices, executive compensation, and sustainability initiatives. This enhances transparency, accountability and trust between companies and their shareholders.

5- Strengthening internal controls and detecting fraudulent activities within institutions. AI algorithms can identify anomalies and potentially fraudulent activity by analyzing data patterns, transaction records, and employee behavior. This enables companies to implement early warning systems, improve internal audit processes, and mitigate the risks of fraud and misconduct.

6- AI can support corporate governance by enhancing ethical behavior and integrity within organizations. AI-powered systems can help monitor and analyze employee behavior, such as detecting conflicts of interest, identifying potential insider trading, or monitoring compliance with codes of conduct. This helps foster a culture of ethics and accountability, ensuring organizations adhere to ethical standards and legal requirements.

8 Conclusion

AI aims to expand and increase the ability and efficiency of humans in the tasks of reshaping nature and governing society as well as raise the efficiency of performance by developing software that can learn automatically to acquire knowledge and experience, then gradually improve learning behavior. AI technologies are increasingly utilized in the field of finance, such as asset management, algorithmic trading, credit underwriting, etc. Integrating AI technologies into investment and financing decisions has shown important results at the level of sustainability through its effects on ESG practices. Accordingly, integrating AI technologies has become crucial in financing, investment, competition, and market entry decisions. However, the positive impact of integrating AI technologies into financing and investment decisions has its own pros and cons in terms of its positive and negative effects on the environment and society in particular that may come along with some challenges in terms of the huge financial resources required to include these technologies, skilled human capital, and prepared institutional frameworks. However, research outcomes in this field indicate the superiority of the positive impact of AI in promoting sustainable growth over the negative one while recommending the necessity of reducing the negative impact on ESG as much as possible.

References

Abbas, S., et al.: Antecedents of trustworthiness of social commerce platforms: a case of rural communities using multi group SEM & MCDM methods. Electron. Commer. Res. Appl. **62**, 101322 (2023)

Abdelazim, H.Y., Wahba, K.: An artificial intelligence approach to portfolio selection and management. Int. J. Finan. Serv. Manag. **1**(2–3), 243–254 (2006)

Adeoye, O.B., Okoye, C.C., Ofodile, O.C., Odeyemi, O., Addy, W.A., Ajayi-Nifise, A.O.: Artificial Intelligence in ESG investing: enhancing portfolio management and performance. Int. J. Sci. Res. Arch. **11**(1), 2194–2205 (2024)

Ahmed, A.D., Salih, M.M., Muhsen, Y.R.: Opinion weight criteria method (OWCM): a new method for weighting criteria with zero inconsistency. IEEE Access (2024)

Ali, J., Hussain, K.N., Alnoor, A., Muhsen, Y.R., Atiyah, A.G.: Benchmarking methodology of banks based on financial sustainability using CRITIC and RAFSI techniques. Decis. Making: Appl. Manag. Eng. **7**(1), 315–341 (2024)

Alnoor, A., Atiyah, A.G., Abbas, S.: Toward digitalization strategic perspective in the European food industry: non-linear nexuses analysis. Asia-Pac. J. Bus. Admin. (2023)

Alnoor, A., Atiyah, A.G., Abbas, S.: Unveiling the determinants of digital strategy from the perspective of entrepreneurial orientation theory: a two-stage SEM-ANN Approach. Glob. J. Flex. Syst. Manag. **25**, 1–18 (2024)

Alnoor, A., Chew, X., Khaw, K.W., Muhsen, Y.R., Sadaa, A.M.: Benchmarking of circular economy behaviors for Iraqi energy companies based on engagement modes with green technology and environmental, social, and governance rating. Environ. Sci. Pollut. Res. **31**(4), 5762–5783 (2024)

Alnoor, A., et al.: How positive and negative electronic word of mouth (eWOM) affects customers' intention to use social commerce? A dual-stage multi group-SEM and ANN analysis. Int. J. Hum. Comput. Interact. **40**(3), 808–837 (2024)

Perrault, A., Fang, F., Sinha, A., Tambe, M.: Artificial intelligence for social impact: learning and planning in the data-to-deployment pipeline. AI Mag. **41**(4), 3–16 (2020). https://doi.org/10.1609/aimag.v41i4.5296

Taeihagh, A.: Governance of artificial intelligence. Policy Soc. **402**, 137–157 (2021). https://doi.org/10.1080/14494035.2021.1928377

Armstrong, A.: Ethics and ESG. Australas. Acc. Bus. Finan. J. **14**(3), 6–17 (2020)

Atiyah, A.G.: Unveiling the quality perception of productivity from the senses of real-time multisensory social interactions strategies in metaverse. In: Al-Emran, M., Ali, J.H., Valeri, M., Alnoor, A., Hussien, Z.A. (eds.) IMDC-IST 2024, vol. 876, pp. 83–93. Springer, Cham (2023). https://doi.org/10.1007/978-3-031-51300-8_6

Atiyah, A.G., Zaidan, R.A.: Barriers to using social commerce. In: Alnoor, A., Wah, K.K., Hassan, A. (eds.) Artificial Neural Networks and Structural Equation Modeling: Marketing and Consumer Research Applications, pp. 115–130. Springer, Singapore (2022). https://doi.org/10.1007/978-981-19-6509-8_7

Atiyah, A.G., Alhasnawi, M., Almasoodi, M.F.: Understanding metaverse adoption strategy from perspective of social presence and support theories: the moderating role of privacy risks. In: Al-Emran, M., Ali, J.H., Valeri, M., Alnoor, A., Hussien, Z.A. (eds.) IMDC-IST 2024, vol. 876, pp. 144–158. Springer, Cham (2023). https://doi.org/10.1007/978-3-031-51300-8_10

Atiyah, A.G., All, N.D.A., Zaidan, A.S., Bayram, G.E.: Understating the social sustainability of metaverse by integrating adoption properties with users' satisfaction. In: Al-Emran, M., Ali, J.H., Valeri, M., Alnoor, A., Hussien, Z.A. (eds.) IMDC-IST 2024, vol. 895, pp. 95–107. Springer, Cham (2023). https://doi.org/10.1007/978-3-031-51716-7_7

Atiyah, A.G., Faris, N.N., Rexhepi, G., Qasim, A.J.: Integrating Ideal Characteristics of Chat-GPT Mechanisms into the Metaverse: Knowledge, Transparency, and Ethics. In: Al-Emran, M., Ali, J.H., Valeri, M., Alnoor, A., Hussien, Z.A. (eds.) IMDC-IST 2024, vol. 895, pp. 131–141. Springer, Cham (2023). https://doi.org/10.1007/978-3-031-51716-7_9

Baburaj, A.: Artificial intelligence v. intuitive decision making: how far can it transform corporate governance? GNLU L. Rev. **8**, 233 (2021)

Belhaj, M., Hachaichi, Y.: Artificial intelligence, machine learning and big data in finance opportunities, challenges, and implications for policy makers (2021)

Bhattacharya, C., Sinha, M.: The role of artificial intelligence in banking for leveraging customer experience. Australas. Acc. Bus. Finan. J. **16**(5), 89–105 (2022)

Boffo, R., Patalano, R.: ESG Investing: Practices, Progress and Challenges. OECD Paris (2020)

Botsari, A., Lang, F.: ESG considerations in venture capital and business angel investment decisions: evidence from two pan-European surveys (No. 2020/63). EIF Working Paper (2020)

Boudway, I.: Waymo's self-driving future looks real now that the hype is fading (2021). https://www.bloomberg.com

Brusseau, J.: AI human impact: toward a model for ethical investing in AI-intensive companies. J. Sustain. Finan. Investment **13**(2), 1030–1057 (2023)

Bussmann, N., Giudici, P., Tanda, A., Yu, E.P.Y.: Explainable machine learning models to identify the key drivers of the implied cost of capital. (2023). Available at SSRN 4173890

Chen, L., et al.: Artificial intelligence-based solutions for climate change: a review. Environ. Chem. Lett. **21**(5), 2525–2557 (2023). https://doi.org/10.1007/s10311-023-01617-y

Chien, F.: The role of corporate governance and environmental and social responsibilities on the achievement of sustainable development goals in Malaysian logistic companies. Econ. Res.-Ekonomska istraživanja **36**(1), 1610–1630 (2023)

Cihon, P., Schuett, J., Baum, S.D.: Corporate governance of artificial intelligence in the public interest. Information **12**(7), 275 (2021)

de Hoyos Guevara, A.J., Dib, V.C.: ESG principles, challenges and opportunities. J. Innovation Sustain. RISUS **13**(4), 18–31 (2022)

de la Barcena Grau, A.M.: Social impact returns. Filling the finance gap with data value. Risk Manag. 125–143 (2021)

De Lucia, C., Pazienza, P., Bartlett, M.: Does good ESG lead to better financial performances by firms? Machine learning and logistic regression models of public enterprises in Europe. Sustainability **12**(13), 5317 (2020)

Del Vitto, A., Marazzina, D., Stocco, D.: ESG ratings explainability through machine learning techniques. Ann. Oper. Res. 1–30 (2023)

Delcker, J.: The man who invented the self-driving car (in 1986). Politico, July (2018)

Delgado-Ceballos, J., Ortiz-De-Mandojana, N., Antolín-López, R., Montiel, I.: Connecting the sustainable development goals to firm-level sustainability and ESG factors: the need for double materiality. BRQ Bus. Res. Q. **26**(1), 2–10 (2023)

Ertel, W.: Introduction to Artificial Intelligence. Springer, Cham (2018)

Folqué, M., Escrig-Olmedo, E., Corzo Santamaria, T.: Sustainable development and financial system: integrating ESG risks through sustainable investment strategies in a climate change context. Sustain. Dev. **29**(5), 876–890 (2021)

Food and Drug Administration. FDA Approves new robotic surgery device. Sci Daily (2000). https://www.sciencedaily.com/releases/2000/07/00071,2070727,2019

Ghodselahi, A., Amirmadhi, A.: Application of artificial intelligence techniques for credit risk evaluation. Int. J. Model. Optim. **1**(3), 243 (2011)

Goodrich, J.: How IBM's deep blue beat world champion chess player Garry Kasparov. IEEE Spectrum (2021)

Guo, T., Jamet, N., Betrix, V., Piquet, L.A., Hauptmann, E.: Esg2risk: a deep learning framework from ESG news to stock volatility prediction (2020). *arXiv preprint* arXiv:2005.02527

Holmes, J., Sacchi, L., Bellazzi, R.: Artificial intelligence in medicine. Ann. R. Coll. Surg. Engl. **86**, 334–338 (2004)

Holzinger, A., Langs, G., Denk, H., Zatloukal, K., Müller, H.: Causability and explainability of artificial intelligence in medicine. Wiley Interdisc. Rev. Data Min. Knowl. Discov. **9**(4), e1312 (2019)

https://www.unpri.org/download?ac=10948

Huang, A.H., Wang, H., Yang, Y.: FinBERT: a large language model for extracting information from financial text. Contemp. Account. Res. **40**(2), 806–841 (2023)

Husin, N.A., Abdulsaeed, A.A., Muhsen, Y.R., Zaidan, A.S., Alnoor, A., Al-mawla, Z.R.: Evaluation of metaverse tools based on privacy model using fuzzy MCDM approach. In: Al-Emran, M., Ali, J.H., Valeri, M., Alnoor, A., Hussien, Z.A. (eds.) IMDC-IST 2024, vol. 895, pp. 1–20. Springer, Cham (2023). https://doi.org/10.1007/978-3-031-51716-7_1

IMF. (2019). Global Financial Stability Report: Lower for Longer

Kaab, A., Sharifi, M., Mobli, H., Nabavi-Pelesaraei, A., Chau, K.W.: Combined life cycle assessment and artificial intelligence for prediction of output energy and environmental impacts of sugarcane production. Sci. Total. Environ. **664**, 1005–1019 (2019)

Kaya, B.C.: The role of artificial intelligence In corporate governance (2022). Available at SSRN 4143846

Khaled, R., Ali, H., Mohamed, E.K.: The sustainable development goals and corporate sustainability performance: mapping, extent and determinants. J. Clean. Prod. **311**, 127599 (2021)

Khan, N., Ahmad, T.: DCarbonX decentralised application: carbon market case study. *arXiv preprint* arXiv:2203.09508.(2022)

Kingo, L.: The UN Global Compact: Finding Solutions to Global Challenges| United Nations. *Un. Org* (2019)

Nevena, K., Amina, H.: Building a sustainable future: ESG business handbook, how environmental, social and governance standards can benefit your business, UNDP (2023)

Li, T.T., Wang, K., Sueyoshi, T., Wang, D.D.: ESG: research progress and future prospects. Sustainability **13**(21), 11663 (2021)

Lim, T.: Environmental, social, and governance (ESG) and artificial intelligence in finance: state-of-the-art and research takeaways. Artif. Intell. Rev. **57**(4), 1–64 (2024). https://doi.org/10.1007/s10462-024-10708-3

Liu, J., et al.: Artificial intelligence in the 21st century. IEEE Access **6**, 34403–34421 (2018)

Lorenzetti, L.: Forget Siri, Amazon Now Brings You Alexa. Fortune, November, 6 (2014)

Mandes, A.: Algorithmic and high-frequency trading strategies: a literature review (2016)

Maree, C., Omlin, C.W.: Balancing profit, risk, and sustainability for portfolio management. In: 2022 IEEE Symposium on Computational Intelligence for Financial Engineering and Economics (CIFEr), pp. 1–8. IEEE (2022)

Meier, R., Danzinger, R.: Personalized portfolio optimization using genetic (AI) algorithms. In: Soldatos, J., Kyriazis, D. (eds.) Big Data and Artificial Intelligence in Digital Finance, pp. 199–214. Springer, Cham (2022). https://doi.org/10.1007/978-3-030-94590-9_11

Chui, M., Harryson, M., Valley, S., Manyika, J., Roberts, R.: Notes from the AI frontier applying AI for social good (2018). discussion paper

Min, H.: Artificial intelligence in supply chain management: theory and applications. Int J Log Res Appl **13**(1), 13–39 (2010)

Minkkinen, M., Niukkanen, A., Mäntymäki, M.: What about investors? ESG analyses as tools for ethics-based AI auditing. AI & Soc. **39**(1), 329–343 (2024)

Modgil, S., Singh, R.K., Hannibal, C.: Artificial intelligence for supply chain resilience: learning from Covid-19. Int. J. Logistics Manag. **33**(4), 1246–1268 (2022)

Mohanty, B., Mishra, S.: Role of artificial intelligence in financial fraud detection. Acad. Market. Stud. J. **27**(S4) (2023)

Muhsen, Y.R., Husin, N.A., Zolkepli, M.B., Manshor, N.: A systematic literature review of fuzzy-weighted zero-inconsistency and fuzzy-decision-by-opinion-score-methods: assessment of the past to inform the future. J. Intell. Fuzzy Syst. **45**(3), 4617–4638 (2023)

Murtaza, S.S., Lak, P., Bener, A., Pischdotchian, A.: How to effectively train IBM Watson: classroom experience. In: 2016 49th Hawaii International Conference on System Sciences (HICSS), pp. 1663–1670. IEEE (2016)

Nabavi-Pelesaraei, A., Rafiee, S., Mohtasebi, S.S., Hosseinzadeh-Bandbafha, H., Chau, K.W.: Integration of artificial intelligence methods and life cycle assessment to predict energy output and environmental impacts of paddy production. Sci. Total. Environ. **631**, 1279–1294 (2018)

Naim, A.: Role of artificial intelligence in business risk management. Am. J. Bus.Manag. Econ. Bank. **1**, 55–66 (2022)

Sachin, N., Mukul, T.: Algorithmic Trading and Strategies (2020). https://www.researchgate.net/publication/345319146_Algorithmic_Trading_and_Strategies

OECD (2022), "Measuring the environmental impacts of artificial intelligence compute and applications: The AI footprint", OECD Digital Economy Papers, No. 341, OECD Publishing, Paris

Ögren, T., Forslund, P.: Screening techniques, sustainability and risk adjusted returns.:-A quantitative study on the Swedish equity funds market (2017)

Pachot, A., Patissier, C.: Towards Sustainable Artificial Intelligence: an overview of environmental protection uses and issues', Green and Low-Carbon Economy [Preprint] (2023). https://doi.org/10.47852/bonviewglce3202608

Powaski, M.C.K., Ordoñez, C.D., Sánchez, L.J.: ESG impact on financial corporate performance and portfolio returns: evidence of Australia and Japan. *Vinculatégica* (2021)

Raman, N., Bang, G., Nourbakhsh, A.: Mapping ESG trends by distant supervision of neural language models. Mach. Learn. Knowl. Extr. **2**(4), 453–468 (2020)

Reig-Mullor, J., Garcia-Bernabeu, A., Pla-Santamaria, D., Vercher-Ferrandiz, M.: Evaluating ESG corporate performance using a new neutrosophic AHP-TOPSIS based approach. Technol. Econ. Dev. Econ. **28**(5), 1242–1266 (2022)

Rouhollahi, Z.: Towards artificial intelligence enabled financial crime detection. *arXiv preprint* arXiv:2105.10866 (2021)

Sadiq, M., Ngo, T.Q., Pantamee, A.A., Khudoykulov, K., Ngan, T.T., Tan, L.P.: The role of environmental social and governance in achieving sustainable development goals: evidence from ASEAN countries. Econ. Res. Ekonomska istraživanja **36**(1), 170–190 (2023)

Sadok, H., Sakka, F., El Maknouzi, M.E.H.: Artificial intelligence and bank credit analysis: a review. Cogent Econ. Finan. **10**(1), 2023262 (2022)

Schaeffer, J., Lake, R., Lu, P., Bryant, M.: Chinook the world man-machine checkers champion. AI Mag. **17**(1), 21 (1996)

Schramade, W.: Integrating ESG into valuation models and investment decisions: the value-driver adjustment approach. J. Sustain. Finan. Investment **6**(2), 95–111 (2016)

Sharma, R., Shishodia, A., Gunasekaran, A., Min, H., Munim, Z.H.: The role of artificial intelligence in supply chain management: mapping the territory. Int. J. Prod. Res. **60**(24), 7527–7550 (2022)

Silvestro, D., Goria, S., Sterner, T., Antonelli, A.: Improving biodiversity protection through artificial intelligence. Nat. Sustain. **5**(5), 415–424 (2022)

Statt, N.: Facebook is using billions of Instagram images to train artificial intelligence algorithms. The Verge (2018)

Stone, W.L.: The history of robotics. In Robotics and Automation Handbook, pp. 8–19. CRC Press (2018)

Svoboda, E.: Artificial intelligence is improving the detection of lung cancer. Nature **587**(7834), S20–S20 (2020)

The Green Dilemma: Can AI Fulfil Its Potential Without Harming the Enviroment?

Torricelli, C., Bertelli, B.: ESG screening strategies and portfolio performance: how do they fare in periods of financial distress? CEFIN WORKING PAPERS (2022)

Lim, T.: Environmental, social, and governance (ESG) and artifcial intelligence in fnance: state of the art and research takeaways. Artif. Intell. Rev. **57**, 76 (2024). https://doi.org/10.1007/s10462-024-10708-3

Tse, T., Esposito, M., Goh, D.: AI-powered ESG: our chance to make a real difference? Digit. Transformation Playbook, **165** (2023)

Turluev, R., Hadjieva, L.: Artificial intelligence in corporate governance systems. In: SHS Web of Conferences, vol. 93, p. 03015s. EDP Sciences (2021)

UN Environment Programme-Finance Initiative. (2004). Who cares wins: The global compact connecting financial markets to a changing world. *Gözden geçirilme tarihi, 21*

Van Duuren, E., Plantinga, A., Scholtens, B.: ESG integration and the investment management process: fundamental investing reinvented. J. Bus. Ethics **138**, 525–533 (2016)

Wallace, R.S.: The Anatomy of ALICE, pp. 181–210. Springer, Netherlands (2009)

Waltersmann, L., Kiemel, S., Stuhlsatz, J., Sauer, A., Miehe, R.: Artificial intelligence applications for increasing resource efficiency in manufacturing companies—a comprehensive review. Sustainability **13**(12), 6689 (2021)

Weizenbaum, J.: ELIZA—a computer program for the study of natural language communication between man and machine. Commun. ACM **9**(1), 36–45 (1966)

Zhang, X., Chen, Y.: An artificial intelligence application in portfolio management. In: International Conference on Transformations and Innovations in Management (ICTIM 2017), pp. 775–793. Atlantis Press (2017)

Žigienė, G., Rybakovas, E., Alzbutas, R.: Artificial intelligence based commercial risk management framework for SMEs. Sustainability **11**(16), 4501 (2019)

Ziolo, M., Bak, I., Cheba, K.: The role of sustainable finance in achieving sustainable development goals: does it work? Technol. Econ. Dev. Econ. **27**(1), 45–70 (2021)

Harnessing Technological Innovation and Artificial Intelligence in Iraqi Commercial Banks to Achieve Sustainability

Mustafa Khudhair Hussein[1], Nahran Qasim Krmln[2], Hakeem Hammood Flayyih[2], and Ruaa Basil Noori[3(✉)]

[1] Imam A'adhum University College, Baghdad, Iraq
[2] Department of Financial and Banking Sciences, College of Administration and Economics, University of Baghdad, Baghdad, Iraq
{nahran.q,hakeem.hmood}@coadec.uobaghdad.edu.iq
[3] Faculty of Economics and Management, Sfax University, Sfax, Tunisia
ruoabasil@yahoo.com

Abstract. The study aimed to measure the mediating role of artificial intelligence (AI) in the relationship between technological innovation (TI) and competitive advantage (CA) in Iraqi commercial banks. The dimensions of artificial intelligence included two variables are ADM and CBSMI. The dimensions of TI included three variables are ATM, MPA, and POS. The dimensions of CA included three variables are Cost, Quality and Market Share. The study sample is employees of Iraqi commercial banks in twelve banks listed on the Iraqi stock market. The sample consisted of 399 employees, including are managers, accountants, auditors, and consultants. We used the Smart-PLS program to measure the results of the questionnaire. The results of the current study showed that ATM, MPA, and POS proxy of the TI influence on ADM and CBSMI proxy of the AI. There is a statistically significant positive relationship between the ATM, MPA, and POS on MPA, ATMs, and points of sale on cost, quality, and market share proxy of the CA. Also, there are the importance of the dimensions of AI as a mediator in the relationship between the dimensions of TI and CA.

Keywords: Technological Innovation · Artificial intelligence · Competitive Advantage · finance technology · Commercial Banks

1 Introduction

Considering technological advancements and continuous changes, it has become necessary to provide modern tools in the banking sector to keep up with modern trends and facilitate service delivery. Employing AI in banks contributes to accelerating and streamlining banking operations and transactions while providing a broad base for customers through digital applications and modern systems (Omoge et al., 2022). However, in comparison to advanced countries, the application of AI in the banking industry in Iraq still requires development and improvement. However, it is considered the future of

A. Alnoor et al. (Eds.): AIRDS 2024, LNNS 1033, pp. 280–296, 2024.
https://doi.org/10.1007/978-3-031-63717-9_18

banking and financial transactions as it offers innovative solutions to enhance efficiency and performance (Sajjad, 2023). Indeed, the CA of a bank in the industry is measured by the market share it achieves, which is determined by the proportion of revenues generated by the company. By implementing AI, banks can increase their market share and improve overall performance (Rabea'Hadi et al., 2023; Ali et al., 2024). AI technologies can enable banks to offer personalized services, streamline processes, automate tasks, and provide predictive analytics, all of which contribute to enhancing their competitiveness in the market and delivering better outcomes (Rabeah et al., 2022). Iraqi banks operating in the sector can develop modern technologies to increase the loyalty of their customers and attract more of them in a reliable and guaranteed manner, through their reliance on AI because of its vital and important role in gaining an AC (Sarhan & Al-Ali, 2022). Commercial banks have an important role in enhancing sector activity through investments and enhancing the economic development of the country (Hadi et al., 2023). Furthermore, On the other hand, Iraqi banks may face a set of challenges regarding compliance with regulations, despite making the necessary adjustments to achieve their advantage (Hamid & Alwan, 2023; Alnoor et al., 2024). The financial and technical innovations achieved by banks can increase customer confidence and their share with competitors in the same sector. Innovation plays a decisive role in strengthening the CA of institutions and increasing their market share. It was found (Jerdea, 2023; Alnoor et al., 2023). Many research and studies have focused on the importance of TI in banks and its role in increasing the number of customers (Raza et al., 2023; Maria, 2023). It is expected that they will gain highly loyal customers through the TIs provided (Kiryakov et al., 2022), which is one of the competitive strategies to enhance a high position among competitors in the same sector. (Pulkka & Cuthbertson, 2023). Aabkhare et al., (2013) highlighted some advantages and disadvantages of using online banking services, which include the fast money transfer service but also the risks of sharing information on multiple devices. The world is shifting towards a cashless economy through online banking facilities, which include services such as online banking, mobile banking, telephone banking, and ATMs. These factors contribute to the variation in a bank's market share (Salman et al., 2023). The banking sector worldwide is rapidly changing its services through innovation, which allows banks to enhance their market share and CA (Sahu & Maity, 2023.; Jihène, 2023). However, banks that are unable to meet the requirements of the modern age may lose their CA and market share compared to other banks (Elroy & Ramesh, 2023). Developing countries like Iraq face challenges in adopting modern tools and technologies to compete in the global banking system (Kumar & Gupta, 2023). Innovative bank financing not only increases market share but also provides cost efficiency, enhancing CA (Niepmann, 2023). The current study adds to the existing literature by being the first study to shed light on TI in enhancing CA through AI as a mediator in banks operating in the Iraqi banking sector. The primary objective of this investigation was to identify the direct role of TI in the CA of commercial banks in Iraq. Furthermore, the study aimed to examine the impact of AI on this relationship in terms of innovative financial practices as a CA to increase market share. The specific objectives of the study were as follows:

First, to study the impact of TI, such as internet & mobile banking and ATM services, on the CA of Iraqi banks.

Second, to analyze the effect of AI in enhancing the CA of banks in the Iraqi banking sector. The following research questions can represent the research problem:

RQ1: How does the use of TI (mobile banking, ATM, and Point-f-Sale (POS) services impact the CA of Iraqi commercial banks?

RQ2: How does AI play a mediating role in the relationship between TI and the CA of Iraqi commercial banks?

The purpose of this study is to measure the mediating role of AI in the relationship between TI and CA, in order to improve the quality of banking services, enhance customer loyalty and achieve their satisfaction with the aim of enhancing the efficiency of banking services activities and achieving competitive advantage for Iraqi banks through Employing TI and AI.

According to World Bank in 2017; The results of the following Fig. 1. it is evident that the Iraqi economy heavily relies on cash transactions, accounting for 77%. Only 23% of adults possess a bank account, which is a low percentage compared to other economies. Iran ranks first, with 94% of adults engaging in banking currency transactions, followed by the United Arab Emirates at 88%. Saudi Arabia ranks third with 72%, and finally, Turkey ranks third with 69%. This indicates that the banking industry in Iraq is weak and lacks sufficient capabilities in modern financial technology-based transactions. This could be attributed to the need for more public trust in the banking industry and financial illiteracy, as only 27% of the adult population possesses financial knowledge. Cash transactions dominate the overall Iraqi banking industry, despite the role banks play in expanding global credit, supporting electronic payments, and providing banking facilities.

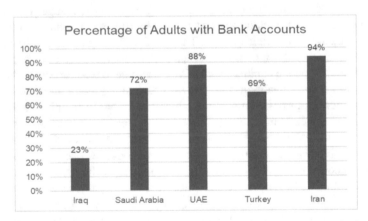

Fig. 1. The proportion of the Iraqi economy's dependence on cash transactions.

The contribution of the banking sector to the gross domestic product (GDP) was 1.94% in 2021. This percentage is modest and weak despite the presence of 74 banking institutions in Iraq, which own 904 branches spread across all provinces of the country. These institutions represent the highest number in the region's economies. Additionally, The World Bank estimates that in 2020, there were 5.63 commercial bank branches per

100000 adults, a figure that is far lower than in nearby nations. As an illustration, Turkey had a ratio of 16.1, Iran had 31.1, and Kuwait had 13.6. As shown in the Fig. 2.

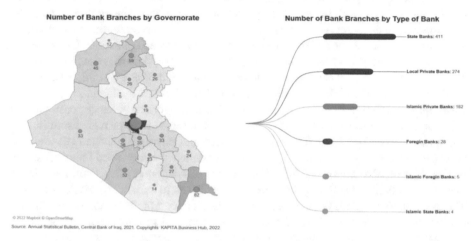

Fig. 2. Number of bank branches by type.

2   Literature Review

2.1   Technological Innovation and AI

Technological innovation is rapidly transforming the banking sector, enabling commercial banks in Iraq to enhance their CA. By adopting AI technologies, banks can streamline operations and offer personalized and responsive experiences to customers (Neama et al., 2023; Alnoor et al., 2024). These advancements in AI allow banks to compete with emerging FinTech players by leveraging digital advantages. AI refers to the intelligence exhibited by machines or software, enabling them to perform tasks traditionally requiring human intelligence (Omoge et al., 2022) The integration of strategic information systems in Iraqi commercial banks has proven to be instrumental in achieving a CA. By employing AI techniques to detect manipulation and fraud, enhance work efficiency, and improve the customer experience of these banks, thus improving their AC (Al-Azzawi & Bedoui, 2023; Atiyah, 2023). On the other hand, employing AI techniques can also reduce operating costs and the accuracy of the decision-making process, and this was confirmed by the results of research and studies that were applied in the Iraqi banking environment during the years 2018–2020 (Al-Sahlani, 2023).

2.2   The Competitive Advantage of Iraqi Commercial Banks and the Role of Technological Innovation

Commercial banks in Iraq contributed to enhancing economic growth. It worked and expanded its activities and customer service through the use of modern technologies and innovations (Hasan et al., 2023). Also, it worked to employ technology by simplifying its activities and increasing its market share by increasing productivity (Rao & Budde, 2015; Nikkeh et al., 2022). In this regard, the study (Kadhem et al., 2021) concluded that employing AI in its work increased its market share, as AI contributed to reducing costs and achieving a different AC over its competitors who did not adopt these technologies, as the use of these technologies in banking operations has a significant The connection to data analysis played an effective role in giving a point of view to management, which contributed to sending a message to customers regarding the bank's ability to meet their needs in the field of banking services (Agarwal et al., 2021; Atiyah and Zaidan, 2022). Developing marketing strategies by providing banking services enhances customer satisfaction (Khanfar & Al-Masry, 2023). These technologies have provided solutions for many innovative digital banking activities, including payment mechanisms and easy access to banking services provided to customers whenever they need them, regardless of time and place (Omarini, 2018). The process of achieving integration between AI methods and banking activities has greatly contributed to enhancing the AC of banks. The use of these methods has enabled banks to achieve comprehensiveness and improve performance (Al-Tamimi et al., 2022). It has been proven that the use of technologies based on AI methods has enhanced customer satisfaction and retention, because AI provides comprehensive and continuous support (Ooi et al., 2023;). The speed of data analysis according to AI identified trends and patterns within large data sets, and this helped enhance risk management strategies, corruption detection techniques, and increase investment opportunities by taking advantage of these innovations (Aziz & Andriansyah, 2023).

2.3 The Rise of Technological Innovation in Iraqi Commercial Banks: An Overview

The process of employing ATMs, points of sale, and mobile applications has a major role in the performance of commercial banks operating in Iraq (Thamer & Mardan, 2023; Hussein et al., 2023), and in the same context, it has also played an important role in increasing bank profits (Turki & Sabbar, 2023), and the process of managing banking services Online access to accounts through mobile applications has an important impact (Makttoof & Khaled, 2018) and thus has a positive impact on improving customer satisfaction and operational efficiency by reducing the need for banking branches and enabling access to them 24 h a day, 7 days a week (Al-Sahlani, 2023; Atiyah et al., 2023)., so that it has become a common aspect in the Iraqi environment, as it provides ease and comfort in the process of deposits, withdrawals, bank statements, financial transfers, and carrying out all activities and needs of customers with complete ease (Mezaal & Alothman, 2022).Additionally, POS services enable merchants to accept card payments effortlessly, promoting cashless transactions in a country where traditional payment methods prevail (Jameel & Alheety, 2022).

2.4 Leveraging Technological Advancements for Competitive Advantage in Iraqi Commercial Banks

The process of TI, represented by ATMs and points of sale, has contributed to raising the efficiency of banking activities and providing safer and more convenient methods in the process of accessing services (Hamid & Alwan, 2023). This is done by providing banking operations and services through the Internet, which leads to an increase in customers' interest in purchasing the services of these banks through the high flexibility achieved by these technologies, and thus is reflected in the flexibility of customers' banking transactions and the increase in the market share of banks (Ammar et al., 2023), as this technical revolution is on the rise. The level of a country like Iraq is an important breakthrough by providing easy, flexible, and fast operations in the financial services provided by these banks through Internet applications, points of sale and ATMs, despite the need to increase the number of branches and workers. MPA has made account access, funds transfer, and bill payments easier than ever before. This TI has increased competition among Iraqi commercial banks as they strive to provide user-friendly mobile applications with advanced features to attract and retain customers (Shirkhodaie, 2023; Murad, 2023). The scale of acceptance networks in developed economies is determined by the number of ATMs and POS devices for bank cards. POS systems are no longer limited to specific industries and have expanded across various sectors. Businesses often need to invest in POS terminals or card readers to accept credit card payments. This trend is driven by the aim of banks, payment institutions, and Payment Service Provider companies to increase the number of card transactions and enhance their platforms by supporting a wide range of channels with digital solutions. They prioritize user-friendly and secure payment platforms that offer easy and quick integration opportunities with merchants. This has led to the Focusing on their core business, by employing these technologies, can enhance AC (Satyajitsinh et al., 2023; Chowdhury, 2022; Atiyah et al., 2023), and point-of-sale devices have an important role in collecting and analyzing data (Ruan et al, 2023). Important and valuable insights into consumer preferences that can be used in implementing and developing marketing strategies (Oriedi et al., 2023). Banks can offer discounts on services and products that are compatible with customers' interests and spending habits (Joy et al., 2022; Abbas et al., 2023). Points of sale are among the necessary things that enhance the AC of banks and enhance performance efficiency (Chen, 2022; Muhsen et al., 2023) by integrating modern digital technology and the decision-making process, which It will reflect positively on staying ahead of competitors in the same sector, banks can capitalize on customer demands and outperform competitors in the sector. Through what is vulnerable to previous research and studies that dealt with the study variables, the following hypotheses can be built:

Ha: There is a relationship between TI (MPA, ATM, and POS) in enhancing AI in Iraqi commercial banks.

Hb: There is a relationship between TI (MPA, ATM, and POS) in enhancing the CA of Iraqi commercial banks.

Hc: AI enhances the relationship between TI (MPA, ATM, and POS) and CA in Iraqi commercial banks.

3 Methodology

The research community is represented by employees in the selected Iraqi commercial banks 12 banks spread in the city of Baghdad based on the volume of activity and spread of the commercial bank out of 44 banks listed in the Iraqi stock exchange. A simple random sample of workers was taken. The total study population consisted of 2894 individuals, and a sample consisting of 399 employees in commercial banks in Baghdad and from all administrative levels (managers, accountants, auditors, consultants), based on the equation (Thompson, 2018); the following Table 1. Shows the distribution of sample members among commercial banks.

Table 1. Distribution of sample members among commercial banks.

Number of Employees	Number of Branches	Bank Name	No
349	13	Iraqi Al-Ahli Bank	1
222	10	Iraqi Commercial Bank	2
245	21	United Investment Bank	3
187	9	Ashur International Investment Bank	4
264	11	Iraqi Union Bank	5
337	12	Economy Investment and Finance Bank	6
180	18	Gulf Commercial Bank	7
307	18	Middle East Bank	8
245	9	Mansour Investment Bank	9
209	7	Mosul Development and Investment Bank	10
200	4	Baghdad Bank	11
149	9	Sumer Commercial Bank	12
2894	204	Total	

To achieve the research objectives, the researcher will seek to use the descriptive analytical method by reviewing the literature related to the subject of the study and collecting and analyzing data by relying on the questionnaire as a tool for collecting the necessary data and information. A questionnaire was designed for this purpose and consists of the following axes: The independent variable is by TI, with the Three dimensions (MPA, ATM, and POS consisting of 15 items. The dependent variable, CA, consists of 15 items Divided into three dimensions: Cost 5, Quality 5, and market share 5. And the mediating variable, AI, consists of 8 items divided into two internal dimensions, which are decision-making. Automated 4 items and the second dimension is automated chat and interaction via social media 4 items; a five-point Likert scale was used to distribute the answers of the respondents, as shown in Table 2.

Table 2. Description of the measurement tool and study variables.

Number of items	Dimension	variables	No
(Raza et al.2017)	MPA (5) items; ATM (5) items; POS (5) items	TI	1
(Rua et al., 2018)	Cost (5); Quality (5); Market share (5)	CA	2
(Kruse et al., 2019)	ADM (4); CBSMI (4)	AI	3

Data on the sample was collected through Google Forms and designed electronically, and the response rate was 100% because it does not accept errors when filling out the questionnaire paragraphs, which contributed to eliminating missing data. The final number of questionnaires that could be analyzed in its final form was 399, and the bank's departments contributed to assisting the researcher and welcomed the idea because it has an important role in enhancing its work and benefiting from the results. Table 3. Shows the sample members according to administrative levels in the Iraqi commercial banks studied.

Table 3. Sample members according to administrative levels.

No	Categories	Sample Size
1	Bank Managers	12
2	Branch and Department Managers	131
3	accountants	112
4	Auditors	98
5	consultants	46
Total		399

To test the proposed hypotheses and analyze data we relied on SMARTPLS version 3 The validity and reliability of the measures were confirmed, and then descriptive statistics and correlation coefficients were presented.

4 Results

Table 4 displays. Demographic characteristics of individuals working in the banking sector.

According to the results in Table .4, the focus on the sample members was on their gender, ages, educational qualifications, and experiences. In terms of gender, the majority were males 57.6%, compared to females 42.4%. In Iraq, men usually hold leadership positions (See, Al-Saadi, 2023). With regard to age, the majority of the sample was 56.4% within the age group from 25 to 45 years. This confirms the presence of young talents working in banks. The bachelor's degree constituted the majority among the

Table 4. Demographic characteristics of sample members.

Percentage %	Number	Variables	Metadata
57.6%	230	male	Sex
42.4%	169	feminine	
19.1%	76	20-less than 25 years old	Age
36.8%	147	25- less than 35 years old	
24.6%	98	35-less than 45 years old	
19.5%	78	45 years and over	
54.1%	216	University education	Qualification
32.5%	130	diploma	
9.1%	36	Master's	
4.3%	17	PhD	
25.3%	101	5-Less than ten years	Experience
32.6%	130	10-Under 15 years	
23.1%	92	15 less than 20 years old	
19%	76	20 years and over	

sample members, accounting for 54.1% compared to other university degrees, diploma, master's, and doctorate. As for job experience, it achieved a percentage of 57.7%, which indicates that experience is of great interest in banks.

We present below the main research model of the study, which measures the dimensions of IT (ATM - MPA - POS) as an independent variable in the influential relationship in the AC of Iraqi banks through AI. The model was built through the PLS-SMART program, as shown in Fig. 3.

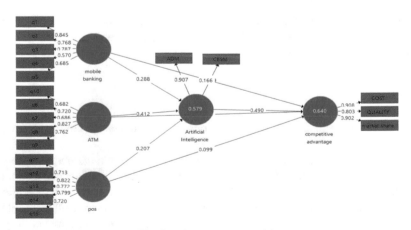

Fig. 3. Measurement model.

Table 5. Outer Loading and convergent validity

Outer Loading			Construct Reliability and Validity		
Construct and their items		factor loading	Cronbach's Alpha	Composite Reliability	(AVE)
AI	ADM	0.907	0.814	0.915	0.843
	CBSM	0.929			
CA	COST	0.908	0.844	0.905	0.762
	QUALITY	0.803			
	market share	0.902			
MPA	Q1	0.845	0.785	0.854	0.543
	Q2	0.768			
	Q3	0.787			
	Q4	0.570			
	Q5	0.685			
ATM	Q6	0.720	0.789	0.856	0.544
	Q7	0.686			
	Q8	0.827			
	Q9	0.762			
	Q10	0.682			
POS	Q11	0.713	0.753	0.834	0.514
	Q12	0.822			
	Q13	0.772			
	Q14	0.799			
	Q15	0.720			

Table 5 show the external loading values for the factors used to measure the structural model of the study. It also includes convergent validity measures such as Cronbach's alpha, composite reliability, and average variance extracted from the study. Cronbach's alpha and composite reliability indicate the internal validity of the constructs, while the average variance extracted indicates the external validity of the constructs. We note that Cronbach's alpha coefficients are all above 0.70, and the average variance extracted is also above 0.50, and this indicates the suitability of the structural model for the study.

The results of Table 6 above indicate that the model has an appropriate level of discriminatory validity, as we note the results of the Fornell–Larcker criterion matrix. It is the correlation of variables with themselves that is greater than their correlation with other variables and thus gives the suitability and validity of the structural model for the current study. After ensuring the suitability of convergent and discriminant validity and reliability, we will apply the structural study model to test the study hypotheses as follows:

Table 6. Discriminant validity of constructs (Fornell–Larcker criterion).

Variables	ATM	AI	CA	MPA	Pos
ATM	0.737				
AI	0.705	0.918			
CA	0.672	0.764	0.873		
MPA	0.637	0.623	0.613	0.737	
Pos	0.527	0.524	0.502	0.348	0.717

Figure 4 displays. The structural equation model of the study, which measures the test of the final estimated models by determining the direct effect of IT in their dimensions, which are each of (ATM - MPA - POS) as an independent variable in the influential relationship in the AC of Iraqi banks, and its indirect effect also through AI as a mediating variable, and also the effect Direct AI in AC.

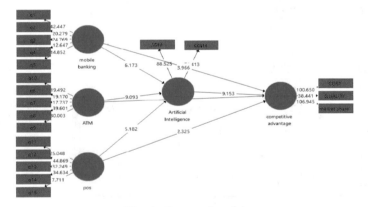

Fig. 4. Structural model

The results of the Table 7. Indicate that there is a direct positive relationship with significant significance between all study variables and that according to the basic rule for the importance of these relationships, the t-values must be 1.96 or greater (Hair et al., 2019). We note that all the relationships between the variables are important according to the following results are It appears that there is a strong positive direct effect of ATM on AI, with a path coefficient of 0.412. The p-value is 0.000, and the T-value is 9.093, which is greater than 1.96, indicating a statistically significant effect. ATM also has a positive direct effect on CA, but it is weaker compared to the previous effect. The path coefficient is 0.169, the p-value is 0.001, and the T-value is 3.288, which is greater than 1.96, further emphasizing the significance and meaningfulness of the relationship. AI has a strong positive direct effect on CA, with a path coefficient of 0.490. The p-value is 0.000, and the T-value is 9.153, which is greater than 1.96. There is a positive direct effect between MPA and AI. The path coefficient is 0.288, the p-value is 0.000, and the

Table 7. Path Coefficients (Direct Effects)

Path	Original Sample	Mean	Standard Deviation	T Statistics	Sig
ATM - > AI	0.412	0.413	0.045	9.093	0.000
ATM - > CA	0.169	0.170	0.051	3.288	0.001
AI - > CA	0.490	0.493	0.054	9.153	0.000
MPA - > AI	0.288	0.286	0.047	6.173	0.000
MPA - > CA	0.166	0.162	0.042	3.966	0.000
POS - > AI	0.207	0.208	0.040	5.182	0.000
POS - > CA	0.099	0.096	0.042	2.325	0.020

T-value is 6.173, which is greater than 1.96. There is a positive direct effect between MPA and CA. The path coefficient is 0.166, the p-value is 0.000, and the T-value is 3.966, which is greater than 1.96. POS (Point of Sale) has a positive direct effect on AI. The path coefficient is 0.207, the p-value is 0.000, and the T-value is 5.182, which is greater than 1.96. There is a positive direct effect of POS on CA, but it is weaker. The path coefficient is 0.099, the p-value is 0.020, and the T-value is 2.325, which is greater than 1.96.

Table 8. Path Coefficients (Specific Indirect Effects)

Path	Original Sample	Mean	Standard Deviation	T Statistics	Sig
ATM - > AI - > CA	0.202	0.204	0.033	6.214	0.000
MPA - > AI - > CA	0.141	0.141	0.026	5.344	0.000
POS - > AI - > CA	0.102	0.103	0.024	4.302	0.000

The results of Table 8 indicate that there is a positive indirect effect for all TI variables represented by (ATM - MPA - POS) on the CA of banks through the mediation of AI and the value of the paths around them were 0.202, 0.141, and 0.102. Respectively, all of them had a moral significance, and the value of P = 0.000, and the value of T was equal to 6.214, 5.344, and 4.302, respectively, which is greater than 1.96, based on (Hair et al., 2019). It indicates the importance of this relationship between the investigated variables.

5   Discussions of Results

The results of the current study showed that TI (MPA, ATMs, and POS) affect AI in Iraqi commercial banks. Therefore, we accept Hypothesis (H1). This result is consistent with the results of the study (Kruse et al., 2019) and the results of the study (Mouna,

2023). There is a positive, statistically significant relationship between the TI of banking operations represented by (MPA, ATMs, and POS) and the CA of Iraqi commercial banks represented by (cost, quality, and market share). Therefore, we accept hypothesis (H2), and these results are consistent with the results of the study (Al-Sahlani, 2023). Banks that pay great attention to financial innovation and invest in the development of modern financial technologies achieve superior performance and maintain strong competitiveness in the banking market. It also agrees with the results of the study by (Neama et al., 2023) that finance technology has a positive and important role in enhancing the performance of private banks in Iraq. These results include improving efficiency, providing innovative banking services and enhancing customer experience. And the results of Muhammad et al., (2021). And the results of a study (Sahu & Maity, 2023). The results also indicated that AI enhances the relationship between TI and CA in Iraqi commercial banks, and therefore, we accept the third hypothesis (H3). This result is consistent with the results of the study (Al-Sahlani. 2023), which indicated that the use of chatbots supported by AI and virtual assistants in commercial banks Iraqiya, like CIBC's virtual assistant powered by IBM Watson Assistant, enhances customer support and reduces operating costs.

6 Conclusions

The study aimed to measure the mediating role of AI in the relationship between TI and AC in Iraqi commercial banks. We found that banks' banking technology contributed to increasing competitiveness. The development of banking technologies, such as MPA, point-of-sale devices, and ATMs, and keeping pace with recent changes in the field of using AI technologies have enhanced these capabilities. Based on this, banks must enhance their capabilities by convincing them of the benefits and uses of modern banking technology. We also concluded that decision-makers in banks must provide support for innovative technologies, including MPA, points of sale, and ATMs, to improve quality, reduce costs, and increase market share. Banks must also work to intensify the training of their employees on modern technology and to use it efficiently and effectively because it achieves a AC and achieves a large market share through its use of technologies and the development of highly efficient information security systems that guarantee the protection of customer data and address potential security threats. Considering the results reached, Iraqi banks must develop easy-to-use and highly secure technologies that are far from complex and compatible with available operating systems that enable their customers to access their accounts easily. And work to enhance cooperation and partnerships between banks and financial technology companies to provide innovative and unusual banking solutions. Providing technical support services to customers around the clock to deal with technical inquiries and problems. The Central Bank of Iraq must encourage the establishment of an efficient and effective banking infrastructure. At the level of legislation and laws, the federal government must improve the business environment and enhance confidence in the banking system to improve the performance of banks.

References

Aabkhare, A.A., Aliloo, B.M.M., Abedini, E.: Advantages and disadvantages of E-banking and commerce. Life Sci. J. **10**(4s) (2013)

Abbas, S., et al.: Antecedents of trustworthiness of social commerce platforms: a case of rural communities using multi group SEM & MCDM methods. Electron. Commer. Res. Appl. **62**, 101322 (2023)

Agarwal, A., Singhal, C., Thomas, R.: AI-powered decision-making for the bank of the future. McKinsey & Company. –2021. –March (2021)

Al-Azzawi, R.Z.A., Bedoui, R.: The impact of modern information technology governance on financial performance (an applied study on a sample of private iraqi commercial banks). Migr. Lett. **20**(S6), 207–215 (2023)

Ali, J., Hussain, K.N., Alnoor, A., Muhsen, Y.R., Atiyah, A.G.: Benchmarking methodology of banks based on financial sustainability using CRITIC and RAFSI techniques. Decis. Making: Appl. Manag. Eng. **7**(1), 315–341 (2024)

Alnoor, A., Atiyah, A.G., Abbas, S.: Toward digitalization strategic perspective in the European food industry: non-linear nexuses analysis. Asia-Pac. J. Bus. Adm. (2023)

Alnoor, A., Atiyah, A.G., Abbas, S.: Unveiling the determinants of digital strategy from the perspective of entrepreneurial orientation theory: a two-stage SEM-ANN approach. Glob. J. Flex. Syst. Manag. **25**, 1–18 (2024)

Alnoor, A., et al.: How positive and negative electronic word of mouth (eWOM) affects customers' intention to use social commerce? A dual-stage multi group-SEM and ANN analysis. Int. J. Hum. Comput. Interact. **40**(3), 808–837 (2024)

Al-Saadi, H.H.F.: The Mediating Role of Audit Quality in The Relationship Between Corporate Governance and Earnings Management: Corporate Social Responsibility as A Moderate Variable. Institut Supérieur de Gestion de Tunis, Université de Tunis (2023)

Al-Sahlani, H.K.: The impact of financial innovation on the performance-evidence from the Iraqi banking system. J. Econ. Finan. Manag. Stud. 06(10) (2023). https://doi.org/10.47191/jefes/v6-i10-35

Al-Tamimi, H.Z.H., Al-Rubaie, A.K.A., Al-Mayali, H.N.H.: The role of marketing with reactionary intelligence in improving the reputation of the bank: an analytical study of the opinions of a sample of iraqi private banks. Middle East Int. J. Soc. Sci. (MEIJSS), 4(3) (2022)

Atiyah, A.G.: Unveiling the quality perception of productivity from the senses of real-time multisensory social interactions strategies in metaverse. In: Al-Emran, M., Ali, J.H., Valeri, M., Alnoor, A., Hussien, Z.A. (eds.). IMDC-IST 2024.LNNS, vol 876, pp. 83–93. Springer, Cham (2023). https://doi.org/10.1007/978-3-031-51300-8_6

Atiyah, A. G., & Zaidan, R. A. (2022). Barriers to using social commerce. In artificial neural networks and structural equation modeling: marketing and consumer research applications (pp. 115–130). Singapore: Springer Nature Singapore

Atiyah, A.G., Alhasnawi, M., Almasoodi, M.F.: Understanding metaverse adoption strategy from perspective of social presence and support theories: the moderating role of privacy risks. In: Al-Emran, M., Ali, J.H., Valeri, M., Alnoor, A., Hussien, Z.A. (eds.) Beyond Reality: Navigating the Power of Metaverse and Its Applications. IMDC-IST 2024. LNNS, vol. 876, pp. 144–158. Springer, Cham (2023). https://doi.org/10.1007/978-3-031-51300-8_10

Atiyah, A.G., All, N.D.A., Zaidan, A.S., Bayram, G.E.: Understating the social sustainability of metaverse by integrating adoption properties with users' satisfaction. In: Al-Emran, M., Ali, J.H., Valeri, M., Alnoor, A., Hussien, Z.A. (eds.) IMDC-IST 2024. LNNS, vol. 895, pp. 95–107. Springer, Cham (2023). https://doi.org/10.1007/978-3-031-51716-7_7

Atiyah, A.G., Faris, N.N., Rexhepi, G., Qasim, A.J.: Integrating ideal characteristics of chat-GPT mechanisms into the metaverse: knowledge, transparency, and ethics. In: Al-Emran, M.,

Ali, J.H., Valeri, M., Alnoor, A., Hussien, Z.A. (eds.) IMDC-IST 2024. LNNS, vol. 895, pp. 131–141. Springer, Cham (2023). https://doi.org/10.1007/978-3-031-51716-7_9

Aziz, L.A.R., Andriansyah, Y.: The role AI in modern banking: an exploration of AI-driven approaches for enhanced fraud prevention, risk management, and regulatory compliance. Rev. Contemp. Bus. Anal. 6(1), 110–132 (2023)

Chen, H.: Prediction and analysis of financial default loan behavior based on machine learning model. Comput. Intell. Neurosci. 2022, 7907210 (2022)

Chowdhury, S.J., Aich, E., Reno, S., Ahmed, M.: Utilizing hyperledger-based private blockchain technology to secure credit card payment system. In: 2022, the 25th International Conference on Computer and Information Technology (ICCIT), pp. 354–359. IEEE (2022)

Monis, E., Pai, R.: Neo banks: a paradigm shift in banking. Int. J. Case Stud. Bus. IT Educ. (IJCSBE), 7(2), 318–332 (2023). https://doi.org/10.47992/ijcsbe.2581.6942.0275

Falih, F.S.: The mediating effect of operational risk management between business attributes and competitive advantage: an exploratory study (survey) in the Iraqi Banking Sector (Doctoral dissertation, Universiti Tun Hussein Onn Malaysia). (2022)

Hadi, A.H., Abdulhameed, G.R., Malik, Y.S., Flayyih, H.H.: The influence of information technology (It) on firm profitability and stock returns. East. Eur. J. Enterp. Technol. 124(13), 87–93 (2023)

Hair, J.F., Anderson, R.E., Babin, B.J., Black, W.C.: Multivariate Data Analysis, 8th edn. Cengage Learning, United Kingdom (2019)

Hamid, N.M., Alwan, I.H.: The positive effects of political change on the banking sector in Iraq after 2003. Russ. Law J. 11(7s), 418–425 (2023)

Hasan, S.I., Saeed, H.S., Al-Abedi, T.K., Flayyih, H.H.: The role of target cost management approach in reducing costs for the achievement of competitive advantage as a mediator: an applied study of the Iraqi electrical industry. Int. J. Econ. Finan. Stud. 15(2), 214–230 (2023)

Hussein, M.K., Al-tekreeti, R.B. N., Hasan, M.F., Flayyih, H.H.: The moderate role of the perceived orientation of information technology in the relationship between human capital and organizational innovation mediating orientations to learning: literature review. Ishtar J. Econ. Bus. Stud. 4(1) (2023). https://doi.org/10.55270/ijebs.v4i1.14

Jameel, A.S., Alheety, A.S.: Customers' perceptions and behavioural intentions regarding mobile banking usage. In: 2022 International Conference on Innovation and Intelligence for Informatics, Computing, and Technologies (3ICT), pp. 189–194. IEEE. (2022)

Jerdea, L.: A bibliometric analysis of four constructs interconnections: innovation, competitive advantage, agility and performance. In: Proceedings of the International Conference on Business Excellence, vol. 17, no. 1, pp. 1213–1226. (2023)

Jihène, B.: The impact of digitalization on the banking sector: evidence from Fintech countries. Asian Econ. Finan. Rev. 13, 269–278 (2023). https://doi.org/10.55493/5002.v13i4.4769

Utamajaya, J.N., Ramadhan, A., Abdurachman, E., Trisetyarso, A., Zarlis, M.: Risk assessment analysis on mobile banking using Cobit 5 framework (2022). https://doi.org/10.1109/iccit5 5355.2022.10118645

Kadhem, L.N.J., Kadhem, A.L.H.A., Dawood, F.S.: The role of AI in achieving ambidextrous performance: a case study in a sample of private banks. J. Acc. Finan. Stud. (JAFS) 16, 61–82 (2021)

Khanfar, I., Almasri, A.: The role of the electronic banking marketing mix elements in enhancing the competitive advantage: a field study on customers of Islamic international arab bank at Amman city/Jordan. In: Alareeni, B.A.M., Elgedawy, I. (eds.) Artificial Intelligence (AI) and Finance. Studies in Systems, Decision and Control, vol. 488, pp. 691–703. Springer, Cham (2023). https://doi.org/10.1007/978-3-031-39158-3_64

Kiryakov, I., Godin, A., James, U. O., Nnamdi, O., Jones, A.O.: Improvements activity of innovation: experience, challenges and opportunities in Nigeria. In: E3S Web of Conferences, vol. 371. EDP Sciences (2023). https://doi.org/10.1051/e3sconf/202337105035

Kruse, L., Wunderlich, N., Beck, R.: Artificial intelligence for the financial services industry: What challenges organizations to succeed. In: A Paper Presented at the Proceedings of the 52nd Hawaii International Conference on System Sciences (2019)

Kumar, J., Gupta, S.S.: Impact of AI towards customer relationships in the Indian banking industry. Gyan Manag. J. **17**(1), 105–115 (2023)

Makttoof, M.N., Khalid, H.: A review of mobile banking adoption factors by customers for Iraqi banks. Int. J. Eng. Technol. **7**(3.20), 629–637 (2018)

Maria, R.: HRM practices and organizational performance: mediation effect of innovation. Southeast Eur. J. Econ. Bus. **18**, 85–99 (2023). https://doi.org/10.2478/jeb-2023-0007

Mezaal, W.C., Alothman, H.A.S.: Troubles and determinants in Using ATMs. An exploratory study of a sample clients opinions of the Iraqi Trade Bank-Dhi Qar branch. Technium Soc. Sci. J. **37**, 713 (2022). https://doi.org/10.47577/tssj.v37i1.2096

Turki, M.D., Sabbar, M.: Studying the role of banking marketing in supporting the competitive advantage of the Iraqi banks. Int. J. Prof. Bus. Rev. Int. J. Prof. Bus. Rev. **8**(4), 13. (2023). https://doi.org/10.26668/businessreview/2023.v8i4.1342

Muhammad, A.A., Mayea, S. S., Madhi, M.S., Thajil, K.M.: The role of creativity and leadership in achieving the competitive advantage of banks. Int. J. Res. Soc. Sci. **12**(4), 889–907 (2022). https://doi.org/10.37648/ijrssh.v12i04.047

Muhsen, Y.R., Husin, N.A., Zolkepli, M.B., Manshor, N.: A systematic literature review of fuzzy-weighted zero-inconsistency and fuzzy-decision-by-opinion-score-methods: assessment of the past to inform the future. J. Intell. Fuzzy Syst. **45**(3), 4617–4638 (2023)

Murad, K.I.: The extent to which financial institutions comply with the US FATCA law—Indonesian J. Law Econ. Rev. **18**(2) (2023)

Neama, N., Abboud, R., Aref, S.: Financial technology and its role in Iraqi banking industry (analyzing study for selected private banks of Iraq). Open J. Bus. Manag. **11**, 1577–1583 (2023). https://doi.org/10.4236/ojbm.2023.114087

Niepmann, F.: Banking across borders with heterogeneous banks. J. Int. Econ. **142**, 103748 (2023)

Nikkeh, N.S., Hasan, S.I., Saeed, H.S., Flayyih, H.H.: The role of costing techniques in reduction of cost and achieving competitive advantage in Iraqi financial institutions. Int. J. Econ. Finan. Stud. **14**(4), 62–79 (2022)

Omarini, A.E.: Fintech and the future of the payment landscape: the mobile wallet ecosystem. A challenge for retail banks? Int. J. Finan. Res. **9**(4), 97–116 (2018)

Omoge, A.P., Gala, P., Horky, A.: Disruptive technology and AI in the banking industry of an emerging market. Int. J. Bank Mark. **40**, 1217–1247 (2022). https://doi.org/10.1108/ijbm-09-2021-0403. (ahead-of-print)

Ooi, K. B., et al.: The potential of Generative AI across disciplines: perspectives and future directions. J. Comput. Inf. Syst. 1–32 (2023)

Oriedi, D., Malanga, K., Musumba, G., Kamau, G., Ollengo, M.: A simulation of the optimal personnel demand for banking hall services. Int. J. Model. Optim. **13**(2), 60–65 (2023)

Pulkka, L., Cuthbertson, R.: The Timing of Innovation (2023). https://doi.org/10.1093/oso/978 0192862617.003.0009

Rabea'Hadi, M., Hasan, M.F., Flayyih, H.H., Hussein, M.K.: Green banking: a literature review on profitability and sustainability implications. Ishtar J. Econ. Bus. Stud. **4**(2), 1–6 (2023). https://doi.org/10.55270/ijebs.v4i2.27

Jasim, R.D., Al-Mashhdani, B.N.A.: Methods of forecasting credit losses in a sample of Iraqi banks-a comparative analysis. J. Econ. Adm. Sci. **28**(132), 174-195 (2022). https://doi.org/10. 33095/jeas.v28i132.2283

Rao, Y.V., Budde, S.R.: Banking technology innovations in India: enhancing customer value and satisfaction. Indian J. Sci. Technol. **8**(33), 1–10 (2015)

Raza, M.A., Naveed, M., Ali, S.: Determinants of Internet banking adoption by banks in Pakistan. Manag. Organ. Stud. **4**(4), 12 (2017). https://doi.org/10.5430/mos.v4n4p12

Raza, S., Laghari, M.K., Junejo, M.A.: The dynamics of innovation on firm performance: empirical analysis of pharmaceutical manufacturing firms in Sindh Pakistan. Pak. J. Hum. Soc. Sci. **11**(2), 1153–1171 (2023). https://doi.org/10.52131/pjhss.2023.1102.0424

Rua, O., França, A., Ortiz, R.F.: Key drivers of SMEs export performance: the mediating effect of competitive advantage. J. Knowl. Manag. **22**(2), 257–279 (2018). https://doi.org/10.1108/JKM-07-2017-0267

Ruan, O., Zhao, W., Tian, Y.: Secure E-transaction system based on smart contract and InterPlanetary file system. In: Second International Conference on Algorithms, Microchips, and Network Applications (AMNA 2023), vol. 12635, pp. 187–196. SPIE (2023)

Sahu, T.N., Maity, S.: Mobile banking a new banking model: an empirical investigation of financial innovation. Int. J. Bus. Innovation Res. **32**(2), 188–206 (2023)

Sajjad, A.: The effect of corporate governance in Islamic banking on the agility of Iraqi banks. J. Risk Finan. Manag. **16**, 292 (2023). https://doi.org/10.3390/jrfm16060292

Salman, M.D., et al.: The impact of engineering anxiety on students: a comprehensive study in the fields of sport, economics, and teaching methods. Revista iberoamericana de psicología del ejercicio y el deporte **18**(3), 326–329 (2023)

Sarhan, S.S., Al-Ali, A.H.: The role of financial structure balance in ensuring the Iraqi bank's financial health an analytical study of a sample of banks listed in Isx-Iq For the Period 2015–2020). TANMIAT AL-RAFIDAIN, 41(136) (2022)

Gohil, S., Pravinbhai, R.B., Ruiwale, A.: A comparative analysis of customer satisfaction over e-banking services of public sector and private sector banks. EPRA Int. J. Multidisc. Res. (IJMR), **9**(2), 313-317 (2023). https://doi.org/10.36713/epra12531

Shirkhodaie, M.: Factors affecting the intention to use electronic banking services via mobile phone in Iraq and solutions to improve them. J. Univ. Babylon, **31**, 45–54 (2023). https://doi.org/10.29196/jubpas.v31i2.4654

Thamer, H., Mardan, Z.A.: The role of information technology applications represented by electronic payment and distribution services in improving the quality of banking performance: a field study on a sample of Iraqi banks. J. Contemp. Issues Bus. Gov. **29**(3), 91–116 (2023)

Turki, M.D., Sabbar, M.: Studying the role of banking marketing in supporting the competitive advantage of the Iraqi banks. Int. J. Prof. Bus. Rev. Int. J. Prof. Bus. Rev. **8**(4), 13 (2023)

Artificial Intelligence and Trends Using in Sustainability Audit: A Bibliometric Analysis

Hakeem Hammood Flayyih[1] ⓘ, Safauldeen Ali Shamukh[2](✉) ⓘ,
Hayder Abdulsattar Jabbar[3] ⓘ, and Hussein Qusay Abbood[1] ⓘ

[1] College of Administration and Economics, University of Baghdad, Baghdad, Iraq
{hakeem.hmood, hussein.hussein1205}@coadec.uobaghdad.edu.iq
[2] Department of Accounting, College of Administration and Economics, Islamic Azad
University, Tehran, Iran
safaadeen1991@gmail.com
[3] Ministry of Agriculture, Baghdad, Iraq

Abstract. The purpose of this study is to give an overview of the evolution of auditing and Artificial Intelligence (AI) research posted in the Scopus database, with the goal of tracking the increase of scientific activities that may direct the way for future studies by exposing gaps in the area. We found 614 studies in this subject, spanning 1976 to 2024. We focused our investigation on 65 auditing and AI research articles published between 2015 and 2024, utilizing bibliometric analysis with VOSviewer. Due to the growing reliance of businesses on AI, along with the rise in the number of businesses sponsoring AI, this article outlines the current research trend in the fields of AI and auditing. It does this by offering a thorough bibliometric analysis of the trend and development of research over the last ten years. This involves looking up nations, scholarly publications, and research-related keywords. The results show that experts in the field of employing AI in auditing are noticing a convergence in the application of this technology across emerging and developed countries. We find, using VOSviewer research, that the decision-making process is the primary reason why audit firms adopt AI. The most frequently published nations and keywords are selected. Our research demonstrates the lack of application of AI by several auditing firms in various nations. Finally, the data analysis indicates several possible research concerns to study in the interaction between auditing and AI, which will serve as a focus for future research. Once again, this report provides a framework for audit companies to focus on AI-related issues.

Keywords: Auditing · Artificial Intelligence · Bibliometric Analysis · VOSviewer

1 Introduction

The term "Artificial Intelligence" piques the interest of both novices and experts. The concept of a man-made machine or the conscious ability to think, learn, and make decisions on its own is absolutely incredible, and before it became a reality as it is now, an attempt to consolidate the ideas of AI was observed through Hollywood films, as we saw the human imagination run wild. Exploring all of the possibilities of AI (Hasan, 2022; Atiyah, 2023; Husin et al., 2023; Al-Enzi et al., 2023). AI in our current era is one of the most important and growing magazines in all areas of life. According to the 2019 IT managers, a survey conducted by Gartner, a leading research and consulting firm, the percentage of companies implementing AI has increased by 270 percent in the past four years. Global spending on AI has crossed the threshold to $97.9 billion in 2023, compared to spending in 2019, which amounted to approximately $37.5 billion (Zemankova, 2019; Muhsen et al., 2023). Numerous industries have seen radical change as a result of the development of AI technology. The service sector is one of these, and it is distinguished by a workforce with specialized skills, a high level of customization, strong client interaction, and knowledge intensity. Although, the individual characteristics of the professional providers sectors are somehow away from advancements in technology. In fact, with the increasing call of AI in the fields of service providing, this environment has changed in the modern day (Yang et al., 2024). New accounting standards must be developed considering the rapid advancement of digital technologies such as blockchain, data analytics, and AI. The regulatory framework in this field needs to be reconditioned because of the huge impact that these technologies have on accounting and reporting procedures (Alyaseri et al., 2023; Shapovalova et al., 2023; Hadi & Flayyih, 2024; Atiyah and Zaidan, 2022; Atiyah et al., 2023). Thus, the accounting and auditing profession faces a serious challenge due to the rapid development in communications and information technologies. This challenge is displayed by the need to provide tools that authorize it to deal with the modern technical environment, as well as the emergence of what is called "Digital Auditing". This is particularly important given that these technologies support the work of accountants and auditors in several ways, including creating knowledge bases, improving outputs, rationalizing, and directing methods for dealing with daily procedures, improving the quality of services, and supporting the audit strategy to reduce audit risks and raise the profitability of accounting and auditing companies (Aljaaidi et al., 2023; Hadi et al., 2023; Flayyih et al., 2024; Atiyah et al., 2023). The first area of business where Information and communications technology (shortly, ICT) tools and procedures have been used is accounting. Even though ICT was initially used for simple accounting systems, financial modeling software has rapidly shown to be quite helpful in the analytical parts of accounting since the 1980 s. Because of the conservative mindset of its practitioners, the acceptance of ICT by the accounting profession has been sluggish. But by the late 1990 s, the industry was compelled to go computerized to stay competitive, increase productivity, and save costs (Omoteso, 2012). Any business's window is its accounting department. We may observe a dramatic shift away from journals and ledgers and toward computer-based formats because of the growing usage of computers in the accounting information system (Ali et al., 2023). Robots (AI) had to be included into the accounting database to keep up with this drastic development. The application of robotics technology in accounting is the application of programs based on specialized systems and other technologies during the process of recording, reporting, and communicating commercial and financial information, as well as in the auditing process, enabling companies to prepare financial reports and communicate information easily and transparently that meets the needs of decision makers (Ashok &

MS, 2019; Alnoor et al., 2024). Currently, many large audit companies are reporting using AI (AI) in their audit and advisory responsibilities. They cite benefits including improved client service, faster data processing, higher accuracy levels, and more in-depth insight into business operations. Considering that AI is a new technology that seeks to replicate human cognitive abilities and judgment, adopters of this technology will have a competitive advantage. As a result, the Big Four audit firms state that they are utilizing their plans to carry out this innovation in a variety of ways, such as creating audit working papers, identifying audit risks, planning the audit process, testing, and analysis, among other uses in the auditing industry (Munoko et al., 2020; Atiyah et al., 2023). With its immense and compelling advantages, AI is currently witnessing what many organizations refer to as the "spring of AI." Naturally, as our reliance on AI grows, auditing will also grow. There is some consensus that, in conjunction with the use of fast machines and huge data, fundamental progress has been made in the last few years, despite the advancement being dotted with false beginnings and overstated claims. Significant investments in AI have been disclosed by numerous auditing firms recently (Issa et al., 2016; Agustí & Orta-Pérez, 2023). Big data has made it possible for enterprises to generate complex and unstructured information, which can make it difficult for audit firms to analyze the data and provide the necessary audited reports on time. In addition to making audits more accurate, AI-powered auditing technology represents a significant advancement in the current auditing landscape. The utilization of AI-assisted auditing approaches in external auditing has the potential to boost audit efficiency, improve financial report quality, guarantee superior audit quality, and assist decision makers in making dependable choices (Hu et al, 2021; Alnoor et al., 2024). With the advent of AI, professional services firms including audit firms should improve their services to adapt to AI. However, traditional ad hoc innovations led by specialists have limitations in incorporating new technology outside their expertise (Goto, 2023). AI systems have brought about a significant change in the auditing process; However, opponents of the AI revolution view this growth as a step backwards, as many auditors will fail to adapt to this new environment and will be left behind (Albawwat & Frijat, 2021; Abass et al., 2023; Alnoor et al., 2023). AI can support a company's internal audit function by providing significant strategic oversight, reducing manual procedures, and enabling an additional value-added audit service (Wassie & Lakatos, 2024). We go over the following research issues to show the problem scenario and investigate the connection between auditing and AI: "1) How are articles in the fields of AI and auditing distributed? 2) Which phrases and fields of study are most frequently used? 3) Which nation is leading in terms of the primary uses of AI in accounting and auditing?". The remainder of this paper is structured as follows. Section 2 describes the literature review. Section 3 describes the research methodology, while Sect. 4 presents the data analysis. Section 5 highlights the findings and discussions surrounding the main findings, followed by Sect. 6 which presents the conclusion.

2 Literature Review

Many academics in their literary studies have proposed the idea of so-called AI in accounting and auditing (Abukhader, 2020; Almufadda & Almezeini, 2022; Estep et al., 2023; Hu et al., 2021; Issa et al., 2016; Kokina & Davenport, 2017; Ali et al., 2024;

Lehner et al., 2022) where points of view important to company success were discussed. Furthermore, models for increasing the use of AI in auditing were established, including a process model. Since then, numerous studies have been undertaken on the advancement of the application of AI in accounting. These studies mostly focused on financial and accounting processes, examining elements that contribute to improve the quality of financial reporting (Balashova et al., 2021; Abbas et al., 2023). Researchers have explored the early conceptual levels of AI in accounting and auditing (Aksoy & Gurol, 2021; Issa et al., 2016; Munoko et al., 2020). Many studies in literature point to different ways of defining and classifying AI in this field. A study (Omoteso, 2012) investigated auditors' usage of AI systems in historical knowledge contributions in order to predict future research and software development trends. The outcomes obtain a specific research gap in evaluating the impact of AI on the design and watching of internal control systems, the effectiveness of audit committees, and the implications of using such systems for small and medium-sized audit firms, supervisory education, auditing public sector institutions, auditor independence, and the performance gap between audit expectations. A study (Munoko et al., 2020) shows the unintended effects that may happen as the uses and advantages of AI in the auditing profession continue to emerge, by looking at the benefits of AI and the ethical issues of adopting this new technology. By reaching two different ethical frameworks, the study obtains a conceptual view of the practical ethical and societal challenges surrounding AI through an examination of past research. According to the report, AI improves efficiencies, provides greater visibility into business operations, and gives adopters a competitive advantage. However, there are significant ethical questions about this technology that, if left unanswered, could offset the promised benefits. A study (Albawwat & Frijat, 2021) investigated whether the simplicity of use, usefulness, and contribution to audit quality varied based on the type of AI system. The study used a questionnaire to collect data from 124 Jordanian auditors who work for local firms. According to the findings, auditors consider supported and augmented AI systems to be simple to use in auditing, however freestanding AI systems are considered complicated. Fedyk et al. (2022) evaluated the impact of AI on auditing quality and efficiency. The Babina and Others scale, published in 2020, was used to assess AI via the lens of intellectual capital. The study analyzed unique data from over 310,000 employment resumes for the 36 largest auditing firms looking for AI workers. The study revealed preliminary findings on the AI workforce in the auditing sector. The survey found that workers in the field of AI tend to be male, relatively young, and generally hold academic degrees. The findings suggest that investing in AI improves audit quality, lowers fees, and eventually replaces human auditors, while the impact on labor takes several years to completely implement. A study of (Goto, 2023) was undertaken in the context of the Big Four audit firms, integrating AI into the external audit service in Japan in the first decade of the twenty-first century. The study highlights the significance of research and development of audit firms' research services in boosting AI adoption in the case firms. The purpose of a study (Aljaaidi et al., 2023) was to determine how using AI apps affected the efficiency of auditing firms and accountants. Thirty-eight Saudi Arabian audit firms made up the final sample. The regression analysis's findings demonstrated that auditing firms who employ AI software view it as a helpful tool for improving the efficiency of both their auditing and accounting departments. It can decrease the time, expense, and effort required for the audit process; give audit firms a competitive edge; assist auditors in determining materiality more accurately; enhance team performance; and elevate continuous auditing above traditional auditing, allowing auditors to choose audit samples with great efficiency. Enhancing the standard of control protocols for client-used electronic files and transactions; helping to manage tasks and operations with increasingly sophisticated and intelligent mechanisms; raising the effectiveness and efficiency of the audit process; raising the effectiveness and efficiency of the process of planning and supervising the audit process; and lowering audit risks and uncertainty. The impact of AI variables and the technology acceptance model on several areas of accounting and auditing processes was the focus of a study (Abdullah & Almaqtari, 2024; Ahmed et al., 2024). In the study, 228 respondents made up the sample. The findings suggest that utilizing deep learning, cloud computing, big data analytics, and AI can enhance accounting and auditing procedures. AI technology enable businesses to make decisions more accurately, efficiently, and with greater efficiency, which improves financial reporting and auditing. A study of (Anh et al., 2024; Alnoor et al., 2024) investigated the impact of technology readiness on AI adoption by accountants and auditors, employing mediating factors such as perceived usefulness and perceived ease of use inside Vietnamese enterprises. Based on 143 questionnaires. The findings revealed a positive correlation between the use of virtual reality and AI by accounting and auditing experts.

3 Methodology

This study examines auditing and AI literature using bibliometric methodologies. We used bibliographic methods to quantitatively analyze bibliographic data. Broadus (1987) developed this strategy to analyze past studies.

Journal articles on auditing and AI were retrieved on March 22, 2024, primarily from the Scopus database, using the following search equation in the "TITLE-ABS-KEY (artificial AND intelligence AND auditing)" approach. It is an important data source for obtaining scientific articles in the literature review at the present time. By relying on the keywords of published articles, the bibliometric study revealed the most important published topics, amounting to 614 articles. This topic was investigated within specific places in the article, which included "Article title, abstract and keywords". Figure 1 presents the roadmap for bibliometric analysis.

VOSviewer was used to examine the bibliographic keywords found in 65 papers. The keywords of the nation or author are one of the elements that are interesting to investigate. A strong relationship can exist between any two elements. Each link has a strength, indicated by a positive numerical number; the higher the value, the stronger the bond. We overlooked a country's co-authorship relationships with other countries.

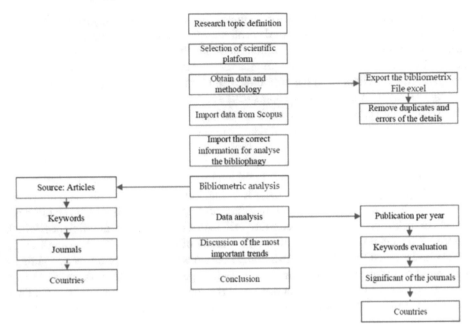

Fig. 1. Workflow in bibliometric analysis (see (Al-Khoury et al., 2022).

4 Results

Over the course of 48 years, 614 research articles were published (see Fig. 2). As of 1976 (Clark, 1976), the earliest known publication date, no additional publications were recorded until 2015. Between 2018 and 2024, interest in the link between AI and auditing has grown significantly. The overall number of publications climbed from six in 2018 to 21 articles, while the number of articles increased to 127 in 2023, and as of the date of this study, the number of indexed articles had reached 38 for the year 2024. Although the yearly growth rate increased by 90% in 2023, it more than doubled (84%) from 2014 and 2024. As a result, the annual figure is likely to grow more. Most papers, however, were closed access and not accessible through the Scopus database. As of 2020, just 38% (207 papers) had been published in open type. As a result, once journal access is granted, the citation score will grow automatically. The results also revealed that the papers used in this study were published in six different languages. The current study focused on papers published in English, the most extensively utilized language, totaling 602 items. Figure 2 displays the annual and cumulative number of research articles in AI, accounting, and auditing indexed in Scopus from 1976 to 2024.

The research contributions of the top ten nations worldwide from 1976 to 2024 are shown in Table 1. It is observed that the United States of America leads the world in terms of research contributions, accounting for 30.27% of all studies. China follows with a rate of 14.41%, and Australia is ranked tenth, accounting for 4.5%.

Because the research includes the words in the study's title, abstract, and primary words, it is highlighted that the cognitive contributions in this field have been spread

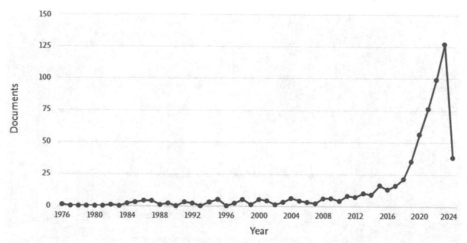

Fig. 2. Annual and cumulative number of research articles in AI and accounting and auditing for the period 1976–2024.

Table 1. Distribution of research articles in AI, accounting and auditing for the ten most published countries for the period 1976–2024.

Precentage	Number	Country	No
30.27%	145	United States	1
14.41%	69	China	2
11.06%	53	India	3
10.44%	50	United Kingdom	4
7.10%	34	Undefined	5
6.89%	33	Germany	6
5.22%	25	Spain	7
5.22%	25	Canada	8
4.80%	23	Italy	9
4.59%	22	Australia	10

among different specializations, as shown Fig. 3. We carried out a thorough inventory to investigate the contributions made by specialized research further. Articles published in specialist journals were chosen based on keywords and open access publications for the years 2015 to 2024.

Journals specialized in administrative, economic, accounting, and financial sciences were selected, as shown in Tables 2 and 3. Which shows that the vast majority of articles published in economic, administrative, financial, and accounting sciences, amounting to 42 articles during the selected period, while the rest of the journals, which constitute a small percentage, were published in other magazines, which numbered 23 articles.

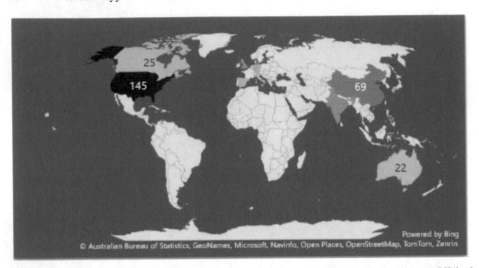

Fig. 3. Distribution of research articles in AI, accounting, and auditing for the ten most published countries for the period 1976–2024.

Table 2. Names of journals that included studies on AI and auditing for the period from 2015 to 2024.

Subject Area	Number of Articles
Business, Management and Accounting	29
Economics, Econometrics and Finance	13
Computer Science	5
Decision Sciences	5
Social Sciences	5
Engineering	3
Pharmacology, Toxicology and Pharmaceutics	2
Arts and Humanities	1
Medicine	1
Nursing	1

Figure 4, created with VOSviewer, illustrates this for us. For the years 2015 through 2024, we prepared this network based on knowledge contributions. We mostly depended on the 428 keywords found in the studies that we chose.

Table 3. Distribution of research articles in AI, accounting and auditing for the ten most published countries for the period 2015–2024.

No	COUNTRY	Number
1	United States	11
2	United Kingdom	4
3	China	4
4	Taiwan	2
5	Australia	3
6	Canada	3
7	Jordan	3
8	Netherlands	3
9	Oman	3
10	Saudi Arabic	3

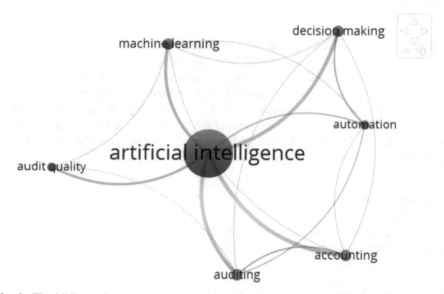

Fig. 4. The bibliometric network is created based on knowledge contributions for the period 2015–2024.

Figure 5, created with VOSviewer, illustrates this for us. For the years 2015 through 2024, we prepared this network based on knowledge contributions. We mostly depended on the 428 keywords found in the studies that we chose.

Following the consideration of using keywords to provide cognitive contributions in data identifiers. We can now clearly see that there are three networks: the primary network, which is blue; there are also networks that are green and red. The quality of

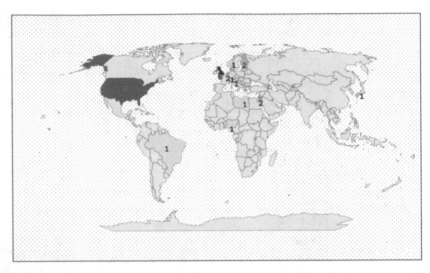

Fig. 5. Distribution of research articles in AI, accounting and auditing for the ten most published countries for the period 2015–2024.

auditing and machine learning in the first place, accounting and auditing in the second place, and decision-making in the third place are found to be clearly correlated with AI. It is evident that automation and audit processes are the main factors that influence decision-making in accounting and auditing, while machine learning and automation are the two ways that accounting is connected to decision-making. On the other hand, there is a route for audit quality that connects machine learning and auditing. The network's primary emphasis was AI, which emphasizes the importance of integrating AI into accounting, auditing, and decision-making processes.

5   Discussions

The business world's accounting cycle is made up of numerous interrelated procedures. But even with the development of new technologies that are helpful at every stage of the process, their integration still appears to be restricted, and their promise is still largely unrealized (Faccia et al., 2019). By employing a thorough research methodology, we can broaden our understanding of earlier contributions to AI. This study used a bibliometric approach to analyze literature pertaining to accounting, auditing, and AI. The findings show that although automation has been used in financial processes since the 1970 s of the previous centuries, published field research on the application of AI in accounting and auditing has been around since the beginning, albeit at a slow pace. With the usage of AI reaching a high recently, the emphasis has shifted to offering precise solutions for the financial reporting process and its reporting quality in relation to audit quality. The number of publications in research has increased significantly during the last five years, yet it cannot be said that this trend is at its highest point. Nonetheless, it can be predicted that the application of AI in accounting and auditing will rise significantly in the upcoming years. This can be attributed to the growth of AI companies as well as the growing number of businesses interested in this technology. Thus, perspective may play a role in how this field of study develops in the future. The consensus among businesses is that implementing AI should aim to increase value creation for the organization. Companies that recognize the importance of AI are seen by the public as not only sharing a competitive advantage, but also aiming to achieve future financial savings. This paper provides evidence of the weak reliance of auditing companies in various countries on resorting to AI.

6 Conclusions

The trend and advancement of research on the application of AI in accounting and auditing were summarized in the current paper based on a database search of 65 articles from Scopus. The study gave a general overview of most of the countries, and the current paper also covered certain topics that were restricted to the author's keywords. This study employed large-scale data analysis and bibliometric analysis to look at scientific publications linked to research variables for the years 2015–2024. According to the findings of this study, research on this area has been rapidly growing since 2015. The statistical results of the data sets revealed that the United States came in top, followed by the United Kingdom, China, and Taiwan in third. The third category featured several Arab countries, including the Sultanate of Oman and the Kingdom of Saudi Arabia, which are among the top ten in the world. We discovered a clear relationship between AI, audit quality, and machine learning through network relationships via VOSviewer, followed by the relationship of AI to accounting and auditing in second place, and decision-making in third place, indicating that decision-making is the primary goal of using AI in accounting. And auditing is mostly dependent on process automation, whereas accounting is related to auditing. We've noticed that AI is the primary focus of analysis, highlighting the key benefit of incorporating AI into accounting, auditing, and decision-making processes. Finally, the data analysis identifies several potential research issues to be investigated regarding the relationship between auditing and AI, which serve as an area for future research. Once again, this study also provides a framework for audit firms to pay attention to AI-related areas.

References

Abass, Z.K., Al-Abedi, T.K., Flayyih, H.H.: Integration between COBIT and COSO for internal control and its reflection on auditing risk with corporate governance as the mediating variable. Int. J. Econ. Finan. Stud. **15**(2), 40–58 (2023). https://doi.org/10.34109/ijefs.202315203

Abbas, S., et al.: Antecedents of trustworthiness of social commerce platforms: a case of rural communities using multi group SEM & MCDM methods. Electron. Commer. Res. Appl. **62**, 101322 (2023)

Abdullah, A.A.H., Almaqtari, F.A.: The impact of artificial intelligence and Industry 4.0 on transforming accounting and auditing practices. J. Open Innov. Technol. Market Complexity **10**(1), 100218 (2024)

Abukhader, S.M.: Extent of artificial intelligence into accounting and auditing work - An analytical attempt of job and duties. Int. J. Bus. Process. Integr. Manag. **10**(2), 125–136 (2020). https://doi.org/10.1504/ijbpim.2020.117165

Agustí, M.A., Orta-Pérez, M.: Big data and artificial intelligence in the fields of accounting and auditing: a bibliometric analysis. Spanish J. Finan. Acc. Revista Española de Financiación y Contabilidad **52**(3), 412–438 (2023)

Ahmed, A.D., Salih, M.M., Muhsen, Y.R.: Opinion weight criteria method (OWCM): a new method for weighting criteria with zero inconsistency. IEEE Access **12**, 5605–5616 (2024)

Aksoy, T., Gurol, B.: Artificial intelligence in computer-aided auditing techniques and technologies (CAATTs) and an application proposal for auditors. In: Aksoy, T., Hacioglu, U. (eds.) Auditing Ecosystem and Strategic Accounting in the Digital Era: Global Approaches and New Opportunities, pp. 361–384. Springer International Publishing, Cham (2021). https://doi.org/10.1007/978-3-030-72628-7_17

Albawwat, I., Frijat, Y.: An analysis of auditors' perceptions towards artificial intelligence and its contribution to audit quality. Accounting **7**(4), 755–762 (2021)

Al-Enzi, S.H.Z., Abbas, S., Abbood, A.A.J., Muhsen, Y.R., Al-Hchaimi, A.A.J., Almosawi, Z.: Exploring research trends of metaverse: a bibliometric analysis. In: Al-Emran, M., Ali, J.H., Valeri, M., Alnoor, A., Hussien, Z.A. (eds.) Beyond Reality: Navigating the Power of Metaverse and Its Applications: Proceedings of 3rd International Multi-Disciplinary Conference - Theme: Integrated Sciences and Technologies (IMDC-IST 2024) Volume 1, pp. 21–34. Springer Nature Switzerland, Cham (2023). https://doi.org/10.1007/978-3-031-51716-7_2

Ali, J., Hussain, K.N., Alnoor, A., Muhsen, Y.R., Atiyah, A.G.: Benchmarking methodology of banks based on financial sustainability using CRITIC and RAFSI techniques. Decis. Mak. Appl. Manage. Eng. **7**(1), 315–341 (2024)

Ali, S.I., Al-taie, B.F.K., Flayyih, H.H.: The effects of negative audit and audit environment on the internal auditor performance: mediating role of auditing process independence. Int. J. Econ. Finan. Stud. **15**(1), 64–79 (2023). https://doi.org/10.34109/ijefs.202315105

Aljaaidi, K., Alwadani, N., Adow, A.: The impact of artificial intelligence applications on the performance of accountants and audit firms in Saudi Arabia. Int. J. Data Netw. Sci. **7**(3), 1165–1178 (2023)

Al-Khoury, A., et al.: Intellectual capital history and trends: a bibliometric analysis using Scopus database. Sustainability **14**(18), 11615 (2022). https://doi.org/10.3390/su141811615

Almufadda, G., Almezeini, N.A.: Artificial intelligence applications in the auditing profession: a literature review. J. Emerg. Technol. Account. **19**(2), 29–42 (2022). https://doi.org/10.2308/JETA-2020-083

Alnoor, A., Atiyah, A.G., Abbas, S.: Toward digitalization strategic perspective in the European food industry: non-linear nexuses analysis. Asia-Pac. J. Bus. Adm. (2023)

Alnoor, A., Atiyah, A.G., Abbas, S.: Unveiling the determinants of digital strategy from the perspective of entrepreneurial orientation theory: a two-stage SEM-ANN approach. Global J. Flex. Syst. Manage. **25**(2), 243–260 (2024). https://doi.org/10.1007/s40171-024-00385-0

Alnoor, A., Chew, X., Khaw, K.W., Muhsen, Y.R., Sadaa, A.M.: Benchmarking of circular economy behaviors for Iraqi energy companies based on engagement modes with green technology and environmental, social, and governance rating. Environ. Sci. Pollut. Res. **31**(4), 5762–5783 (2024)

Alnoor, A., et al.: How positive and negative electronic word of mouth (eWOM) affects customers' intention to use social commerce? A dual-stage multi group-SEM and ANN analysis. Int. J. Hum. Comput. Interact. **40**(3), 808–837 (2024)

Alyaseri, N.H.A., et al.: Exploring the modeling of socio-technical systems in the fields of sport, engineering and economics. Revista iberoamericana de psicología del ejercicio y el deporte **18**(3), 338–341 (2023)

Anh, N.T.M., et al.: The effect of technology readiness on adopting artificial intelligence in accounting and auditing in Vietnam. J. Risk Finan. Manage. **17**(1), 27 (2024)

Ashok, M.L., MS, D.: Emerging trends in accounting: an analysis of impact of robotics in accounting, reporting and auditing of business and financial information. Int. J. Bus. Analytics Intell. **7**(2), 28–34 (2019).

Atiyah, A.G.: Unveiling the quality perception of productivity from the senses of real-time multisensory social interactions strategies in metaverse. In: Al-Emran, M., Ali, J.H., Valeri, M., Alnoor, A., Hussien, Z.A. (eds.) International Multi-Disciplinary Conference-Integrated Sciences and Technologies, pp. 83–93. Springer Nature Switzerland, Cham (2023). https://doi. org/10.1007/978-3-031-51300-8_6

Atiyah, A.G., Zaidan, R.A.: Barriers to using social commerce. In: Alnoor, A., Wah, K.K., Hassan, A. (eds.) Artificial Neural Networks and Structural Equation Modeling: Marketing and Consumer Research Applications, pp. 115–130. Springer Nature Singapore, Singapore (2022). https://doi.org/10.1007/978-981-19-6509-8_7

Atiyah, A.G., Alhasnawi, M., Almasoodi, M.F.: Understanding metaverse adoption strategy from perspective of social presence and support theories: the moderating role of privacy risks. In: Al-Emran, M., Ali, J.H., Valeri, M., Alnoor, A., Hussien, Z.A. (eds.) International Multi-Disciplinary Conference-Integrated Sciences and Technologies, pp. 144–158. Springer Nature Switzerland, Cham (2023). https://doi.org/10.1007/978-3-031-51300-8_10

Atiyah, A.G., All, N.D.A., Zaidan, A.S., Bayram, G.E.: Understating the social sustainability of metaverse by integrating adoption properties with users' satisfaction. In: Al-Emran, M., Ali, J.H., Valeri, M., Alnoor, A., Hussien, Z.A. (eds.) International Multi-Disciplinary Conference-Integrated Sciences and Technologies, pp. 95–107. Springer Nature Switzerland, Cham (2023). https://doi.org/10.1007/978-3-031-51716-7_7

Atiyah, A.G., Faris, N.N., Rexhepi, G., Qasim, A.J.: Integrating ideal characteristics of chat-GPT mechanisms into the metaverse: knowledge, transparency, and ethics. In: Al-Emran, M., Ali, J.H., Valeri, M., Alnoor, A., Hussien, Z.A. (eds.) International Multi-Disciplinary Conference-Integrated Sciences and Technologies, pp. 131–141. Springer Nature Switzerland, Cham (2023). https://doi.org/10.1007/978-3-031-51716-7_9

Balashova, N.N., Vardanyan, S.A., Volodina, M.V., Ishkina, N.A., Koshkarev, I.A.: Prospects of the use of artificial intelligence and automatization systems in accounting and auditing in the realities of the digital economy. In Adv. Res. Russ. Bus. Manage. **2021**, 527–534 (2021)

Broadus, R.N.: Toward a definition of "bibliometrics." Scientometrics **12**, 373–379 (1987)

Clark, I.A.: Relational data dictionary implementation. In: Hasselmeier, H., Spurth, W.G. (eds.) IBM 1975. LNCS, vol. 39, pp. 279–290. Springer, Heidelberg (1976). https://doi.org/10.1007/3-540-07612-3_12

Estep, C., Griffith, E.E., MacKenzie, N.L.: How do financial executives respond to the use of artificial intelligence in financial reporting and auditing? Rev. Acc. Stud. (2023). https://doi. org/10.1007/s11142-023-09771-y

Faccia, A., Al Naqbi, M.Y.K., Lootah, S.A.: Integrated cloud financial accounting cycle: how artificial intelligence, blockchain, and XBRL will change the accounting, fiscal and auditing practices. In: Proceedings of the 2019 3rd International Conference on Cloud and Big Data Computing, pp. 31–37 (2019).

Fedyk, A., Hodson, J., Khimich, N., Fedyk, T.: Is artificial intelligence improving the audit process? Rev. Acc. Stud. **27**(3), 938–985 (2022)

Flayyih, H.H., Hadi, H.A., Al-Shiblawi, G.A.K., Khiari, W.: The effect of audit team and audit committee performance on the quality of audit. J. Gov. Regul. **13**(2), 59–67 (2024). https:// doi.org/10.22495/jgrv13i2art5

Goto, M.: Anticipatory innovation of professional services: The case of auditing and artificial intelligence. Res. Policy **52**(8), 104828 (2023)

Hadi, A.H., Abdulhameed, G.R., Malik, Y.S., Flayyih, H.H.: The influence of information technology (IT) on firm profitability and stock returns. Eastern Eur. J. Enterp. Technol. **4**(13 (124)), 87–93 (2023). https://doi.org/10.15587/1729-4061.2023.286212

Hadi, H.A., Flayyih, H.H.: Analysis of the accounting financial performance of private listed banks in the emerging market for the period 2010–2022. Corp. Bus. Strategy Rev. **5**(1), 8–15 (2024). https://doi.org/10.22495/cbsrv5i1art1

Hasan, A.R.: Artificial intelligence (AI) in accounting & auditing: a literature review. Open J. Bus. Manage. **10**, 440–465 (2022). https://doi.org/10.4236/ojbm.2022.101026

Kuang-Hua, H., Chen, F.H., Hsu, M.F., Tzeng, G.H.: Identifying key factors for adopting artificial intelligence-enabled auditing techniques by joint utilization of fuzzy-rough set theory and MRDM technique. Technol. Econ. Dev. Econ. **27**(2), 459–492 (2020). https://doi.org/10.3846/tede.2020.13181

Husin, N.A., Abdulsaeed, A.A., Muhsen, Y.R., Zaidan, A.S., Alnoor, A., Al-mawla, Z.R.: Evaluation of metaverse tools based on privacy model using fuzzy MCDM approach. In: Al-Emran, M., Ali, J.H., Valeri, M., Alnoor, A., Hussien, Z.A. (eds.) Beyond Reality: Navigating the Power of Metaverse and Its Applications: Proceedings of 3rd International Multi-Disciplinary Conference - Theme: Integrated Sciences and Technologies (IMDC-IST 2024) Volume 1, pp. 1–20. Springer Nature Switzerland, Cham (2023). https://doi.org/10.1007/978-3-031-51716-7_1

Issa, H., Sun, T., Vasarhelyi, M.A.: Research ideas for artificial intelligence in auditing: the formalization of audit and workforce supplementation. J. Emerg. Technol. Account. **13**(2), 1–20 (2016). https://doi.org/10.2308/jeta-10511

Kokina, J., Davenport, T.H.: The emergence of artificial intelligence: how automation is changing auditing. J. Emerg. Technol. Account. **14**(1), 115–122 (2017). https://doi.org/10.2308/jeta-51730

Lehner, O.M., Ittonen, K., Silvola, H., Ström, E., Wührleitner, A.: Artificial intelligence based decision-making in accounting and auditing: ethical challenges and normative thinking. Account. Auditing Accountability J. **35**(9), 109–135 (2022). https://doi.org/10.1108/AAAJ-09-2020-4934

Muhsen, Y.R., Husin, N.A., Zolkepli, M.B., Manshor, N.: A systematic literature review of fuzzy-weighted zero-inconsistency and fuzzy-decision-by-opinion-score-methods: assessment of the past to inform the future. J. Intell. Fuzzy Syst. **45**(3), 4617–4638 (2023)

Munoko, I., Brown-Liburd, H.L., Vasarhelyi, M.: The ethical implications of using artificial intelligence in auditing. J. Bus. Ethics **167**(2), 209–234 (2020)

Munoko, I., Brown-Liburd, H.L., Vasarhelyi, M.: The ethical implications of using artificial intelligence in auditing. J. Bus. Ethics **167**(2), 209–234 (2020). https://doi.org/10.1007/s10551-019-04407-1

Omoteso, K.: The application of artificial intelligence in auditing: looking back to the future. Expert Syst. Appl. **39**(9), 8490–8495 (2012). https://doi.org/10.1016/j.eswa.2012.01.098

Shapovalova, A., Kuzmenko, O., Polishchuk, O., Larikova, T., Myronchuk, Z.: Modernization of the national accounting and auditing system using digital transformation tools. Financ. Credit Act. Probl. Theory Pract. **4**(51) (2023).

Wassie, F.A., Lakatos, L.P.: Artificial intelligence and the future of the internal audit function. Humanit. Soc. Sci. Commun. **11**(1), 1–13 (2024)

Yang, J., Blount, Y., Amrollahi, A.: Artificial intelligence adoption in a professional service industry: a multiple case study. Technol. Forecast. Soc. Chang. **201**, 123251 (2024)

Zemankova, A.: Artificial intelligence in audit and accounting: development, current trends, opportunities and threats-literature review. In: 2019 International Conference on Control, Artificial Intelligence, Robotics & Optimization (ICCAIRO), pp. 148–154. IEEE (2019). https://doi.org/10.1109/ICCAIRO47923.2019.00031

Using Artificial Intelligence to Predict the Financial Gearing's Capability to Achieve Financial Sustainability

Mohmmed Jasim Mohmmed[✉]

College of Administration and Economics, Basrah University, Basrah, Iraq
mohammed.jassim@uobasrah.edu.iq

Abstract. The current study aims to use artificial intelligence to predict financial sustainability based on one of the most crucial artificial intelligence techniques, namely multi-layer neural networks, for a sample of commercial banks listed on the Iraqi Stock Exchange for the period (2013–2022). This is accomplished by adopting the gearing ratio (leverage) within the financial structure as a proxy for measuring financial gearing. Here, three indicators of the financial and economic dimension of financial sustainability were adopted: the market value added index, the earnings per share index, and the credit index, as a proxy for measuring financial sustainability. In order to achieve the goal of the study, the researcher relied on a main hypothesis, which is "the ability of artificial layered neural networks to predict financial sustainability through financial gearing ratios". The study reveals several conclusions, the most important of which is the use of neural networks to predict the values of the growth rate in market value added, the earnings per share index, and the credit index. The study gave relatively accurate results, and the experimental results of the study varied among the three approved indicators of financial sustainability. The study also recommended, within the future studies section, the necessity of comparing the results of different models, as is the case in time series models, to determine the pros and cons of modern models, especially those relying on artificial intelligence.

Keywords: Artificial Intelligence · Neural Networks · Financial Gearing (leverage) · Financial Sustainability · Market Value Added · Earnings per Share · Credit Index

1 Introduction

The debate of the accuracy and superiority of modern models in prediction, including artificial intelligence models (neural networks), is one of the areas of research and experimental studies due to the existence of intellectual and experimental dialectics on the one hand and the other hand, financial gearing is one of the important concepts in financial management. Many studies have been conducted and dealt with it by tackling more than one aspect, including its impact on some variables or its determinants and the variables that affect it, starting with the perspectives of Modigliani & Miller. Financial

© The Author(s), under exclusive license to Springer Nature Switzerland AG 2024
A. Alnoor et al. (Eds.): AIRDS 2024, LNNS 1033, pp. 311–327, 2024.
https://doi.org/10.1007/978-3-031-63717-9_20

gearing is an important variable that is related to both the value of the company and the cost of capital, in addition to the savings achieved within the tax shield by protecting the company's income from taxes by subtracting the cost of financial borrowing (financial gearing) from the income realized before the imposition of taxes. In addition to being one of the aspects of the financing decision in companies, and according to the propositions of economic and financial theory and the results of most empirical studies, it represents the basis of financial sustainability for companies, especially in its economic and financial dimensions represented by the three indicators (added market value, earnings per share, credit). On the other hand, it is more than an intellectual debate about the interaction of indicators of financial sustainability, especially in its economic and financial dimensions.

2 Literature Review

2.1 Financial Gearing

Financial gearing occurs when a company is financed at least partially by borrowing or by fixed-rate financing sources. The higher the borrowing ratio compared to the financing provided by ordinary shareholders, the higher the level of financial gearing (Alhasnawi et al., 2023; Ali et al., 2024). Thus, the term financial gearing is applied to any borrowing or use of financial instruments that results in an exaggeration of the effect on profits or losses and is often used to describe the ratio of debt to equity in companies. The higher the debt-to-equity ratio, the greater the impact of financial gearing on the company's profits (Alhasnawi et al., 2023). The concept of financial gearing has been revealed in the literature on corporate finance and financial management, as the term financial gearing is applied to describe the use of certain fixed costs (acting as gearing) that affect the company's performance, i.e., its profitability. For a company, gearing is a fixed financing cost, and in this sense, a distinction must be made between three types of gearing, namely financial gearing, operating gearing, and combined gearing (Bobinaite, 2015:30). DOFL (degree of financial gearing) is measured by dividing the change in EPS by the change in EBIT; thus DOFL is defined as the percentage change in EPS that results from the change in EBIT (Tunji et al., 2015:70).

As for the operating gearing, which is related to the cost structure, it indicates the extent to which fixed costs are used in the company's cost structure (Ross et al., 2022). The cost structure in companies includes two types of costs: variable and fixed costs. Fixed costs are distinguished from variable costs in that they do not change with the change in the volume of production (van Horne & Wachowicz, 2008:445). A firm's fixed costs are also measured as a percentage of its total costs, as a firm with a higher fixed cost will have higher operating gearing compared to a firm with a higher variable cost. Thus, this means that a small change in sales volume leads to a large change in EBIT and ROIC (return on invested capital), i.e., companies with high operating gearing are highly sensitive to changes in sales. Accordingly, we conclude that operating gearing is a function of fixed costs, and DOPL is measured by dividing the change in PBIT (profit before interest and taxes) to the change in sales.

On the other hand, total or combined gearing refers to the gearing that combines the effect of financial risk and business risk, i.e., between financial gearing and operating gearing. Thus, the degree of financial gearing can be combined with the degree of operating gearing to show the overall effect of gearing for a given change of sales in the earnings available to ordinary shareholders. So, combined gearing refers to the benefits and risks of gearing at a fixed level. Competing firms choose a high level of combined gearing, while conservative firms choose a lower level of combined gearing. The degree of total gearing is measured by multiplying the degree of financial gearing (DOFL) by the degree of operating gearing (DOPL) (Tunji et al., 2015:70).

Management influences equity through financial gearing. In contrast to profit margin and asset turnover ratio, some companies prefer a certain level of financial gearing, as it is not something that management necessarily wants to increase even when doing so increases return on equity. The challenge here is to strike a balance between the benefits and costs of financial gearing. More financial gearing is not necessarily preferable, while some companies have a great deal of freedom in choosing the amount of financial gearing. However, there are economic and institutional restrictions on this, as the nature of the company's work and its assets affect the financial gearing. Companies with stable and highly predictable OCF can safely use more financial gearing than companies facing a high degree of uncertainty. In addition, companies such as banks, which usually have diversified portfolios of readily salable liquid assets, can safely use more financial gearing (Robert et al., 2023:48).

2.2 Financial Sustainability

It refers to the ability of companies to continue with current and future policies without causing the debt ratio to rise continuously. A business that has achieved financial sustainability is a business that allows selling a product or service at a price that covers expenses in addition to making a profit. Also, financial sustainability is an interaction of three main dimensions, which are the environmental, social and economic dimensions; each dimension has a set of indicators (Stankeviciene & Nikonorovaa, 2014:1190; Alnoor et al., 2023). The International Monetary Fund also defined it as the situation in which the borrower can continue to service his/her debt without the need to make any adjustments in future expenditures and revenues. The term sustainability is used to include all aspects of sustainability, including environmental responsibility, social responsibility, corporate citizenship, and green marketing (Roghanian et al., 2012:550; Alnoor et al., 2024). Achieving financial sustainability is also related to the ability of companies to achieve financial stability. Sound financial practices are considered the first and foremost element of financial sustainability. In addition, companies seek to achieve financial sustainability in order to be able to borrow to cover their requirements. However, the distinction between financial sustainability and financial self-sufficiency is important during the life of the company, as self-sufficiency is limited to managing financial operations independently without outside help. As for financial sustainability, it is the ability to repeat performance in the future and, thus, allow continuity and provision of services (Chikaza, 2015:8; Alnoor et al., 2024). The emergence of the concept of banking sustainability, which is one of the new topics in financial concepts, has a role in measuring the success or failure of banks through banking sustainability indicators (Atiyah, 2023). On the other hand, the results of a study revealed that the financial sustainability of banks is positively and significantly driven by the intensity or size of loans. The inefficiency of bank management and the portfolio of loans at risk have an adverse and important impact on financial sustainability (Tehulu, 2014:152). Thus, the economic and financial dimension is one of the important and main dimensions that are related to the financial and banking fields. Therefore, we will touch upon the most important indicators of this dimension according to the following:

2.2.1 MVA Indicator

The added market value is one of the criteria marketed by the American investment company Stern & Steward Co. With the economic value added as a criterion for performance evaluation. In general, market value added is referred to as the difference between the current market value of a company and the capital contributed by investors. If the MVA is positive, the company has an added value, and if it is negative, the company has destroyed the value (Atiyah and Zaidan, 2022). The added market value does not only reflect the wealth of the shareholders, but also the market's assessment of the present net value of the company as a whole, as it represents the difference between the current market value and the employed capital or the book value, that is, it represents the acquired wealth of the shareholders (Ahmed et al., 2024; Alnoor et al., 2024). It is also defined as the difference between the company's market share price and the nominal value of the shares (Satwiko & Agusto, 2021:80). It is considered one of the relatively recent measures that show the success and ability of companies to invest the resources available to the company. In addition, it is possible to distinguish between added (EVA) and (MVA), given that the first (EVA) represents a pattern and approach for evaluating internal performance, while the second (MVA) presents an evaluation of external performance. The added market value is affected by a group of main factors, some of which affect the increase and some decrease, such as growth rates, invested capital intensity, and profit margin value. In addition, it is measured in two ways: the first is the difference between the market value of the company's shares and the book value of the equity, and the second is measured by deducting the future economic added value of the company, and, therefore, the added market value is the market standard for the value sought by the company's financial management. (Brealey et al., 2023:865; Husin et al., 2023; Al-Enzi et al., 2023).

2.2.2 EPS Indicator

Earnings per share is the most common and widely used individual criterion for evaluating financial performance. EPS is one of the numbers that is widely used in providing shareholders with information about the profits achieved by the share. It is used as an indicator of earned income for ordinary shares, so companies disclose data and information that lead to an accurate calculation (Kieso et al., 2013:899). In addition, a distinction must be made between two types of earnings per share: basic earnings per share and diluted earnings per share. Management can resort to accounting procedures to maximize return on share rather than shareholder value. This is confirmed by some studies that there is a problem of unprecedented reporting and profit manipulation to the point of including increasing fake profits, with reference to Enron Corporation, where large companies have become trapped in the vicious circle of EPS management to the extent that they have resorted to deceptive accounting fraud to hide a lot of debt which they were using to fund EPS (Wet, 2013:266).

2.2.3 Credit Indicator

Banks and financial institutions take into account a range of factors when providing loans, including relying on objective and real data devoid of personal impressions. This does not necessarily mean that the work of banks is easy, as banks are tasked with providing the largest possible number of loans to achieve material gains and, at the same time, avoid customers defaulting on payments, which is the biggest risk they may face. Therefore, banks allocate two different managements to achieve this difficult equation, which are credit management and credit risk management. The credit indicator is also an important indicator of sustainability due to banks financing the economic sectors through granting credit, including cash credit (bank loans). The greater the volume of bank loans, the more it leads to achieving economic well-being. On the other hand, the credit indicator in Iraq formed the lowest averages among groups of countries in the world, including low-income countries. It appears that the sources of credit financing, including deposits, were not the determining factor for cash credit.

3 Hypothesis Development

Companies strive to achieve a set of goals, including survival, growth, and continuity (sustainability) across various dimensions, particularly the financial and economic aspects. This is accomplished through making a series of decisions known as financial decisions, which revolve around three core decisions. Financing decisions focus on companies obtaining a low-cost financial structure (a mix of funding sources) with a low (WACC). Financial gearing constitutes an important part of the company's value and, therefore, its sustainability. According to financial theory and the results of most experimental studies, there assumed to be a positive relationship between financial gearing (within an acceptable range without reaching high financial risk) and sustainability, especially financially, through its economic and financial dimensions. Experimental studies have adopted various programs, models, or different standard methods. The current study has relied on one of the modern prediction methods, namely artificial neural network software, to predict the capability of financial gearing to achieve financial sustainability. Thus, the study is based on a main hypothesis, which is "the ability of artificial neural networks to predict financial sustainability through financial gearing".

4 Methodology

4.1 Sample Size and Measurements

Ten banks were selected from the commercial banks to provide the required data for the period (2013–2022), representing approximately 27% of the banks listed in the market. The purpose was to either approve or reject the study hypothesis. The data included a set of indicators representing the study variables, namely the percentage growth rate in

Market Value Added, Earnings per Share index, and Credit index. It is not necessary to have a large number of input variables in the neural network field as too many variables may confuse the network (Bjorklund & Uhlin, 2017:8; Abbas et al., 2023; Muhsen et al., 2023; Atiyah et al., 2023; Atiyah et al., 2023). Some preliminary tests were conducted to ensure the validity and acceptability of the data through stability and normal distribution test. Some descriptive statistical indicators for the variables were calculated according to Table 1.

Table 1. Descriptive Statistics for Variables.

S.D	Mean	Max	Min	N	Variable
.39089	.6018	2.06	.07	110	Market Value Added AMVIND
.07075	.0613	.35	.00	110	Earnings per share EPSIND
.16860	.2092	.71	.00	110	Credit Indicator CREIND
.16824	.5720	.90	.24	110	Financial Gearing GEAIND

4.2 Statistical Input (Artificial Neural Networks)

One of the modern computational techniques is Artificial Neural Networks (ANNs), which are highly important due to their ability to mimic human brain behavior and are a type of machine learning algorithm (Kim & Won, 2018:26; Atiyah et al., 2023). The method of analysis using Artificial Neural Networks is one of the techniques of artificial intelligence and involves simulating the human mind to distinguish between different things and recognize patterns using the machine through self-learning. This process mimics the human mind's ability to recognize different things by leveraging accumulated experiences to achieve accurate and optimal results. The Artificial Neural Network consists of three main parts as follows:

1. Input Layer: This layer feeds the neural network with information from outside the network.
2. Hidden Layer: This is the second part of the Artificial Neural Network and is located between the input layer and the output layer. This layer receives information from the input layer through interconnections and processes the information, which is then sent to the output layer.
3. Output Layer: This layer receives information from the hidden layer and consists of processing units that produce final results. Figure 1 illustrates a schematic diagram of the Artificial Neural Network.

There are several types of Artificial Neural Networks based on the number of layers, such as single-layer networks and multi-layer networks. Additionally, network types differ based on their structure and the connections between their components, such as feedforward networks and feedback networks. Moreover, neural network types can be distinguished based on the learning algorithm used, whereas networks using gradient-based learning or non-gradient-based learning algorithms can be identified.

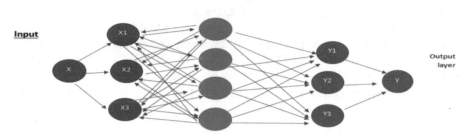

Fig. 1. Key Components of Artificial Neural Network.

5 Results

Description of the Artificial Neural Network

Three artificial neural networks were used along with MATLAB software to analyze and predict the relationship between financial gearing and MVA, earnings per share index, and credit index. Below are the three artificial neural networks that were built for each dependent variable.

5.1 The Artificial Neural Network for the Percentage Growth in Market Value Added Variable

The artificial neural network for the dependent variable, Percentage Growth in Market Value Added, includes the independent variable Financial Gearing. 70% of the observations were allocated for training the artificial neural network, while 15% were dedicated to testing the network and assessing its output accuracy. Additionally, 15% of the total observations were reserved to verify the accuracy of the neural network results. The purpose of building an artificial neural network for the relationship between financial gearing and the Percentage Growth in the Market Value Added variable is to determine the importance of financial gearing in predicting the three variables. By executing the neural network, statistical indicators were obtained, including the Mean Squared Error for the difference between the predicted value and the actual value, as well as the correlation value between the actual values of the input data and the outputs.

Table 2. Sum of Squared Errors and Correlation Coefficient for the Artificial Neural Network of Financial Gearing Model and Percentage Growth in Market Value Added.

R	MSE	Sample Type
0.522	0.01914	Training
0.567	0.02162	Validation
0.342	0.03371	Testing

From Table 2, it is observed that the average squared error values for the difference between the actual value and the predicted value by the neural network are similar for the training, validation, and testing samples, reaching 0.01914, 0.02162, and 0.03371, respectively. Additionally, there is convergence in the correlation coefficient between inputs and outputs, with values of 0.522 and 0.567, respectively. However, the correlation coefficient value for the testing sample is 0.342, which is the weakest. To ensure the effectiveness and accuracy of the neural network, a repeatability plot of the error was generated.

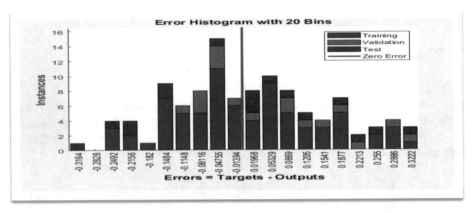

Fig. 2. Repeatability Plot of Errors for the Artificial Neural Network between Financial Gearing and Percentage Growth in Market Value Added.

Figure 2 shows that the errors are symmetrical around zero, indicating that they follow a normal distribution, especially for the training, validation, and testing samples. This means that the network is trained appropriately. From Fig. 3, which represents the quality of fitting the relationship between the GEAIND and MVA, we observe a strong correlation between the target values and the values of the dependent variable for the samples of training, validation, and testing, as well as for the overall data, as this is evident from the alignment between the estimated values during the neural network process and the actual values.

Figure 4 confirms the alignment between the actual values and the values predicted by the neural network.

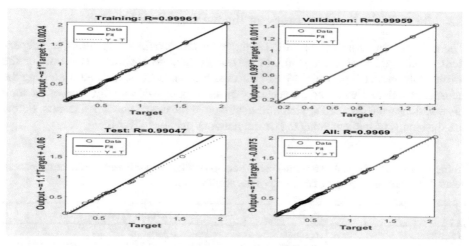

Fig. 3. The regression curve (quality of relationship fitting) between GEAIND and MVA.

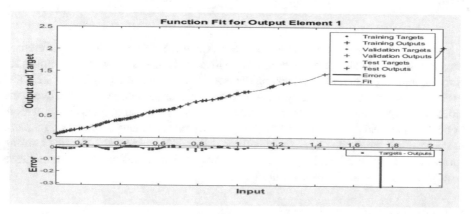

Fig. 4. The regression curve between GEAIND and MVA, along with the error curve.

The above Fig. 4 indicates that the error values between the actual and predicted values are approaching zero, indicating the accuracy of the results of the artificial neural network in predicting the values of the dependent variable (percentage growth in market value added).

From the results of the neural network tests, it became clear that the size of the training sample was able - through the results of the test sample - to reflect results that were close to the true values of the variable of growth in market value, and this was confirmed by the verification sample. Therefore, this applies to most of the experimental results that confirmed the accuracy of neural networks' predictions, and, therefore, the change in the growth of market value in a positive way gives a good perception regarding the possibility of the company's sustainability.

5.2   The Artificial Neural Network for the Earnings-Per-Share Variable

The artificial neural network for the earnings per share variable includes the independent variable financial gearing. 70% of the observations were allocated for training the artificial neural network, while 15% of the observations were designated for testing the neural network and assessing the accuracy of its outputs. Additionally, 15% of the total observations were reserved to verify the accuracy of the neural network's results. Below are the results of implementing the neural network using MATLAB software.

Table 3. The sum of squares errors and the correlation coefficient for the artificial neural network related to the financial gearing and earnings per share model.

R	MSE	Sample Type
0.472	0.00369	Training
0.737	0.00222	Validation
0.647	0.00537	Testing

From Table 3, we observe that the average squared error values for the differences between the actual values and the predicted values by the neural network are similar for both the training, validation, and testing datasets, with values of 0.00369, 0.00222, and 0.00537 respectively. Additionally, we notice that the correlation coefficient between the input and output values in the training dataset is 0.472, indicating a weak correlation. In contrast, the validation dataset where 0.737, which represents a strong correlation. In the testing dataset, the correlation coefficient is 0.647, which also indicates a strong correlation. To ensure effectiveness and accuracy of the neural network, the iterative error plot was generated, as shown in Fig. 5.

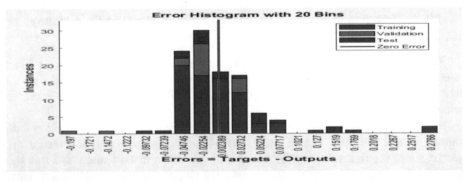

Fig. 5. The iterative plot of errors for the artificial neural network between financial gearing and earnings per share.

Figure 5 shows that the errors between the estimated values and the actual values follow a somewhat normal distribution as they are symmetrical around zero (for the training, validation, and testing samples). From the iterative error plot, we can conclude that the network is trained appropriately. In Fig. 6, which represents the quality of the relationship between the GEAIND and EPSIND, we notice a moderately strong correlation between the estimated values and the actual values for the training sample, with some deviation in the alignment vector of the estimated data from the actual data. Besides, we observe a strong correlation between the estimated values and the actual values during the testing phase, with the deviation between the vectors becoming smaller in this phase. Regarding the relationship between the actual value and the estimated value for the testing sample, the Figure shows a moderate correlation, while the deviation between the vectors indicates a larger deviation than in the other phases.

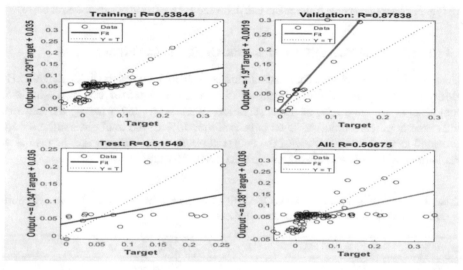

Fig. 6. Regression curve (relationship quality) between the financial gearing variable and the earnings per share.

Figure 7 indicates a relatively weak match between the actual values and the predicted values obtained through the neural network.

The error vector curve depicting the difference between estimated values and actual values shows that the errors fall within the range of (0.2, −0.2), which is acceptable.

Although the error vector curve is acceptable with the values obtained from the test sample results, there is a deviation in the matching vector of the estimated data from the real data. This applies to the fact that what banks may achieve in certain profitability, which is reflected in earnings per share, is not a strong indicator of achieving sustainability. This, in turn, applies to the financial idea in that achieving earnings per share is not the ultimate goal that companies seek to achieve but rather the goal of maximizing market value.

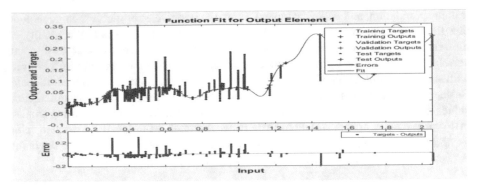

Fig. 7. Regression curve between the GEAIND and EPSIND, along with the error curve.

5.3   The Artificial Neural Network for the Credit Index Variable

The artificial neural network for the credit index variable includes the independent **variable** of financial gearing. 70% of the observations were allocated for training the artificial neural network, while 15% of the observations were set aside for testing the network and assessing the accuracy of its outputs. Another 15% of the total observations were allocated to verify the accuracy of the neural network results. Below are the results of implementing the neural network using MATLAB software.

Table 4. A Sum of Squared Errors and Correlation Coefficient for the Artificial Neural Network Specific to the Financial Gearing and Credit Index Model.

R	MSE	Sample Type
0.588	0.0185	Training
0.712	0.0358	Validation
0.777	0.0362	Testing

From Table 4, we observe that the average squared error values for the difference between the actual value and the estimated value predicted by the neural network are similar for both the validation and test samples, reaching 0.0358 and 0.0362, respectively. The average squared error for the training sample was smaller than the rest of the samples, at 0.0185. Additionally, we notice a convergence in the correlation coefficient between inputs and outputs, with values of 0.712 and 0.777, respectively. However, the correlation coefficient value for the training sample was the weakest at 0.588. To ensure the effectiveness and accuracy of the neural network, a repetitive error plot was created.

Figure 8 shows that the errors between the estimated values and the actual values follow a normal distribution, being symmetrical around zero (for the training, validation, and testing samples). Through the iterative error chart, we can conclude that the network is appropriately trained. In Fig. 9, which represents the quality of the relationship between the CREIND and GEAIND, we observe a moderately strong correlation between the

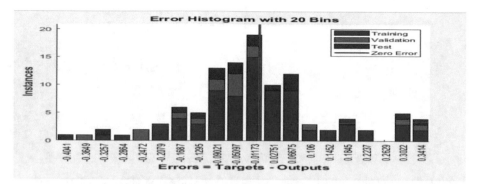

Fig. 8. The iterative error chart for the artificial neural network between financial sustainability and the credit index.

estimated values and the actual values for the training sample, with a relatively minor deviation in the matching vector for the estimated data from the real data. We also note that the correlation between the estimated values and the actual values is weak during the validation stage, with a significant deviation between the vectors at this stage. Regarding the relationship between the actual values and the estimated values for the test sample, the Figure indicates a weak correlation, as evidenced by the significant deviation between the vectors.

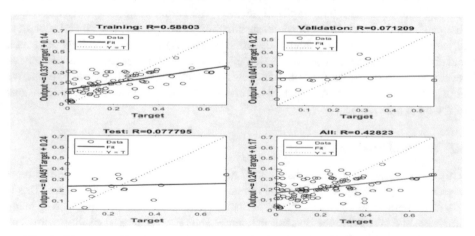

Fig. 9. The regression curve (relationship quality) between the CREIND and GEAIND.

Figure 10 below shows a relatively weak match between the actual values and the values predicted by the neural network.

The error vector curve for the differences between the estimated values and the actual values shows that the errors fall within the range (0.5, −0.5), which is acceptable.

The acceptability of error values for predicting financial sustainability through the granted credit index does not give definitive results, but there is variations in the accuracy of the results of each of these indicators. Therefore, the large credit values granted by

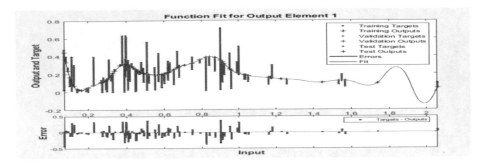

Fig. 10. The regression curve between the CREIND and GEAIND, along with the error curve.

these banks may not have a role in achieving sustainability and growth, which depends on the degree of risk of this credit.

6   Implications and Future Studies

6.1   Theoretical Implications

The current study addresses two aspects of inquiry. The first aspect concerns the accuracy and superiority of modern models in prediction, including neural networks, which has sparked both theoretical and empirical debates. The second aspect explores the role of financial gearing in achieving financial sustainability through its financial and economic dimensions, as indicated by three indicators: market value added, earnings per share, and credit index. The first aspect of the inquiry was adopted to answer the second aspect, revealing that the modern predictive models employed in the study exhibited a relative ability to predict financial sustainability, with variations observed among the dimensions of the three indicators.

6.2   Practical Implications

Artificial intelligence, through its adopted algorithms, has become prominent in most aspects of life, particularly in financial domains and decision-making processes. This is due to its significant ability to mimic human cognition and learn by abstracting many processes and computations that trained neural cells have mastered dealing with. The connections of information between neural cells vary, making the most relevant information stronger in synaptic connections, while the less relevant information has weaker synaptic connections. Unlike digital computers that operate sequentially, neural networks excel in the high distribution of tasks, providing high capacity for the required functions.

6.3   Future Studies

Today's world is rapidly moving towards the programming and automation of most processes and decisions, with artificial intelligence taking over a significant space. This has led many experts to explore the potential of utilizing and benefiting from this advancement to reduce human effort and achieve more accurate and realistic results. Therefore, this study aims to contribute experimental evidence to the achievements of previous researchers by applying it to a sample of banks and variables that have occupied the financial arena. Consequently, future studies can adopt other methods or additional companies and variables to generalize and confirm the results obtained by previous studies by tracking the progress of these studies. The study focused on the financial and economic dimensions of financial sustainability, paving the way for further research on other dimensions or indicators.

7 Conclusion

Predicting future values remains an experimental process, especially when it comes to financial sustainability. The financial sector has witnessed the emergence and utilization of numerous models and programming methods that have managed to reduce the immense variety of measurement methods through several abstraction processes. It's not always necessary for modern models to be the most accurate or for models with the lowest MSE to be the best. Using neural networks to predict values (such as the growth rate in market value added, earnings per share index, and credit index) using the financial gearing variable has yielded relatively accurate results. According to the prediction results, it became evident that the artificial neural network's predictions for the relationship between financial gearing and market value added were more accurate than its predictions for the relationship between earnings per share and financial gearing, as well as its predictions for the credit index and financial gearing relationship.

References

Abbas, S., et al.: Antecedents of trustworthiness of social commerce platforms: a case of rural communities using multi group SEM & MCDM methods. Electron. Commer. Res. Appl. **62**, 101322 (2023)

Ahmed, A.D., Salih, M.M., Muhsen, Y.R.: Opinion weight criteria method (OWCM): a new method for weighting criteria with zero inconsistency. IEEE Access 1–1(2024)

Al-Enzi, S.H.Z., Abbas, S., Abbood, A.A., Muhsen, Y.R., Al-Hchaimi, A.A.J., Almosawi, Z.: Exploring research trends of metaverse: a bibliometric analysis. In: Al-Emran, M., Ali, J.H., Valeri, M., Alnoor, A., Hussien, Z.A. (eds.) Beyond Reality: Navigating the Power of Metaverse and Its Applications: Proceedings of 3rd International Multi-Disciplinary Conference - Theme: Integrated Sciences and Technologies (IMDC-IST 2024) Volume 1, pp. 21–34. Springer Nature Switzerland, Cham (2023). https://doi.org/10.1007/978-3-031-51716-7_2

Alhasnawi, M.Y., Said, R.M., Daud, Z.M., Muhammad, H.: Enhancing managerial performance through budget participation: insights from a two-stage A PLS-SEM and artificial neural network approach (ANN). J. Open Innov. Technol. Mark. Complex. **9**(4), 100161 (2023)

Alhasnawi, M., Said, R.M., Daud, Z.M., Muhamad, H.: Budget participation and managerial performance: bridging the gap through budget goal clarity. Adv. Soc. Sci. Res. J. **10**(9), 187–200 (2023). https://doi.org/10.14738/assrj.109.15539

Ali, J., Hussain, K.N., Alnoor, A., Muhsen, Y.R., Atiyah, A.G.: Benchmarking methodology of banks based on financial sustainability using CRITIC and RAFSI techniques. Decis. Mak. Appl. Manage. Eng. **7**(1), 315–341 (2024)

Alnoor, A., Atiyah, A.G., Abbas, S.: Toward digitalization strategic perspective in the European food industry: non-linear nexuses analysis. Asia Pac. J. Bus. Adm. (2023)

Alnoor, A., Atiyah, A.G., Abbas, S.: Unveiling the determinants of digital strategy from the perspective of entrepreneurial orientation theory: a two-stage SEM-ANN approach. Global J. Flex. Syst. Manage. **25**(2), 243–260 (2024). https://doi.org/10.1007/s40171-024-00385-0

Alnoor, A., Chew, X., Khaw, K.W., Muhsen, Y.R., Sadaa, A.M.: Benchmarking of circular economy behaviors for Iraqi energy companies based on engagement modes with green technology and environmental, social, and governance rating. Environ. Sci. Pollut. Res. **31**(4), 5762–5783 (2024)

Alnoor, A., et al.: How positive and negative electronic word of mouth (eWOM) affects customers' intention to use social commerce? A dual-stage multi group-SEM and ANN analysis. Int. J. Hum. Comput. Interact. **40**(3), 808–837 (2024)

Atiyah, A.G.: Unveiling the quality perception of productivity from the senses of real-time multisensory social interactions strategies in metaverse. In: Al-Emran, M., Ali, J.H., Valeri, M., Alnoor, A., Hussien, Z.A. (eds.) Beyond Reality: Navigating the Power of Metaverse and Its Applications: Proceedings of 3rd International Multi-Disciplinary Conference - Theme: Integrated Sciences and Technologies (IMDC-IST 2024) Volume 2, pp. 83–93. Springer Nature Switzerland, Cham (2023). https://doi.org/10.1007/978-3-031-51300-8_6

Atiyah, A.G., Zaidan, R.A.: Barriers to using social commerce. In: Alnoor, A., Wah, K.K., Hassan, A. (eds.) Artificial Neural Networks And Structural Equation Modeling: Marketing and Consumer Research Applications, pp. 115–130. Springer Nature Singapore, Singapore (2022). https://doi.org/10.1007/978-981-19-6509-8_7

Atiyah, A.G., Alhasnawi, M., Almasoodi, M.F.: Understanding metaverse adoption strategy from perspective of social presence and support theories: the moderating role of privacy risks. In: Al-Emran, M., Ali, J.H., Valeri, M., Alnoor, A., Hussien, Z.A. (eds.) Beyond Reality: Navigating the Power of Metaverse and Its Applications: Proceedings of 3rd International Multi-Disciplinary Conference - Theme: Integrated Sciences and Technologies (IMDC-IST 2024) Volume 2, pp. 144–158. Springer Nature Switzerland, Cham (2023). https://doi.org/10.1007/978-3-031-51300-8_10

Atiyah, A.G., All, N.D.A., Zaidan, A.S., Bayram, G.E.: Understating the social sustainability of metaverse by integrating adoption properties with users' satisfaction. In: Al-Emran, M., Ali, J.H., Valeri, M., Alnoor, A., Hussien, Z.A. (eds.) Beyond Reality: Navigating the Power of Metaverse and Its Applications: Proceedings of 3rd International Multi-Disciplinary Conference - Theme: Integrated Sciences and Technologies (IMDC-IST 2024) Volume 1, pp. 95–107. Springer Nature Switzerland, Cham (2023). https://doi.org/10.1007/978-3-031-51716-7_7

Atiyah, A.G., Faris, N.N., Rexhepi, G., Qasim, A.J.: Integrating ideal characteristics of chatGPT mechanisms into the metaverse: knowledge, transparency, and ethics. In: Al-Emran, M., Ali, J.H., Valeri, M., Alnoor, A., Hussien, Z.A. (eds.) Beyond Reality: Navigating the Power of Metaverse and Its Applications: Proceedings of 3rd International Multi-Disciplinary Conference - Theme: Integrated Sciences and Technologies (IMDC-IST 2024) Volume 1, pp. 131–141. Springer Nature Switzerland, Cham (2023). https://doi.org/10.1007/978-3-031-51716-7_9

Bobinaite, V.: Financial gearing and its determinants in companies producing electricity from wind resources in Latvia. Econ. Bus. **27**(1), 29–39 (2015)

Brealey, R.A., Myers, S.C., Marcus, A.J.: Fundamentals of Corporate Finance. McGraw-Hill (2023)

Chikaza, Z.: Analysis of financial sustainability and outreach of microfinance institutions (MFIs) in Zimbabwe: case study of harare. Doctoral dissertation, Stellenbosch: Stellenbosch University (2015)

de Wet, J.: Earnings per share as a measure of financial performance: does it obscure more than it reveals? Corp. Ownership Control **10**, 265–275 (2013)

Husin, N.A., Abdulsaeed, A.A., Muhsen, Y.R., Zaidan, A.S., Alnoor, A., Al-mawla, Z.R.: Evaluation of metaverse tools based on privacy model using fuzzy MCDM approach. In: Al-Emran, M., Ali, J.H., Valeri, M., Alnoor, A., Hussien, Z.A. (eds.) Beyond Reality: Navigating the Power of Metaverse and Its Applications: Proceedings of 3rd International Multi-Disciplinary Conference - Theme: Integrated Sciences and Technologies (IMDC-IST 2024) Volume 1, pp. 1–20. Springer Nature Switzerland, Cham (2023). https://doi.org/10.1007/978-3-031-51716-7_1

Kieso, D.E., Weygandt, J.J., Warfield, T.D.: Intermediate Accounting. Wiley (2013)

Kim, H.Y., Won, C.H.: Forecasting the volatility of stock price index: a hybrid model integrating LSTM with multiple GARCH-type models. Expert Syst. Appl. **103**, 25–37 (2018)

Muhsen, Y.R., Husin, N.A., Zolkepli, M.B., Manshor, N.: A systematic literature review of fuzzy-weighted zero-inconsistency and fuzzy-decision-by-opinion-score-methods: assessment of the past to inform the future. J. Intell. Fuzzy Syst. **45**(3), 4617–4638 (2023)

Robert, C.H., Jennifer, L.C., Todd, M.: Analysis for Financial Management. McGraw-Hill (2023)

Roghanian, P., Rasli, A., Gheysari, H.: Productivity through effectiveness and efficiency in the banking industry. Procedia Soc. Behav. Sci. **40**, 550–556 (2012)

Ross, S.A., Westerfield, R.W., Jordan, B.D.: Fundamentals of Corporate Finance. McGraw-Hill (2022)

Satwiko, R., Agusto, V.: Economic value added, market value added, dan kinerja keuangan terhadap return saham. Media Bisnis **13**(1), 77–88 (2021)

Siyanbola, T.T., Olaoye, S.A., Olurin, O.T.: Impact of gearing on performance of companies. Arab. J. Bus. Manage. Rev. **3**(1), 68–80 (2015)

Stankeviciene, J., Nikonorova, M.: Sustainable value creation in commercial banks during financial crisis. Procedia Soc. Behav. Sci. **110**, 1197–1208 (2014)

Tehulu, T.A.: Determinants of financial sustainability of microfinance institutions in East Africa. Eur. J. Bus. Manage. **5**(17), 152–158 (2013)

Van Horne, J.C., Wachowicz Jr, J.M.: Fundamentals of Finance Management. YD Williams, Moscow (2008)

Towards Digital Sustainability: Integrating Canonical Correlation with Artificial Neural Network

Ali Naser Hussein[✉]

College of Administration and Economics, Basrah University, Basrah, Iraq
ali.hussien@uobasrah.edu.iq

Abstract. The study aims to use an artificial neural network to analyze the canonical correlation by determining the optimal weights, as well as comparing the results of the correct correlation and the canonical correlation using the artificial neural network. And comparing the results of the correlation between two sets of variables using the canonical correlation and the legal correlation using the artificial neural network to show the extent to which the results of the artificial neural networks match the results of the statistical model to adopt the artificial neural network model as another means of conducting the canonical correlation analysis. The study concluded that the group represented by (the age of the husband, the age of the wife, the educational level of the husband, the educational level of the wife, the husband's occupation, the wife's occupation, the standard of living of the family, the presence of kinship between the spouses, independent housing, the husband's smoking, the husband's consumption of alcoholic beverages) had a clear and significant effect on the variables of the second group, represented by (the variable of the duration of marriage, the variable of the number of children). The first function also proved its high significance according to the value of the canonical correlation coefficient, which helps in making the decision that the first function is the function that can be relied upon in interpreting the relationship between the two components, meaning that the function the first is reliable in analyzing the relationship between the first and second group of variables, as well as using it in prediction. The study also concluded that the best neural network model is the third model, as it has the lowest value for the criteria for comparing error measures and the highest coefficient of determination, meaning that the explanatory power of this model reached 97% of what can be relied upon in building canonical variables and prediction.

Keywords: Artificial Neural Networks · canonical correlation Analysis · Back Propagation Algorithm · Hidden · Input

1 Introduction

In light of the tremendous technological progress in all the various fields of life, research in the various sciences is no longer sufficient by simply presenting problems, studying phenomena, identifying causes, drawing conclusions, and making decisions superficially

A. Alnoor et al. (Eds.): AIRDS 2024, LNNS 1033, pp. 328–341, 2024.
https://doi.org/10.1007/978-3-031-63717-9_21

and abstractly far from the method of objectivity and measurement. The general trend in such research has become to use quantitative measurement methods and statistical approaches to classify scientific phenomena, highlight their characteristics, and analyze the interrelationships between these phenomena on an objective basis (XinYing et al., 2024; Muhsen et al., 2023).

In the current study, the artificial neural network model and the canonical correlation were used together. Canonical correlation analysis (CCA) is a statistical method used to find linear relationships between two groups of variables. On the other hand, artificial neural networks (ANNs) are a type of machine learning algorithm that can learn complex relationships between inputs and outputs through the use of interconnected nodes or neurons. Hybridization of CCA and ANNs involves incorporating CCA into the training process of the ANN model (Zubaidi et al., 2021; Zubaidi et al., 2020). This involves using CCA to transform the input variables into a new set of variables that are most closely related to the output variables, and then training the ANN model on this transformed data. His approach has been studied and applied to the phenomenon of divorce in the province of Basra, where it is clear the importance of linking the use of an artificial neural network with the canonical correlation to study the phenomenon of divorce in all respects depending on the extent of the use of artificial neural networks in the process of internal auditing of the phenomenon of divorce under study and interest. Overall, the hybridization of CCAs and ANNs is an exciting area of research with promising applications in a wide range of areas of life (Chew et al., 2023).

A study (Shu & Ouarda, 2007) used canonical correlation analysis (CCA) and artificial neural networks (ANNs) to obtain estimates of the amount of flood optimization at unpressurized sites.The proposed methods were applied to 151 watersheds in the province of Quebec, Canada.ANN models were used to determine the functional relationships between flood amounts and physiographic variables in the CCA area.Two models of artificial neural networks have been developed which are single ANN model and group ANN model.Two evaluation procedures were applied, namely the Jackknife validation procedure and the split sample validation procedure, and to evaluate the performance of the proposed models, the results of the proposed models were compared with the original CCA model, the basic kriging model, and the original ANN models.The results showed that the ANN models based on CCA provide better estimation than the original ANN models, and have the best performance among all models in terms of prediction accuracy.Also, group ANN approaches provide better generalizability than individual ANN models (Shu and Ouarda, 2007; Atiyah, 2023; Alnoor et al., 2024).

The study (Mohamed et al., 2013) aims to apply canonical correlation analysis with artificial neural networks to EEG data sets that track the state of the brain under different subjects, as such data needs careful analysis when it consists of a series of different subjects. The network's capabilities were demonstrated on EEG data to determine subject dependence in terms of correlation and then compared its effectiveness to that of cosine reference signals (Hossain et al., 2013; Alnoor et al., 2023). (Mumtaz et al. 2016) Proposed a new method based on canonical correlation analysis (CCA) and artificial neural network (ANN) to detect epileptic seizures from EEG signals. CCA was applied to the EEG signals and feature vectors corresponding to the Eigen values were obtained. It used Eigen values as inputs into an artificial neural network (ANN) model that has been widely explored for neural networks to classify the occurrence of epileptic seizures and epileptic seizures. It was concluded that the Eigen values obtained using CCA are the best detector for epileptic seizures and provide average classification accuracy and very high sensitivity (Soomro et al., 2016; Ali et al., 2024).

(Ari et al., 2018) Based on a proposed method SVCCA, developed a projection weighted CCA (canonical correlation analysis) as a tool for understanding neural networks, as the basic method was improved, showing how to distinguish between signal and noise, and then applied this technique for comparison to a group of CNN networks, and this Evidence that generalized networks converge with more similar representations than other networks, wider networks converge with more similar solutions than narrow networks, and that trained networks with identical topologies that have different learning rates converge into distinct clusters with diverse representations. The study also showed the representative dynamics of RNNs, across both training and sequential time steps. It was found that RNNs converge in a bottom-up pattern over the course of training and that the hidden state is highly variable over the course of sequencing, even when linear transformations are accounted for. Together, these results provide insights. New about the function of CNNs and RNNs, and demonstrates the benefit of using CCA to understand representations (Morcos et al., 2018; Ahmed et al., 2024; Al-Enzi et al., 2023; Alnoor et al., 2024).

2 The Neural Network Model to the Canonical Correlation

Multi-layer neural networks (MLPNNs) are used to classify the inputs and MLPNNs having two or more layers are often placed in a feed-forward neural network because of its fast operation, no implementation limitations. As mentioned earlier, there are three networking layers for MLPNN, which are the input layer, the hidden layer, and the output layer. As the amount of input neurons is matched with the characteristics and features chosen to organize the expected output categories, and this is what the output layers function, and the amount of output neurons depends on the amount of expected output categories. The system is fully compatible with intermediate layers, which can be combined to enhance network efficiency (Hsieh, 2001; Hsieh, 2021; Naghsh et al., 2010; Abbas et al., 2023; Atiyah and Zaidan, 2022).

An MLPNN has many hidden layers, but an MLPNN with single hidden layers is ideal as a classifier because a hidden layer with a large number of neurons is more prone to processing delays and a small number of neurons can cause classification errors. Certainly, the MPLNN model, which has insufficient or abundant hidden neurons, leads to serious problems of overfitting and generalization, so finding the required amount is done by trial and error.

Correct Correlation Analysis (CCA) is applied to the divorce data to extract the feature vectors, and then the extracted feature vectors are combined to train the Multilayer Neural Network (MLPNNs). In order to avoid over fitting the MPLNN classifiers, the MLPNN model is trained by the back propagation algorithm, and the iteration is performed to the method used a 10-fold validation to generalize the results of the workbook. The MPLNN classifier is trained on the characteristics of factors leading to divorce. MPLNN compares features extracted using CCA with features trained on the basis of characteristic vectors corresponding to the largest Eigen values associated with the energy of the maximum and minimum signals. If the extracted features contain the characteristics of the causes of divorce, the trained classifier alerts their occurrence.

Three networks (NNs) were used, with thecanonical correlation cca, to form the (NLCCA) model. The double network in Fig. 1 shows the inputs Y'S, X'S, which are the canonical variables (V, U). At the start, the inputs represent X'S periodically, and here the network indicates the hidden layer for the inputs (hx), in the same way the inputs for the variables (Y'S) are also periodic, and (hy) represents the hidden layer of the inputs, then the values are extracted the linear components Ui and Vi, which represent the canonical variables.

In the second network NN, h^u is a hidden layer of (u) that represents the output layer (x'), and for the third network NN, h^v represents the hidden layer of the output layer Y'. As shown in Fig. 1.

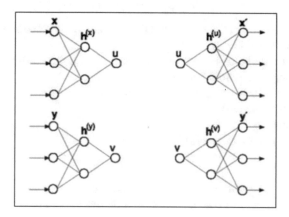

Fig. 1. NLCCA (Ubeyli, 2009; Husin et al., 2023)

3 Statistical Analysis

In this paragraph, we will discuss the analysis of the canonical correlation of the variables of the phenomenon of divorce in Basra and identify the most correlated variables, and we will also use one of the most important methods of artificial intelligence represented by the artificial neural network in the analysis of the correct correlation by determining the optimal weights, as well as the comparison between the results of the canonical correlation and the canonical correlation using the artificial neural network.

The data of the study were obtained from the Basra Court of Appeal regarding the phenomenon of divorce in Basra Governorate during the year 2020 AD, as the data was divided into two groups, as follows:

The first group:
x_1: the age of the husband
x_2: the age of the wife

X_1 and X_2: the age variable of the husband and wife, respectively
It is classified into three categories as follows:
The first category: (less than 20) years $= 1$
The second category: (20–40) years $= 2$
Third category: (40 or more) $= 3$
X_3: The educational level of the spouse
X_4: The educational level of the wife

X_3 and X4: the educational level variable for the husband and wife, respectively, and classified into three categories (Postgraduate $= 1$, Bachelor's $= 2$ &otherwise $= 3$).

X_5: Husband's profession
X_6: Wife's profession

X_5 and X_6: the variable of the occupation of the husband and wife, respectively: it is a qualitative variable and it is (employee $= 0$ &otherwise $= 1$).

X_7: the standard of living of the family and divided into three categories (below good $= 1$, good $= 2$, & very good $= 3$)
X_8: Husband and wife belong to the same family: it is a qualitative variable and it is (The couple belong to the same family $= 1$ &otherwise $= 0$)
X_9: independent housing variable: it is a qualitative variable and it is (independent housing $= 1$ &otherwise $= 0$)
X_{10}: Husband smoking
X_{11}: The husband's abuse of alcoholic beverages

The second group:

Y_1: the variable of the duration of marriage: it is a quantitative variable.
Y_2: the number of children variable: it is a quantitative variable.

The R programming language program was used to analyze the data using the "Canonical Correlation" method. The following results were obtained:
This stage is considered the first step in the canonical correlation analysis, where each function consists of two components, one of which represents the first group of original variables (Y's), and the second represents the second group (X's). The group (Y's) consists of only two variables, so we will have two groups of correlation functions in this study, two of the functions have been derived, each consisting of two components.

The first function has the highest canonical correlation coefficient (Rc), which reached(0.725108), which is a strong canonical correlation, while the value of (χ^2)calculated was (125.0690) with a degree of freedom (22), and we also note that the p-value is equal to (0.000), which is less than (0.05), which means that it is significant, then followed by the second function With a correlation coefficient (0.354970), and the calculated value of $(\chi^2)$was (19.1244), with degrees of freedom (10). We also note that the p-value is equal to (0.000), which is less than (0.05), which means that it is significant. This means that there is a statistically significant relationship between the two groups of the study, that is, the variables of the first group represented by (the age of the husband, the age of the wife, the educational level of the husband, the educational level of the wife, the husband's profession, the wife's profession, the living standard of the family, the presence of kinship between the spouses, independent housing Smoking for the husband, the husband's consumption of alcoholic beverages) have a clear and significant effect on the variables of the second group, represented by (the variable of the duration of marriage, the variable of the number of children).

Which also confirms these results, the values of "Lambda Prime" which are present in the last column of this table and are similar to R^2 in the multiple correlation, but their interpretation is the opposite of the interpretation of R^2, meaning that the values, when close to zero indicate a high correlation, while the value close to one indicates low correlation. The purpose of this analysis is to study the structural relationship between the first and second group of variables and its not to present any causal relationships or to identify any dependent or independent variables, but there is in fact a reciprocal relationship. There are two functions that have been extracted, and we take into account the variance in the original variables entirely, in both groups, the abundance index reached (13.55%) of the compounds of the first group compared to the compounds of the second group, meaning that the compounds of the first group explain about (13.55%) of the variation in the compounds of the second group, while the latter explains (45.63%) of the variation in the compounds of the first group.

Then comes the stage of determining which of the variables can be relied upon in interpreting the relationship between the two sets of variables, according to the previous three criteria. With regard to the first function, we find that the value of the correlation coefficient amounted to (0.725108), which gives it a high significance, as its significance was statistically proven at a significant level (0.05).

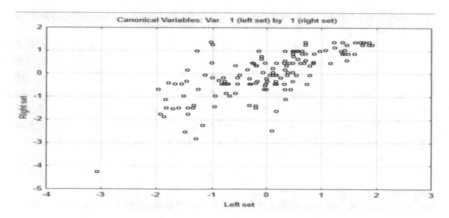

Fig. 2. The relationship between the components of the first function

Figure 2 represents the relationship between the two components of the first function, where the horizontal axis represents (the left set) of the first group of variables, and the vertical axis represents (the right set) of the component of the second group of variables. The figure shows the strength of the relationship between the two components of the function. As for the third criterion, which is the index of abundance, the amount of variance extracted from the original variables of the first group combined, through the first component of the first group, amounted to (0.826205), i.e. (82.62%), while the amount of variance extracted from the variables was The original value of the first group combined, through the first component (0.173795), i.e. (17.37%), while the remaining amount of variation in each of the variables of the first and second groups, which amounted to (17.38%, 82.63%), respectively, is distributed among the remaining functions.

It is also clear that the abundance index (Reddncy) of the first component of the first group of variables contributed to the interpretation of (0.434404), i.e. 43.44% of the variation of the original variables of the second group, while the first component of the second group of variables contributed to the interpretation of (0.021899), i.e. % 2.18 of the variances of the original variables for the first set of variables (Muhsen et al., 2023; Atiyah et al., 2023).

The characteristic values of the correct correlation for a number of (13) variables divided into two groups), so we find that we have (2) distinct values and the first value amounted to (0.525782), while the second value amounted to (0.126004). The variable $x2$ (wife's age) had the greatest impact on the variables of the second group, as its synthetic coefficient was (-0.912759). The second variable in terms of the strength of influence in the second group (Y) is the variable $x1$ (the age of the husband), as its synthetic coefficient was (-0.836624). The variable $X5$ (husband's profession) is the third factor in terms of the strength of influence, as its synthetic coefficient was (0.582802). The $X9$ variable (independent housing variable) is the fourth factor in terms of the strength of influence in the second group, as its synthetic coefficient was (-0.543373). The variable $y2$ (the number of children variable) had the greatest impact on the variables of the first group, as its synthetic coefficient was (-0.920102). The second variable in terms of the strength of influence in the first group (X) is the variable ($y1$), the variable of the duration of marriage, as its synthetic coefficient was (-0.897676).

One of the things that encourage the use of neural networks in the treatment of canonical correlation, as neural networks can be used to process non-linear time series, in addition to that it is scientifically useful to see this modern method and compare it with classical methods to identify its capabilities and efficiency in conducting canonical correlation. A computer program (see appendix..) has been designed to obtain the values of the canonical correlations, and the first step in using the program is to specify the inputs for the neural network, as the inputs are the number of independent variables, whose number is (11) variable, and to determine the number of hidden nodes, which is determined by During the training, which includes conducting many computer experiments.

In this study, the amount of error $E_{tolerance}$ was fixed to be 0.01 and the number of training times ($N_{train} = 500 * N_{pts}$) and $N_{output} = 1$. Applying equation, we find that the number of hidden nodes should be (3) for the data and according to the number of training times. And since the data used in this study amounted to (150) observations, (13%) of them were considered test observations for the purpose of testing, and 15% of the observations were considered as the validation group while the rest of the observations were used for training and estimating the value of Mean Square Error (MSE). All observations were used, and the test was carried out using the back-error propagation network, and the network was trained using several hidden neurons to choose the best one, and their number was between 2–8. Table (3–11) shows the amount of absolute error AE, which represents the difference between the output values of the network and the desired output, as shown in Table 1.

Table 1. Choosing the number of hidden neurons.

Validation	Training	The number of hidden neurons
The absolute value of the error (AE)	The absolute value of the error (AE)	
385.59	445.58	2
*370.787	449.417	3
386.016	445.17	4
376.701	450.29	5
378.109	441.83	6
372.412	446.919	7
373.252	446.07	8

In this test, the number of training times was fixed as 5000 repetitions, the MSE value was 0.01, and the weights were between [−5, 5]. Thus, it was considered that the best number of neurons in the hidden layers is (3) neurons, because the value of the absolute error of legitimacy in this case is the least. After selecting the best number of neurons, the second test was performed to choose the best number of iterations that makes the absolute error value as low as possible. Table 2 shows the results of this test.

Table 2. Choosing the number of repetitions

Validation	Training	Number of Iteration
The absolute value of the error	The absolute value of the error	
421.209	424.417	1000
421.328	425.663	2000
421.093	425.756	3000
421.794	425.378	4000
419.20	423.050	5000
*414.642	412.353	10000
434.293	413.737	20000
453.972	371.977	50000
472.181	354.591	100000

In this test, a neural network with error backpropagation was used with the number of (3) neurons and MSE = 0.01. After conducting the test, the number of repetitions was chosen to be (10000) repetitions to obtain the least error. A computer program designed in the R language was used to find the values of the canonical correlation and calculate the statistical parameters (weights), and a neural network with back propagation of error was built for the original observations and was used (3) hidden nodes in the hidden layer, where the training begins with random weights using an average Learning of 0.01 and 10,000 repetitions and three neurons in the hidden layers. The initial weights of the network for the three layers were determined as random values by the program, as shown in Table 3.

Table 3. Represents the initial weights of the artificial neural network

$W_{(1)i}$	h1	h2	h3
Bias	−0.69842	1.945851	−1.19862
W(1)1	−0.27595	0.800914	0.639492
W (1)2	1.114649	1.165253	2.430227
W(1)3	0.550044	0.358856	−0.55722
W(1)4	1.236676	−0.60856	0.844904
W(1)5	0.139098	−0.20224	−0.7822
W(1)6	0.410275	−0.27325	1.110711
W(1)7	−0.55846	−0.4687	0.249825
W(1)8	0.605371	0.704167	1.651915
W(1)9	−0.50633	−1.19736	−1.45897
W(1)10	−1.42057	0.866366	−0.0513
W(1)11	0.127993	0.864153	−0.52693
$W_{(2)i}$	Y1	Y2	
Bias	−0.19726	−1.08758	
W(2)1	−0.62958	1.484031	
W (2)2	−0.83384	−1.18621	
W(2)3	0.578722	0.101079	

After training the network using the back propagation algorithm with the learning rate and iteration for this process, we will obtain the weights of the network, which lead to the canonical variables u,v. The value of the canonical correlation coefficient after the training process has reached (0.99454). The network weights can be clarified for the canonical correlation of the variables of the first and second groups, as in shown in Table 4.

Table 4. Represents the weights of the canonical network after the training process

$W_{(i)}$	W_{11}	W_{12}	W_{13}
Bias	0.514757	15.12491	−15.2786
W1	0.262473	−0.68948	−0.17793
W 2	0.219373	−1.44004	13.39607
W3	−0.32169	0.179885	7.230664
W4	0.134093	−4.17486	−12.6935
W5	−0.14888	−2.27509	−8.76954
W6	−0.04057	0.46352	10.71123
W7	0.129349	1.573945	3.745104
W8	0.933922	4.451711	−19.0567
W9	−0.06109	−9.14546	−7.47316
W10	−0.07823	−0.26341	8.3691
W11	0.154543	2.79689	5.938075
$W_{(1)i}$	Y1	Y2	
Bias	−0.56691	−1.07959	
W1	3.374869	3.626358	
W 2	−2.19049	−1.68036	
W3	1.52905	1.207536	

The weights of the canonical network can be illustrated in the form of an architectural structure of the artificial neural network, as shown in Fig. 3.

To determine the best model among these models, RMSE, MAPE, MABE and R^2 statistical measures will be calculated, acceptable models are indicated and the best model is determined by comparing the statistical errors rates associated with all models, and the best model with the lowest error rates value is shown. As shown in Table 5.

We note from Table 5 that the best neural network model is the third model, as it has the lowest value of comparison criteria, error measures, and the highest determination coefficient, meaning that the explanatory power of this model amounted to 97% of what can be relied upon in building canonical variables and prediction. And to complete the interpretation of the canonical variables, the correlations between the canonical variables and the original variables have been calculated, and these correlations are sometimes called canonical Variable Loadings, and they are dealt with and interpreted as is the case in factorial analysis. For the purpose of comparing the results, we will use the mean square error criterion, the root means square error criterion, and the mean absolute error criterion for each of the two components resulting from the canonical correlation, as shown in Table 6.

We note that the canonical correlation using the method of artificial neural networks is much better than the normal canonical correlation method based on comparison criteria,

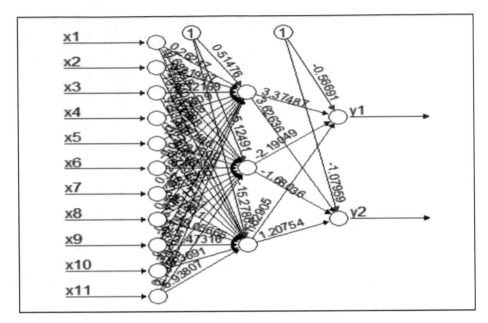

Fig. 3. Neural network architecture

Table 5. Comparison of neural network models

Model	RMSE	M APE	MABE	R2
1	0.945	0.977	3.256	0.9767
2	1.353	1.184	4.431	0.9709
3	0.843	0.910	3.268	0.9769

Table 6. Comparison between the normal canonical correlation and the canonical correlation

standard	Ordinary correlation		Correlation of neural networks	
	The first vehicle	The second vehicle	The first vehicle	The second vehicle
MSE	0.3482245	0.337773	0.029541	0.098732
RMSE	0.5901055	0.115713	0.025648	0.051143
MAE	0.4356959	0.511120	0.002799	0.061586

as neural networks are considered one of the most important intelligent techniques that depend on training and learning, as they give very accurate weights or coefficients.

4 Conclusions

The variables of the first group (the age of the husband, the age of the wife, the educational level of the husband, the educational level of the wife, the husband's occupation, the wife's occupation, the standard of living of the family, the existence of kinship between the spouses, independent housing, smoking for the husband, the husband's consumption of alcoholic beverages) have a clear and significant effect On the variables of the second group, represented by (the variable of the duration of marriage, the variable of the number of children). The index of abundance of compounds of the first group for the compounds of the second group was 13.55% total redundancy, meaning that the compounds of the first group explain about (13.55%) of the variance in the compounds of the second group, while the latter explains (45.63%) from of the variance in the compounds of the first group. The value of the correlation coefficient in the first function amounted to (0.725108), which gives it a high significance, as its significance was statistically proven at a significant level (0.05). The strength of the relationship between the two components of the first function becomes clear, which helps in making the decision that the first function is the function that can be relied upon in explaining the relationship between the two components. So, the first function is the function that can be relied upon in analyzing the relationship between the first and second group of variables, as well as using it in prediction. It was concluded that the variables of the first group, which have a strong influence in the second group (Y), are primarily the wife's age (X_2), then the husband's age (X_1), followed by the husband's occupation variable (X_5), then the independent housing variable (X_9). The variable y_2 (the variable of the number of children) had the greatest impact on the variables of the first group, followed by the variable y_1 (the variable of the duration of marriage). The best number of neurons in the hidden layers is (3) neurons, because the value of the absolute error of legitimacy in this case is the lowest. Three different neural network models with different weights were obtained and compared to choose the best one using statistical measures of errors, as the best model that has the lowest error value is chosen. The best neural network model is the third model, as it has the lowest value for comparison criteria of error measures and the highest determination coefficient, meaning that the explanatory power of this model reached 97% of what can be relied upon in building correct variables and prediction.

References

Abbas, S., et al.: Antecedents of trustworthiness of social commerce platforms: a case of rural communities using multi group SEM & MCDM methods. Electron. Commer. Res. Appl. **62**, 101322 (2023)

Ahmed, A.D., Salih, M.M., Muhsen, Y.R.: Opinion Weight criteria method (OWCM): a new method for weighting criteria with zero inconsistency. IEEE Access **12** (2024)

Al-Enzi, S.H.Z., Abbas, S., Abbood, A.A., Muhsen, Y.R., Al-Hchaimi, A.A.J., Almosawi, Z.: Exploring research trends of metaverse: a bibliometric analysis. In: Al-Emran, M., Ali, J.H., Valeri, M., Alnoor, A., Hussien, Z.A. (eds.) Beyond Reality: Navigating the Power of Metaverse and Its Applications: Proceedings of 3rd International Multi-Disciplinary Conference - Theme: Integrated Sciences and Technologies (IMDC-IST 2024) Volume 1, pp. 21–34. Springer Nature Switzerland, Cham (2023). https://doi.org/10.1007/978-3-031-51716-7_2

Ali, J., Hussain, K.N., Alnoor, A., Muhsen, Y.R., Atiyah, A.G.: Benchmarking methodology of banks based on financial sustainability using CRITIC and RAFSI techniques. Decis. Mak. Appl. Manage. Eng. **7**(1), 315–341 (2024)

Alnoor, A., Atiyah, A.G., Abbas, S.: Toward digitalization strategic perspective in the European food industry: non-linear nexuses analysis. Asia Pac. J. Bus. Adm. (2023)

Alnoor, A., Atiyah, A.G., Abbas, S.: Unveiling the determinants of digital strategy from the perspective of entrepreneurial orientation theory: a two-stage SEM-ANN approach. Global J. Flex. Syst. Manage. **25**(2), 243–260 (2024). https://doi.org/10.1007/s40171-024-00385-0

Alnoor, A., et al.: How positive and negative electronic word of mouth (eWOM) affects customers' intention to use social commerce? A dual-stage multi group-SEM and ANN analysis. Int. J. Hum. Comput. Interact. **40**(3), 808–837 (2024)

Atiyah, A.G.: Unveiling the quality perception of productivity from the senses of real-time multisensory social interactions strategies in metaverse. In: Al-Emran, M., Ali, J.H., Valeri, M., Alnoor, A., Hussien, Z.A. (eds.) Beyond Reality: Navigating the Power of Metaverse and Its Applications: Proceedings of 3rd International Multi-Disciplinary Conference - Theme: Integrated Sciences and Technologies (IMDC-IST 2024) Volume 2, pp. 83–93. Springer Nature Switzerland, Cham (2023). https://doi.org/10.1007/978-3-031-51300-8_6

Atiyah, A.G., Zaidan, R.A.: Barriers to using social commerce. In: Alnoor, A., Wah, K.K., Hassan, A. (eds.) Artificial Neural Networks and Structural Equation Modeling: Marketing and Consumer Research Applications, pp. 115–130. Springer Nature Singapore, Singapore (2022). https://doi.org/10.1007/978-981-19-6509-8_7

Atiyah, A.G., Alhasnawi, M., Almasoodi, M.F.: Understanding metaverse adoption strategy from perspective of social presence and support theories: the moderating role of privacy risks. In: Al-Emran, M., Ali, J.H., Valeri, M., Alnoor, A., Hussien, Z.A. (eds.) Beyond Reality: Navigating the Power of Metaverse and Its Applications: Proceedings of 3rd International Multi-Disciplinary Conference - Theme: Integrated Sciences and Technologies (IMDC-IST 2024) Volume 2, pp. 144–158. Springer Nature Switzerland, Cham (2023). https://doi.org/10.1007/978-3-031-51300-8_10

Chew, X., Khaw, K.W., Alnoor, A., Ferasso, M., Al Halbusi, H., Muhsen, Y.R.: Circular economy of medical waste: novel intelligent medical waste management framework based on extension linear diophantine fuzzy FDOSM and neural network approach. Environ. Sci. Pollut. Res. **30**(21), 60473–60499 (2023)

Hossain, M.Z., Rabin, M.J.A., Uddin, A.F.M.N., Shahjahan, M.: Canonical correlation analysis with neural network for inter subject variability realization of EEG data. In: International Conference on Informatics, Electronics and Vision (ICIEV), pp. 1–5 (2013)

Hsieh, W.: Nonlinear canonical correlation analysis of the tropical pacific climate varability using neural approach. J. Clim. **14**(12), 25–29 (2021)

Hsieh, W., Wu, A.: Nonlinear canonical correlation analysis of the tropical climate variability using neural approach. J. Clim. **14**(12), 2528–2539 (2001)

Husin, N.A., Abdulsaeed, A.A., Muhsen, Y.R., Zaidan, A.S., Alnoor, A., Al-mawla, Z.R.: Evaluation of metaverse tools based on privacy model using fuzzy MCDM approach. In: Al-Emran, M., Ali, J.H., Valeri, M., Alnoor, A., Hussien, Z.A. (eds.) Beyond Reality: Navigating the Power of Metaverse and Its Applications: Proceedings of 3rd International Multi-Disciplinary Conference - Theme: Integrated Sciences and Technologies (IMDC-IST 2024) Volume 1, pp. 1–20. Springer Nature Switzerland, Cham (2023). https://doi.org/10.1007/978-3-031-51716-7_1

Morcos, A.S., Raghu, M., Bengio, S.: Insights on representational similarity in neural networks with canonical correlation. In: 32nd Conference on Neural Information Processing Systems (NIPS 2018), Montréal, Canada (2018)

Muhsen, Y.R., Husin, N.A., Zolkepli, M.B., Manshor, N.: A systematic literature review of fuzzy-weighted zero-inconsistency and fuzzy-decision-by-opinion-score-methods: assessment of the past to inform the future. J. Intell. Fuzzy Syst. **45**(3), 4617–4638 (2023)

Muhsen, Y.R., Husin, N.A., Zolkepli, M.B., Manshor, N., Al-Hchaimi, A.A.J., Ridha, H.M.: Enhancing NoC-based MPSoC performance: a predictive approach with ANN and guaranteed convergence arithmetic optimization algorithm. IEEE Access **11**, 90143–90157 (2023). https://doi.org/10.1109/ACCESS.2023.3305669

Naghsh-Nilchi, A.R., Aghashahi, M.: Epilepsy seizure detection using Eigen-system spectral estimation and multiple layer perceptron neural network. Biomed. Signal Process. Control **5**(2), 147–157 (2010). https://doi.org/10.1016/j.bspc.2010.01.004

Shu, C., Ouarda, T.B.M.J.: Flood frequency analysis at ungauged sites using artificial neural networks in canonical correlation analysis physiographic space. Water Resour. Res. J. 43(7) (2007)

Soomro, M.H., Musavi, S.H.A., Pandey, B.: Canonical correlation analysis and neural network (CCA-NN) based method to detect epileptic seizures from EEG signals. Int. J. Bio Sci. Bio Technol. **8**(4), 11–20 (2016). https://doi.org/10.14257/ijbsbt.2016.8.4.02

Ubeyli, E.D.: Combined neural network model employing wavelet coefficients for EEG signals classification. Digit. Signal Process. J. **19**, 297–308 (2009)

XinYing, C., Tiberius, V., Alnoor, A., Camilleri, M., Khaw, K.W.: The dark side of metaverse: a multi-perspective of deviant behaviors from PLS-SEM and fsQCA findings. Int. J. Hum. Comput. Interact. 1–21 (2024)

Zubaidi, S.L., et al.: Simulating a stochastic signal of urban water demand by a novel combination of data analytic and machine learning techniques. IOP Conf. Ser. Mater. Sci. Eng. **1058**(1), 012066 (2021). https://doi.org/10.1088/1757-899X/1058/1/012066

Zubaidi, S.L., Al-Bugharbee, H., Muhsin, Y.R., Hashim, K., Alkhaddar, R.: Forecasting of monthly stochastic signal of urban water demand: Baghdad as a case study. IOP Conf. Ser. Mater. Sci. Eng. **888**(1), 012018 (2020). https://doi.org/10.1088/1757-899X/888/1/012018

Artificial Intelligence Simulation of Ant Colony and Decision Tree in Terms Sustainability

Asmaa Ayoob Yaqoob and Waleed Meiya Rodeen[✉]

Department of Statistics, College of Administration and Economics, University of Basrah,
Basrah, Iraq
{asmaa.yaqoob,Waleed.rodeen}@uobasrah.edu.iq

Abstract. Complain AI with disease diagnosis holds huge potential to improve the sustainability of healthcare systems by improving efficiency, optimizing resources, and developing in-person and remote care initiatives. In this research, artificial intelligence simulation of ant colony optimization was used to achieve the advantage of community communication in ants. In addition, the ant colony algorithm (ACO) was compared with the decision tree algorithm (DT), the logistic model, and neural networks. The study was applied to thalassemia, which is one of the diseases widely spread around the world, which causes the breakdown of red blood cells in the body. Therefore, the aim of this research is to classify the incidence of this disease or not, taking into account the effect of some variables on the probability of infection, in order to reduce deaths due to this disease and achieve one of the goals of sustainable development, which is reaching the standard of good health. The research results indicate that the ant colony artificial intelligence algorithm is superior in classification operations compared to classical methods such as logistic regression and neural networks. The decision tree also concluded that the genetic factor had the greatest influence on the probability of contracting the disease among the rest of the factors included in the study, which were gender, blood type, genetic factor, and place of residence.

Keywords: Artificial Intelligence · Swarm · Data Mining · Classification · Decision Tree · Ant Colony Optimization · Thalassemia

1   Introduction

Thalassemia is one of the hereditary blood diseases that has a wide spread in various countries of the world, especially the Mediterranean countries, so the disease is sometimes called Mediterranean anemia, as the disease causes the breakdown of red blood cells, and thalassemia is of two types, alpha and beta, and this disease There are many serious complications that may be fatal in most cases, especially of the beta type (Abdulwahab, 2008). Due to the extreme importance of the disease and the high costs of treating it, it was necessary to identify the most important reasons that lead to a person being infected with it and classify the presence of infection or not. This is done through artificial intelligence algorithms as a method. Advanced and modern data mining. We notice increased interest in data science in terms of analysis and processing. Methods of data mining and extracting information from huge amounts of data have increased.

© The Author(s), under exclusive license to Springer Nature Switzerland AG 2024
A. Alnoor et al. (Eds.): AIRDS 2024, LNNS 1033, pp. 342–351, 2024.
https://doi.org/10.1007/978-3-031-63717-9_22

Data classification is one of the most used things in data mining tasks and has a fundamental role in the decision-making process, classification means assigning an object to a pre-defined category according to its characteristics. Classification is usually used in various fields, especially speech recognition, pattern recognition, medical diagnosis, etc. That Many types of classification methods have emerged, including linear and non-linear such as logistic, nearest neighbor classifiers, support vector, neural networks, decision trees, and many others. Recently, new classification algorithms have appeared in artificial intelligence, and one of these algorithms is the ant colony optimization algorithm(ACO), which is one of the developed swarm intelligence algorithms and relies in its method on simulating the intelligence of real ants in thinking to find the optimal path when searching for food (Neagoe et al., 2010). In addition, the decision tree algorithm is one of the important algorithms that works to make the appropriate decision and has a parallel value to artificial intelligence algorithms, as it is considered one of the algorithms subject to supervision. From the above, some algorithms will be applied in this paper and compared between them for the purpose of classifying the incidence of thalassemia, taking into account some of the factors affecting the occurrence of the disease as inputs to the algorithms.

2 Literature Review

There are many studies on the use of the ACO and DT algorithms, some of which will be mentioned in relation to the subject of the study on the data classification mechanism. (Silva et al., 2002; Atiyah et al., 2023) studied a new methodology for improving logistics operations using the ACO algorithm. The proposed methodology was applied using simulation to prove its efficiency in choosing the appropriate parameters for the ACO algorithm. It was then applied to real data about scheduling and distributing goods (Atiyah et al., 2023). The results of the study reached: The ACO algorithm provides good scheduling of logistics operations. (Martens et al., 2007; Alnoor et al., 2023) gave an overview of the classical approach based on C4.5 classification techniques, comparison with the ACO algorithm, and developed a new method based on the characteristics of the Ant-miner algorithm, which deals with the problem in a dynamic way by including the time interval. In (Nejad et al., 2008; Ali et al., 2024) study, an algorithm was proposed to discover irregular classification rule sets based on the characteristics of the Ant-Miner type ACO algorithm, by creating a tree graph consisting of a group of nodes representing the basic operators and workers of the algorithm (Atiyah and Zaidan, 2022). The Ant-Miner results were compared with Tree-Miner proposed, Tree-Miner algorithm, gave better and more comprehensive predictive results and accuracy (Neagoe et al., 2010; Abbas et al., 2023) The ACO algorithm was used to estimate a logistic regression model in evaluating the quality of red wine, and the results were compared with multiple linear regression, and it was found that ACO had better MAE and MSE values. (Zhang & Zhao, 2014; Al-Enzi et al., 2023) proposed the generalized ant colony algorithm (GACO) for logistical model in order to overcome the shortcomings of classical ACO such as the long time to search for a path. The results showed that GACO worked to reduce transportation costs and achieve optimal solutions to the vehicle routing problem. In a study conducted by (Dabbagh & Bashi, 2018; Alnoor et al., 2024), a system was developed based on the characteristics of ACO in order to detect and classify intrusions into computer networks, A variety of ACO algorithms were used and compared between them, and the study concluded that Ant-Miner is superior in the field of network security. (Habash, 2022; Husin et al., 2023) explained the decision tree algorithms and hybridized them with the genetic algorithm (GA) and the ACO algorithm. Taking the advantage of inheriting good traits from the GA and the community outreach advantage from the ACO algorithm, the study was applied in classifying heart failure disease by taking into account the effect of some factors on the incidence of the disease. In this research, the effect of some factors on thalassemia infection was studied, in addition to classification processes, using one of the most important swarm intelligence algorithms, which is the Ant Colony Optimization (ACO) algorithm with decision tree construction (Muhsen et al., 2023).

3 Materials and Methods

In this section, everything related to the materials and methods that were followed in the classification process, represented by both decision trees and the ant colony algorithm, will be clarified.

3.1 Artificial Intelligence for Ant Colony

Artificial intelligence Swarm algorithms are considered one of the most important algorithms used nowadays, which have good and accurate results at the level of prediction and classification, what is meant by swarm is colonies based on the collective intelligence method (Atiyah et al., 2023). One of the most important of these algorithms is the ant colony algorithm, the idea of which was inspired by insects such as ants and termites, where one can look at the intelligent behavior of a swarm of ants when they communicate with each other through a single network for the purpose of reaching the best paths, through the secretion of a substance called pheromone, where the appropriate path is chosen in which the proportion of pheromone is higher than the rest of the paths, so the ants have the ability to discover the shortest distances (Martens et al., 2007; Atiyah, 2023; Alnoor et al., 2024). Artificial intelligence has employed artificial ants that mimic in their work the behavior of real ants in the mechanism of searching for food, that is, the behavior of shortest distances in order to reach food. The goal of this process is to reach optimal solutions and improve performance for various issues. This algorithm works to find and build solutions repeatedly and gives the amount of pheromone to the paths corresponding to the solutions, choosing the optimal path depends on two parameters: the amount of pheromone and the heuristic values. The amount of pheromone depends on the number of ants involved in making the algorithm, when the ant reaches a place and point, the decision is made and determined (Alnoor et al., 2024). The path with the highest pheromone increases the amount of pheromone and then evaluates the path and creates positive feedback as a process of self-motivation, Ultimately, all the ants converge on the shorter path (Daly & Shen, 2009).

The mechanism of artificial intelligence in simulating the style of real ants is through employing m artificial ants and n nodes at random. Each ant works to take a round repeatedly by applying the potential nearest neighbor method, and the ant adjusts the pheromone level on all the edges that it visited by applying Local updating rule. After all the ants have completed their rounds, the pheromone level is adjusted by applying the global updating rule that favors the edges associated with the best round found.

Each ant k has a set of characteristics in order to fully determine randomness, and these characteristics can be summarized as follows (Daly & Shen, 2009):

1. Memory: Every ant has a very strong memory in which all information about the paths taken is stored.
2. Starting point: Every ant has a starting point from which it starts and a set of conditions that must be met for the purpose of completion.
3. Termination procedures: If one of the termination criteria is met, the ant stores that, otherwise it moves to the neighboring node.
4. Decision: The ant chooses the neighboring node or point according to the rules of probability and according to the percentage of phoromone present. The choice is often random but biased towards the pheromone value.
5. Updating the amount of pheromone: Each ant updates the amount of pheromone as it passes through the beginning or during the return line, and the basic principle remains to increase the amount of pheromone in good places and reduce it in bad places or paths.

Figure 1 shows the path of the ants between the nest and the food source (Seidlova & Pozivil, 2005). We notice from the figure that the ants follow the path between the nest and the food center, and when a threshold appears on the path, the ants choose the direction of travel randomly and with equal probability, and after that, the pheromone is applied more quickly on the path the best and the shortest.

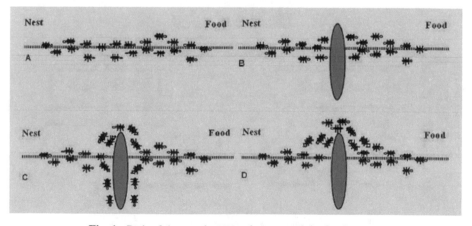

Fig. 1. Path of the ants between the nest and the food source.

The ants continue the process of moving between the nodes until all the nodes are visited, and after all the ants have set their solutions, the appropriate nodes are kept in determining the most appropriate path, as the optimal path is directly proportional to the amount of pheromone used. In other words, the best rules for walking are those in which update larger quantities of pheromone (Neagoe et al., 2010).

4 Decision Tree Algorithm

The classification methodology is considered one of the most important techniques used in various fields because of the effective role it provides in dealing with a large amount of data and information. One of the most important algorithms for this purpose is the decision tree algorithm through which classifications are established for different groups through training sets and is considered one of the supervised algorithm (Jijo & Abdulazez, 2021). In data mining, decision trees are predictive models that can be used in classification and refer to a hierarchical model of decisions, as the decision maker works to use them to determine the optimal strategies that achieve goals (Rokash & Maimon, 2015).

The shape of the decision tree is a hierarchical structure resembling a flow chart. Each decision node works on a feature and each branch represents the result of this decision. The tree begins with a single node known as the root node and then branches into multiple nodes until it reaches the leaves that represent the final decision. The data included in the study is divided into a training and validation data set, where the training data is used to build the tree model and the validation data is used to determine the appropriate tree size (Song and Lu, 2015). Figure 2 below shows the architectural structure of the decision tree (Jijo and Abdulazez, 2021). We notice from the figure that nodes and branches are included in the decision tree, as each node requires one or more properties or features. The tree is divided into three stages: model building (learning), then the evaluation form and finally the classification form.

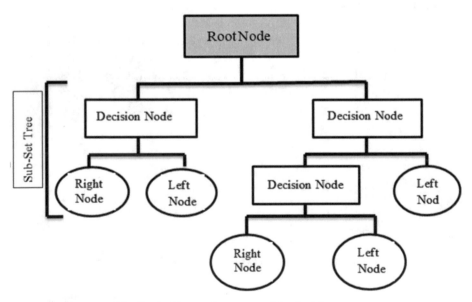

Fig. 2. Architectural structure of the decision tree.

The decision tree bases its calculations in the classification stage on entropy values to measure the randomness of the data in order to determine whether a node should be divided or not, it is known that the entropy value lies between 0 and 1 (Jijo & Abdulazez, 2021; Liu et al., 2013). The information is obtained as one of the measures used in segmentation, which gives the amount of knowledge about the value of the random variable. This is the opposite of the value of entropy, as the higher the value of the measure, the better it is in the classification process. For more details about decision trees in classification, please refer to (Rokach & Mimon, 2015).

5 Application

In this section, the algorithms included in the study and which were explained previously will be applied, as the results were obtained using the SPSS(V.28) program and the MATLAB(V.2021) programming language.

One of the most important criteria for sustainable development is achieving good health, and the most important of them is reducing the number of deaths due to some non-communicable diseases such as cancer, thalassemia, and others. In this study a sample size of (95) was drawn from the blood transfusion center of Al-Faw hospital in the city of Basra, Iraq. The sample included a group of patients who needed blood transfusions, and the sample contained patients suffering from thalassemia and others who did not suffer from it, group of variables was taken to demonstrate their role in thalassemia. The study variables were represented as follows:

Y: Diagnosis, 0: thalassemia is not found, 1: thalassemia is found.
X_1: The patient's gender is male or female.
X_2: Type of blood (A,B,AB,...etc)
X_3: Genetic factor, if one of the parents or other family members is injury
X_4: Place of residence: city center or countryside.

In this paragraph, the ACO algorithm will be applied and a decision will be made about the effect of variables and the classification of thalassemia. As a first step, the best number of nodes in the ACO algorithm was chosen using (7000) ants. It was shown from the results in Table 1 that the number of nodes (25) gave the lowest execution time, but the results of both MSE and MAE were better when the number of nodes was increased to (50), thus the number of nodes was taken into account when calculating the rest of the results.

Table 1. The number of nodes in ACO algorithm, the values of MSE and MAE and the execution time in seconds, by using 7000 ants.

Number of nodes	MSE	MAE	Time in seconds
25	0.386	0.371	0.17
50	0.341	0.336	0.14
75	0.418	0.421	0.28
100	0.464	0.473	1.23

To demonstrate the quality of the ACO algorithm in classification, the classification matching rates of some classical methods (Logistic, N.N were compared with artificial intelligence methods. From Table 2 and Fig. 3, we note that the ACO algorithm gave the best classification quality is (89.86%) with the lowest error values (MSE = 0.341 and MAE = 0.336).

Table 2. Comparison between classification algorithms.

Algorithm	Classification quality %	MSE	MAE
DT	87.42	0.397	0.368
Logistic	87.33	0.392	0.373
Perceptron, N.N	86.2	0.481	0.425
ACO	89.86	0.341	0.336

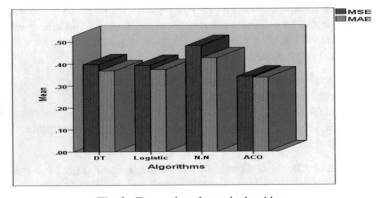

Fig. 3. Error values for each algorithm.

Figure 4 shows the ROC curve for the quality of thalassemia classification, where the further the lines are from the baseline indicates that the classification is good.

The decision tree for classifying thalassemia data. We note that the decision reached is that there is a significant influence of the genetic factor on the incidence of thalassemia. Studies have shown that if one of the parents is injury with thalassemia, there is a high probability that one of the children will be injury. In addition, the place of residence has a clear impact on the probability of injury, and this may be due to the difference in the standard of living or cultural level of those living in the city center and those living in the countryside.

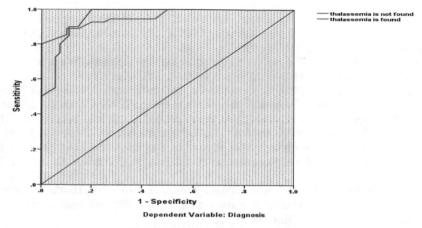

Fig. 4. Roc Curve for thalassemia.

6 Conclusions and Recommendations

In this research, the effect of some factors on thalassemia infection was studied, in addition to classification processes, using one of the most important of artificial intelligence algorithms, which is the Ant Colony Optimization (ACO) algorithm with decision tree construction. The results of the analysis showed that the ACO algorithm gave better results in classification when compared with some other methods and algorithms (DT, Logistic, Perceptron N.N), as the ACO algorithm had the lowest values for both MSE and MAE. The results showed that the genetic factor has a significant impact on the probability of contracting the disease, and the place of residence also has a clear effect, as people who live in the city center were less susceptible to infection compared to people who live in the countryside. In conclusion, we recommend increasing studies on thalassemia due to its extreme importance and increasing scientific research by applying and developing the most important statistical and programming methods in classification processes.

References

Abbas, S., et al.: Antecedents of trustworthiness of social commerce platforms: a case of rural communities using multi group SEM & MCDM methods. Electron. Commer. Res. Appl. **62**, 101322 (2023)

AbdulWahab, B.E.: Factors affecting hereditary anemia (thalassemia). Al- Nahrain Univ. J. **11**(2), 64–74 (2008)

Ahmed, A.D., Salih, M.M., Muhsen, Y R.: Opinion weight criteria method (OWCM): a new method for weighting criteria with zero inconsistency. IEEE Access 1-1 (2024)

Al-Enzi, S.H.Z., Abbas, S., Abbood, A.A., Muhsen, Y.R., Al-Hchaimi, A.A.J., Almosawi, Z.: Exploring research trends of metaverse: a bibliometric analysis. In: Al-Emran, M., Ali, J.H., Valeri, M., Alnoor, A., Hussien, Z.A. (eds.) Beyond Reality: Navigating the Power of Metaverse and Its Applications: Proceedings of 3rd International Multi-Disciplinary Conference - Theme: Integrated Sciences and Technologies (IMDC-IST 2024) Volume 1, pp. 21–34. Springer Nature Switzerland, Cham (2023). https://doi.org/10.1007/978-3-031-51716-7_2

Ali, J., Hussain, K.N., Alnoor, A., Muhsen, Y.R., Atiyah, A.G.: Benchmarking methodology of banks based on financial sustainability using CRITIC and RAFSI techniques. Decis. Mak. Appl. Manage. Eng. **7**(1), 315–341 (2024)

Alnoor, A., Atiyah, A.G., Abbas, S.: Toward digitalization strategic perspective in the European food industry: non-linear nexuses analysis. Asia Pac. J. Bus. Adm. (2023)

Alnoor, A., Atiyah, A.G., Abbas, S.: Unveiling the determinants of digital strategy from the perspective of entrepreneurial orientation theory: a two-stage SEM-ANN approach. Global J. Flex. Syst. Manage. **25**(2), 243–260 (2024). https://doi.org/10.1007/s40171-024-00385-0

Alnoor, A., Chew, X., Khaw, K.W., Muhsen, Y.R., Sadaa, A.M.: Benchmarking of circular economy behaviors for Iraqi energy companies based on engagement modes with green technology and environmental, social, and governance rating. Environ. Sci. Pollut. Res. **31**(4), 5762–5783 (2024)

Alnoor, A., et al.: How positive and negative electronic word of mouth (eWOM) affects customers' intention to use social commerce? A dual-stage multi group-SEM and ANN analysis. Int. J. Hum. Comput. Interact. **40**(3), 808–837 (2024)

Atiyah, A.G.: Unveiling the quality perception of productivity from the senses of real-time multisensory social interactions strategies in metaverse. In: Al-Emran, M., Ali, J.H., Valeri, M., Alnoor, A., Hussien, Z.A. (eds.) Beyond Reality: Navigating the Power of Metaverse and Its Applications: Proceedings of 3rd International Multi-Disciplinary Conference - Theme: Integrated Sciences and Technologies (IMDC-IST 2024) Volume 2, pp. 83–93. Springer Nature Switzerland, Cham (2023). https://doi.org/10.1007/978-3-031-51300-8_6

Atiyah, A.G., Zaidan, R.A.: Barriers to using social commerce. In: Alnoor, A., Wah, K.K., Hassan, A. (eds.) Artificial Neural Networks and Structural Equation Modeling: Marketing and Consumer Research Applications, pp. 115–130. Springer Nature Singapore, Singapore (2022). https://doi.org/10.1007/978-981-19-6509-8_7

Atiyah, A.G., Alhasnawi, M., Almasoodi, M.F.: Understanding metaverse adoption strategy from perspective of social presence and support theories: the moderating role of privacy risks. In: Al-Emran, M., Ali, J.H., Valeri, M., Alnoor, A., Hussien, Z.A. (eds.) Beyond Reality: Navigating the Power of Metaverse and Its Applications: Proceedings of 3rd International Multi-Disciplinary Conference - Theme: Integrated Sciences and Technologies (IMDC-IST 2024) Volume 2, pp. 144–158. Springer Nature Switzerland, Cham (2023). https://doi.org/10.1007/978-3-031-51300-8_10

Atiyah, A.G., All, N.D.A., Zaidan, A.S., Bayram, G.E.: Understating the social sustainability of metaverse by integrating adoption properties with users' satisfaction. In: Al-Emran, M., Ali, J.H., Valeri, M., Alnoor, A., Hussien, Z.A. (eds.) Beyond Reality: Navigating the Power of Metaverse and Its Applications: Proceedings of 3rd International Multi-Disciplinary Conference - Theme: Integrated Sciences and Technologies (IMDC-IST 2024) Volume 1, pp. 95–107. Springer Nature Switzerland, Cham (2023). https://doi.org/10.1007/978-3-031-51716-7_7

Atiyah, A.G., Faris, N.N., Rexhepi, G., Qasim, A.J.: Integrating ideal characteristics of chatGPT mechanisms into the metaverse: knowledge, transparency, and ethics. In: Al-Emran, M., Ali, J.H., Valeri, M., Alnoor, A., Hussien, Z.A. (eds.) Beyond Reality: Navigating the Power of Metaverse and Its Applications: Proceedings of 3rd International Multi-Disciplinary Conference - Theme: Integrated Sciences and Technologies (IMDC-IST 2024) Volume 1, pp. 131–141. Springer Nature Switzerland, Cham (2023). https://doi.org/10.1007/978-3-031-51716-7_9

Bilal, W.H.: Improved ant colony system algorithm to solve the vehicle routing problem. Tishreen Univ. J. Res. Sci. Stud. Basic Sci. Ser. **36**(5), 193–207 (2014)

Dabbagh, N.B., Bashi, M.S.: Using the ANT algorithm to detect and classify intrusions in computer networks. J. Educ. Sci. **27**(2), 146–168 (2018)

Daly, R., Shen, Q.: Learning bayesian network equivalence classes with ant colony optimization. J. Artif. Intell. Res. **35**, 391–447 (2009)

Habash, H.A.: Developing a new methodology to construct decision trees by using a hybrid method based on genetic algorithms and ant colony optimization. Al-Baath Univ. J. **44**(1), 33–56 (2022)

Husin, N.A., Abdulsaeed, A.A., Muhsen, Y.R., Zaidan, A.S., Alnoor, A., Al-mawla, Z.R.: Evaluation of metaverse tools based on privacy model using fuzzy MCDM approach. In: Al-Emran, M., Ali, J.H., Valeri, M., Alnoor, A., Hussien, Z.A. (eds.) Beyond Reality: Navigating the Power of Metaverse and Its Applications: Proceedings of 3rd International Multi-Disciplinary Conference - Theme: Integrated Sciences and Technologies (IMDC-IST 2024) Volume 1, pp. 1–20. Springer Nature Switzerland, Cham (2023). https://doi.org/10.1007/978-3-031-51716-7_1

Liu, Y., Hu, L., Yan, F., Zhang, B.: Information gain with weight based decision tree for the employment forecasting of undergraduates. In: IEEE International Conference on Green Computing and Communications and IEEE Internet of Things and IEEE Cyber, Physical and Social Computing, Beijing, China, pp. 2210–2213 (2013). https://doi.org/10.1109/GreenCom-iThings-CPSCom.2013.417

Martens, D., De Backer, M., Haesen, R., Vanthienen, J., Snoeck, M., Baesens, B.: Classification with ant colony optimization. IEEE Trans. Evol. Comput. **11**(5), 651–665 (2007). https://doi.org/10.1109/TEVC.2006.890229

Muhsen, Y.R., Husin, N.A., Zolkepli, M.B., Manshor, N.: A systematic literature review of fuzzy-weighted zero-inconsistency and fuzzy-decision-by-opinion-score-methods: assessment of the past to inform the future. J. Intell. Fuzzy Syst. **45**(3), 4617–4638 (2023)

Neagoe, V.E., Neghina, C.E., Neghina. M.: Ant colony optimization for logistic regression and its application to wine quality assessment mathematical models for engineering science. In: Proceedings of the 2010 International Conference on Mathematical Models for Engineering Science, pp. 195–200 (2010)

Nejad, N.Z., Bakhtiary, A.H., Analoui, M.: Classification using unstructured rules and ant colony optimization. In: Proceedings of the International Multi Conference of Engineers and Computer Scientists 2008. Vol I IMECS 2008, 19–21 March, Hong Kong (2008)

Seidlová, R., Poživil, J.: Implementation of Ant Colony Algorithms In Matlab. Institute of Chemical Technology, Department of Computing and Control Engineering Technická 5, Prague 6, 166 28, Czech Republic(2005)

Silva, C.A., Runkler, T.A., Sousa, J.M., Palm, R.: Ant colonies as logistic processes optimizers. In: Dorigo, M., Di Caro, G., Sampels, M. (eds.) Ant Algorithms: Third International Workshop, ANTS 2002 Brussels, Belgium, September 12–14, 2002 Proceedings, pp. 76–87. Springer Berlin Heidelberg, Berlin, Heidelberg (2002). https://doi.org/10.1007/3-540-45724-0_7

Song, Y.Y., Lu, Y.: Decision tree methods: applications for classification and prediction. Shanghai Arch. Psychiatry **27**(2), 130–135 (2015)

Zhang, D., Zhao, H.: Research of generalized ant colony algorithm for logistics and distribution. Appl. Mech. Mater. **513–517**, 691–694 (2014). https://doi.org/10.4028/www.scientific.net/AMM.513-517.691

The Role of Artificial Intelligence in Improving Sustainable Audit Quality

Abdulhussein Tofeeq Shibli[1], Nahla Ghalib Abdul Rahma[2],
and Jalil Ibrahim Salih[3(✉)]

[1] Economic and Administration College, University of Basra, Basra, Iraq
[2] Technical Institute of Basra, Southern Technical University, Basra, Iraq
nahla.jalee@stu.edu.iq
[3] Southern Technical University, Technical College of Administration, Basra, Iraq
Jalil.ibrahim@stu.edu.iq

Abstract. The purpose of this study is to determine the significance of artificial intelligence (AI) contribution to upgrading and improving Audit Quality (AQ) throughout the audit process (AP). In addition, AI has led to the reduction of traditional auditing (TA) procedures, which are no longer able to keep up with rapid corporate innovations. To meet the research aims, was employed the descriptive analytical approach, conducting a field study. The study population comprises of 320 external auditors in Iraq, with (260) questionnaires issued to the research sample. The research tool was provided to them electronically, and (232) surveys were returned, yielding a response rate of 89%. The statistical software SPSS V. 20 is used to analyze the data, and the study comes to the conclusion that using AI technology by the auditor increases the AQ and decreases the amount of time spent on it. There is a connection between the auditor's use of AI and the auditing process because of its dependence on conventional auditing methods.

Keywords: Artificial intelligence · Audit Quality · Audit Process · Sustainability

1 Introduction

There are two distinct eras in the historical development of AI. The first era, which is represented by ancient history, is when the ideas of autonomous machines and mechanical gadgets with restricted capabilities first appeared. The second era encompasses modern history, starting with the post-World War II development of modern computers. In the modern day, sophisticated computer programs have been created to tackle challenging intellectual issues, and tools with broad applications across other domains have also been built (Krakowski et al., 2023; Al-Janabi et al., 2024; Atiyah: et al., 2023). The Dartmouth conference founded the expansion of the modern field of AI research in 1956, marking the beginning of AI, and one of the first, most famous and enduring projects on AI and the law focused on AI in tax law (Hadi et al., 2023; Groumpos, 2023; Atiyah et al., 2023). In 1977, a knowledge-based system was developed by L. Thorne McCarty and called it "Taxman", and after the widespread use of the Internet and calculators, AI technology spread greatly (Kieslich et al, 2022). Towards the middle of the

20th century, a few scientists started investigating a novel method for creating intelligent machines. After the invention of the digital computer, a device that could mimic human computational thought processes was created (Gašević et al., 2023; Alnoor et al., 2024). Expert systems are a type of AI program that simulates knowledge and analytical skills. Their commercial success in the early 1980s gave rise to a new wave of interest in AI research. The market for AI research saw earnings of over a billion dollars by 1985, and the AI research certificate was granted in 1987. AI has its drawbacks, but in the early 21st century, more especially, in the 1990s, it made significant strides and was applied to a wide range of tasks, including data mining, medical diagnosis, logistics, and many other operations (Qasaimeh & Jaradeh, 2022; Salman et al., 2023; Alnoor et al., 2023; Atiyah, 2023; Atiyah and Zaidan, 2022; Atiyah et al., 2023; Ali et al., 2024). The business environment of all kinds has changed dramatically in the modern era, particularly since the 1980s (Al-taee, & Flayyih, 2022; Abass et al., 2023; Alnoor et al., 2024). This change has been driven primarily by the use of expert systems in business process implementation, which is regarded as one of the applications of AI programs. AI research has also been increasingly prominent in this era. It mimics the user's knowledge and analytical abilities. AI programs have become more widespread in the 1990s and early 21st centuries, assisting businesses of all stripes in providing and utilizing financial and non-financial data and information, as well as extracting it quickly and accurately (Huynh et al., 2023; Abbas et al., 2023). In the middle of the twentieth century, a small number of scientists began to explore a new approach to building intelligent machines. The digital computer was invented, and then a machine was invented that could simulate the process of human computational thinking. In 1956, the Dartmouth conference was the one that confirmed the breadth of the modern field regarding AI research (Gašević et al, 2023; Muhsen et al., 2023; Ahmed et al., 2024).

The commercial success of expert systems, a type of AI software that simulates knowledge and analytical skills, in the early 1980s marked a new dawn for AI research. Although the AI research certificate faced difficulties in 1987, AI made significant progress in the early 21st century, particularly in the 1990s. Since then, AI has been applied to a wide range of tasks, including data mining, medical diagnosis, logistical operations, and more (Qasaimeh & Jaradeh, 2022; Alnoor et al., 2024). According to the International Federation of Certified Public Accountants, information technology proficiency is a must for accountants and auditors. This means that to use modern technologies in the accounting and auditing fields, particularly in tax accounting procedures, administrative accounting, finance, auditing, and other areas, they must improve their scientific and practical ability to recognize and understand the requirements for doing so (Maseer et al., 2022; Agustí & Orta, 2023; Husin et al., 2023). Auditing companies are making significant investments in advanced information technology to enhance the efficiency and effectiveness of the AP, and specialists estimate that the Big4 auditing companies spend $250 million annually on AI (Fedyk et al, 2022; Al-Enzi et al., 2023). According to (Munoko et al, 2020), when auditors rely more and more on AI, they may make a few basic assumptions. Firstly, they may assume that these AI systems are always accurate. Secondly, they may assume that AI systems always operate within the necessary constraints. Lastly, they may assume that deviations from the desired restrictions will be discovered and corrected. These assumptions may not always be true, which could have negative ethical, legal, or financial ramifications. By 2025, it is expected that AI would handle 30% of audit procedures performed by auditors. The use of information technology in the field of auditing is one of the contemporary issues that has emerged as a result of auditors' awareness of the importance of such use because it assists auditors in selecting audit samples, thus increasing the credibility of the inspection results in the audit and assisting the auditor in performing calculations to obtain more accurate and faster results (Seethamraju & Hecimovic, 2023).

The concept of expert systems has spread with the requirements for carrying out accounting and auditing work and the many uses of electronic transactions, especially in the banking sector, which has become dominated by the nature of electronic work in all transactions with bank customers (Rehman, 2022; Flayyih & Khiari, 2023). AI seeks to understand the nature of human intelligence by developing computer programs capable of simulating human behavior, which is distinguished by intelligence and the ability to process processes electronically and provide users with the data and information they require to make various decisions quickly. One of the reasons that encouraged the use of expert systems in the field of accounting and auditing is the large number of problems faced by the external auditor, which have a large number of alternative solutions that are difficult to sort out to determine the best of them, requiring the auditor's experience in knowing the best alternative and using expert systems (Dalwai et al. al, 2022). In light of the development of computers, the Internet, and sophisticated AI technologies, AI is seen as a technological outcome that will contribute to human progress in the future. Specifically, AI aims to alter the work practices of conventional external auditors (Zhang & Vasarhelyi, 2022; Atiyah et al., 2020). AI's vast analytical capabilities have drastically changed the auditing function by enabling quick calculations, precise analysis, and the extremely efficient completion of many audit tasks. These capabilities support and improve the performance levels and abilities of auditors (Baghdasaryan, 2022). With the increasing importance of the role of AI in audit operations due to new methods and technologies that contribute to modernizing the AP and changing the role played by the auditor, through its impact on the way auditors perform their work, how they collect audit evidence, and how they deal with, analyze and interpret data, which affects the AQ (Boer et al., 2023; Atiyah, 2022).

Despite the great importance of studying the topic of AI in enhancing the quality of external auditing, and within the limits of researchers' knowledge, there is a scarcity of writing on such a topic, especially in Iraq. Given the significance of auditors and their pursuit of high AQ, as well as the fact that they are subject to a variety of internal and external factors and constraints, we should be aware of the ways in which AI can improve AQ. Consequently, the study aims to close a research gap in this area by identifying the role of AI in improving the quality of external auditing. Given the foregoing, the problem of the study can be expressed in light of the following questions: Does AI help to improve and enhance the AQ? What impact does AI have on auditors and the auditing profession? Does AI help auditors reduce their reliance on TA methods? So, Research aims to Recognizing the value of AI in increasing and improving AQ during the AP. Determine the extent to which AI helps to reduce ineffective TA methods and identifying the favorable relationship between the auditing process and the technological advancement of information systems in general, and AI specifically.

2 Literature Review and Hypothesis Development

Thomas Hobbes first suggested the concept of AI in Europe in the early seventeenth century. Hobbes claimed that mechanical terms and symbols, including numbers, graphs, calculations, and statistics, could be used to understand human behavior (Al-Aroud, 2020). To assist its users in making decisions and producing customized reports that satisfy the demands of the business, AI can be utilized to streamline the process of auditing and evaluating financial data and documentation. AI is also employed to assist auditors in decision-making, pattern recognition, and prediction (Bose et al., 2023; Atiyah, 2023). Since AI has grown more adept at spotting fraud and errors in financial statements, its potential use could reduce the need for external auditors and possibly eliminate the need for the company to hire outside auditors to audit its books. This could result in the loss of significant jobs in the auditing industry as well as a decline in the caliber of financial audits (Albawwat & Frijat, 2021; Zaidan et al., 2023).

AI is considered one of the most prominent modern applications of information systems, as it represents one of the most important modern sciences that arose due to the convergence of the technical revolution in the field of computer science and automatic control on the one hand, and logic, mathematics, languages, and psychology on the other hand. In this regard, AI represents an umbrella for many technologies that allow machines to mimic human intelligence (Rodgers et al, 2023; AL-Fatlawey et al., 2021). AI, as defined by Kondratenko (2023), is "that science that includes all algorithms and theoretical and applied methods that are concerned with automating the decision-making process on behalf of humans, whether completely or with the participation of humans with the ability to adapt and predict". AI is also defined as "one of the branches of computer science and engineering, concerned with developing computers capable of working, learning, and thinking independently. AI helps in analyzing large amounts of data, as well as making decisions on its own. There are several ways to use intelligence." artificial in companies" (Zemankova, 2019).

As for (Rashwan & Alhelou, 2020), AI is defined as "an application of computer science and engineering systems with intelligent computers that are capable of displaying human traits to learn, think and act independently, and have the ability to analyze big data and make high-quality decisions". One of the most significant contemporary technologies that will enable the auditor to effect change and satisfy stakeholders and clients is AI. The utilization of AI systems yields numerous benefits, chief among them being the attainment of efficiency and the reduction of time spent on diverse tasks. In addition to the conventional advantages, using AI systems in auditing can offer additional advantages (Pimenov et al., 2023; Chew et al., 2023). Cazazian (2022) pointed out that the use of information technologies in the field of auditing is not a new thing. For example, computer-assisted auditing methods are among the supporting methods in the field of auditing at the present time and have been used for long periods. However, the introduction of modern technologies such as AI and the large volume of data were among the factors that audit offices consider when planning audit operations. According to (Albawwat & Frijat, 2021), auditing firms are presently investigating and evaluating the use of machine learning technologies in auditing processes.

Argus, a machine learning program that can understand contracts for sales and rentals, was employed by Delitte and Touche. Algorithms that enable establishing the primary contract criteria were used in the programming of Argus. Not only that, but it also shows values that deviate greatly from the statistical population. This helps auditors save a ton of time when performing audit tasks. Auditors are affected by AI in many ways, as they are affected by all the changes that occur in their clients' environment. Therefore, these clients adopt modern and advanced technologies that will bring about changes in the AP, starting from the planning stage of the applied process, through field work, and ending with the auditor's report. On the other hand, auditors are directly affected by the need to adopt AI technologies to be able to perform their work in line with technological developments and customer expectations and improve the accuracy and quality of services provided by auditors (Lehner et al, 2022).

According to (Hasan, 2021), the auditing profession will not be able to survive if it does not adjust to the changes that surround audit clients. The most significant of these changes are linked to contemporary technological advancements, which means that an auditor cannot examine the enormous amounts of data that clients have without the use of technologies. Modern technology, and that the auditor will not be able to plan the AP without taking into account the risks arising from the adoption of modern technology by his clients. The significance of AI methods is emphasized by their exceptional capacity to scrutinize the complete statistical population, irrespective of its magnitude. This facilitates the auditor in detecting atypical operations that prove challenging to uncover through sample-based examination (Rashwan & Alhelou, 2020). By implementing AI, the auditor can be directed toward high-risk regions more successfully and documents can be processed more quickly utilizing content analysis and natural language processing than with the conventional applied approach (Munoko et al., 2020).

The efficiency of the auditing process is one of the most significant advantages of utilizing AI in auditing, since it enables the auditor to achieve the highest levels of assurance with the least amount of time and effort. The machine audits contracts on the auditor's behalf in record time, saving him from putting in long hours of work. In record time, this enables the auditor to save time and devote more attention to more crucial tasks that are impossible to complete with electronic devices, like better understanding client demands and corresponding with them (Rodrigues et al., 2023). AI systems increases the auditor's independence and the independence of audit operations, which is reflected in the auditor's effectiveness and quality of performance in identifying and minimizing errors quickly and finishing the AP more quickly, affordably, and effectively (Zemankova, 2019). According to (Goo et al., 2016), AI has a variety of tools. The decision tree assists the auditor in expressing his assessment of the company's continuity, the supporting beam technique assists the auditor in determining the degree to which businesses use discretionary accruals to manage profits, and neural networks Along with "estimating the risks of administrative control and predicting the auditor's fees, and assisting him in logical analysis when planning the AP and saving time and effort," it aids the auditor in uncovering substantial misrepresentation and management fraud (Faúndez et al., 2020) state that although AI is still in its early stages, it has made remarkable and continuous progress, and that AI methods are one of the most important factors that the auditor considers when planning its work with the aim of exploiting the increasing amount of huge data available to it.

The study (Fedyk et al., 2020) aimed to identify the mechanism of the impact of AI on AQ, and the study sample represented a number of audit companies, numbering 36 companies. This is to determine the extent to which audit companies employ a number of workers in the field of AI, and the study concluded that AI is a central function within the audit company. Investing in AI helps improve AQ and reduces audit fees, which in turn leads to reducing the number of auditors in the AP. The study (Gultom et al, 2021) indicated an examination of AI to improve the AQ with the efficiency and doubts of auditors regarding customer satisfaction. To achieve the objectives of the study, wes relied on a questionnaire to collect data. The statistical program SPSS was used as the main tool in analyzing the data. The study population was represented by a number of employees in Indonesian companies listed on the Jakarta Stock Exchange, numbering 229 respondents. One of the most important findings of this study is that AI has a positive impact on customer satisfaction and has positive effects in improving AQ, and that the efficiency of auditors has a positive impact on customer satisfaction. Regarding the study (Asiri at el, 2020), its goal is to determine how AI affects AQ. We used a questionnaire form to collect data for the study. Ninety questionnaires were included in the SPSS Analysis of Statistical, and the study sample consisted of several audit offices across the Kingdom. The study found that AI applications are used by audit offices in Saudi Arabia, with an application rate of 87.92% and a high AQ rate of 90.22%.

The goal of the study (Albawwat & Frijat, 2021) was to determine how much the independent, supported, and augmented AI systems contributed to the AP and how much they improved the audit's quality. The questionnaire form was a key component of we's investigation. The questionnaires suitable for Analysis of Statistical amounted to 124 forms. The research sample represented auditors in Jordan and were distributed electronically. We used the statistical program SPSS to analyze the collected data. The study concluded that the external auditor believes that the enhanced AI systems and the supported AI systems are easy to use by the auditor and useful for the AP. On the contrary, he believes that independent AI systems are complex to use and not useful for the auditing process. In addition, there is a big difference in the contribution of these types to the AQ. Supported AI solutions contribute the most to AQ. While the study (Saad, 2022) intends to assess the interaction between AI and the auditor and the auditing profession, it also investigates the influence of employing AI in increasing AQ. The study sample included a number of auditors from the State of Palestine. In his study, we relied on the questionnaire form, as there were 104 questionnaires eligible for analysis. We analyzed the data using the statistical program SPSS, and the study indicated that there is a favorable association between AI, auditors, and the auditing profession. Furthermore, there is a positive correlation between AI and increased audit efficiency.

The study (Oluwasegun et al, 2023) used a questionnaire to assess the impact of AI on AQ. There were 641 questionnaires appropriate for Analysis of Statistical. The study sample consisted of several firm managers, accountants, and auditors. According to the survey, participants of the study sample agree that AI has a substantial and beneficial impact on AQ and helps supply reliability and correctness of financial data while also preparing it on time. The study (Noordin et al, 2022) aimed to identify the extent to which AI contributes to influencing the AQ in UAE auditing companies. It also aims to identify the extent to which AI contributes to influencing the AQ among local and international auditing firms. The study sample represented several employees of local auditing companies, numbering 22 companies, and a number of employees of international auditing companies, numbering 41 companies. Wes relied on the statistical program SPSS to analyze the data. The study found that there is an impact of AI on AQ, as well as a non-significant difference in the contribution of AI to AQ between local and international audit firms. Based on literary contributions; We developed the following hypotheses to appropriately deal with the research variables and provide answers to the research questions:

H1: AI in auditing does not enhance or improve AQ.
H2: AI doesn't help auditors rely less on conventional auditing methods.
H3: AI in the AP is not associated with any statistically significant or meaningful association.

3   Methodology

3.1   Sample Size and Measurements

To achieve the aims this study, we are adopting the approach of descriptive analytical, through which it attempts to describe the phenomenon under investigation, analyze its data, the relationship between its components, the opinions surrounding it, the processes it includes, and the effects it produces. The research technique, which aims to identify the function of AI in improving the AQ in audit offices, can be elucidated. The research community consists of a number of auditors working in audit offices in Iraq, numbering 320 auditors. The number of questionnaires distributed to the research sample reached (260), and the research tool was distributed to them electronically, and (232) question-naires were returned; thus the response rate is (approximately 89%). This percentage is considered acceptable for generalizing its results to the community of research.

3.2   Statistical Approach

The following is a set of methods used in data analysis (SPSS V. 20); Many relevant SPSS (Statistical Package for Social Sciences) statistical methods have been implemented utilizing the Statistical Packages for the Social Sciences to meet the study objectives and analyze the data obtained.

4   Results

According to this hypothesis, "The application of AI does not contribute to improving and enhancing the AQ." After examining the questionnaire responses of the research sample members, the arithmetic means and standard deviations for this hypothesis are found, and the results are shown in Table 1.

It is clear from the table above that the arithmetic averages for the third hypothesis ranged between (4.52–3.30), and the arithmetic average for all paragraphs is (4.50). As for the standard deviation, it reached (0.67), and we conclude from this that paragraph

No. (8), which states (AI systems contributes to the appropriateness and adequacy of collecting audit evidence) is characterized by a level of high importance, as its arithmetic mean reached (4.52).

The research hypotheses are tested through the data of the actual responses to the survey of the research group, which is represented by a number of auditors, using the Z test with a significance level of (0.05). The research hypotheses will be verified according to the sequence in which they were presented in the research, as follows:

Testing the second hypothesis, which states the following (the application of AI does not contribute to improving and enhancing the AQ). To determine the validity or error of this hypothesis within the context of the Analysis of Statistical results, a Z test on the arithmetic mean was performed, the results of which are provided in Table 1.

Table 1. Z test results

Arithmetic Mean	Sample size	Z		Test Results
4.2	232	Calculated	Tabular	First Hypothesis
		26.707	1.96	First Hypothesis rejected

Analysis of Statistical reveals that the calculated (Z) value is (26.707), which exceeds the tabulated value of 1.96. As a result, the null hypothesis is rejected, and the alternative hypothesis is accepted, stating that (AI contributes to improving and enhancing AQ).

Testing the second hypothesis, which states the following (AI does not contribute to reducing the auditor's reliance on TA methods). To determine the validity or error of this hypothesis within the context of the Analysis of Statistical results, a Z test on the arithmetic mean was performed, the results of which are provided in Table 2.

Table 2. Z test results

Arithmetic Mean	Sample size	Z		Test Results
4.27	232	Calculated	Tabular	Second Hypothesis
		26.87	1.96	Second Hypothesis rejected

Analysis of Statistical revealed that the calculated (Z) has a value of (26.87), which is greater than its tabulated value of (1.96), rejecting the null hypothesis and accepting the alternative hypothesis, which states that (AI contributes to reducing the auditor's reliance on TA methods).

which states the following (there is no statistically significant and significant relationship between the application of AI and the auditing process), and to prove the validity or error of this hypothesis within the framework of the results of the Analysis of Statistical, a Z- test was conducted on the arithmetic mean, the results of which presented in Table 3.

Table 3. Z test results

Arithmetic Mean	Sample size	Z		Test Results
4.50	232	Calculated	Tabular	Third Hypothesis
		34.10	1.96	Third Hypothesis rejected

Analysis of Statistical revealed that the estimated (Z) value (34.10) is bigger than the tabulated value (1.96). As a result, the null hypothesis is rejected, and the alternative hypothesis is accepted, stating that (there is a statistically significant and substantial association between AI and the auditing process).

5 Conclusion

The application of AI enables auditors to store and evaluate massive amounts of data while also providing corroborating evidence, which is represented in the audit report and decreases human errors. In the present day, information technology, particularly AI, is extremely important in the field of auditing because it clarifies the distinction between traditional audits and auditing utilizing AI. AI by the auditor in the AP leads to improving and enhancing the AQ. Reliance on AI technologies in the AP causes the auditor to rely less on conventional audit methods. There is a relationship between the auditing process practiced by the auditor and the application of AI. Auditors' awareness of the importance of AI systems during the provision of audit services, and its role in achieving the AQ process for audit companies in the information technology environment, should be enhanced. To promote the use of information technology by auditors in their auditing operations, all businesses operating in different sectors should switch from manual to computerized accounting systems. It is recommended that professional and supervisory organizations provide auditors with training on AI systems in all phases of the AP, including planning, conducting analytical procedures, and carrying out and documenting audit operations. The auditor's reliance on AI systems in the AP leads to reducing his/her reliance on traditional audit methods. Expansion of AI-related scientific study and efforts by researchers to assess AI accurately.

References

Abass, Z.K., Al-Abedi, T.K., Flayyih, H.H.: Integration between cobit and coso for internal control and its reflection on auditing risk with corporate governance as the mediating variable. Int. J. Econ. Financ. Stud. **15**(2), 40–58 (2023)

Abbas, S., et al.: Antecedents of trustworthiness of social commerce platforms: a case of rural communities using multi group SEM & MCDM methods. Electron. Commer. Res. Appl. **62**, 101322 (2023)

Agustí, M.A., Orta-Pérez, M.: Big data and artificial intelligence in the fields of accounting and auditing: a bibliometric analysis. Span. J. Financ. Account./Revista Española de Financiación y Contabilidad **52**(3), 412–438 (2023)

Ahmed, A.D., Salih, M.M., Muhsen, Y.R.: Opinion weight criteria method (OWCM): a new method for weighting criteria with zero inconsistency. IEEE Access (2024)

Al-Aroud, S.F.: The impact of artificial intelligence technologies on audit evidence. Acad. Account. Financ. Stud. J. **24**, 1–11 (2020)

Albawwat, I., Frijat, Y.: An analysis of auditors' perceptions towards artificial intelligence and its contribution to audit quality. Accounting **7**(4), 755–762 (2021)

Al-Enzi, S.H.Z., Abbas, S., Abbood, A.A., Muhsen, Y.R., Al-Hchaimi, A.A.J., Almosawi, Z.: Exploring research trends of metaverse: a bibliometric analysis. In: Al-Emran, M., Ali, J.H., Valeri, M., Alnoor, A., Hussien, Z.A. (eds.) Beyond Reality: Navigating the Power of Metaverse and Its Applications. IMDC-IST 2024. LNNS, vol. 895, pp. 21–34. Springer, Cham (2023). https://doi.org/10.1007/978-3-031-51716-7_2

AL-Fatlawey, M.H., Brias, A.K., Atiyah, A.G.: The role of strategic behavior in achievement the organizational excellence analytical research of the manager's views of Ur state company at Thi-Qar governorate. J. Adm. Econ. **10**(37) (2021)

Ali, J., Hussain, K.N., Alnoor, A., Muhsen, Y.R., Atiyah, A.G.: Benchmarking methodology of banks based on financial sustainability using CRITIC and RAFSI techniques. Decis. Mak. Appl. Manag. Eng. **7**(1), 315–341 (2024)

Al-Janabi, A.S.H., Almado, A.A.G., Mhaibes, H.A., Flayyih, H.H.: The role of strategic agility in promoting organizational excellence: a descriptive analytical study. Corp. Bus. Strategy Rev. **5**(2), 129–138 (2024). https://doi.org/10.22495/cbsrv5i2art11

Alnoor, A., Atiyah, A.G., Abbas, S.: Toward digitalization strategic perspective in the European food industry: non-linear nexuses analysis. Asia-Pac. J. Bus. Adm. (2023)

Alnoor, A., Atiyah, A.G., Abbas, S.: Unveiling the determinants of digital strategy from the perspective of entrepreneurial orientation theory: a two-stage SEM-ANN approach. Glob. J. Flex. Syst. Manag. 1–18 (2024)

Alnoor, A., Chew, X., Khaw, K.W., Muhsen, Y.R., Sadaa, A.M.: Benchmarking of circular economy behaviors for Iraqi energy companies based on engagement modes with green technology and environmental, social, and governance rating. Environ. Sci. Pollut. Res. **31**(4), 5762–5783 (2024)

Alnoor, A., et al.: How positive and negative electronic word of mouth (eWOM) affects customers' intention to use social commerce? A dual-stage multi group-SEM and ANN analysis. Int. J. Hum. –Comput. Interact. **40**(3), 808–837 (2024)

Al-taee, S.H.H., Flayyih, H.H.: The impact of the audit committee and audit team characteristics on the audit quality: mediating impact of effective audit process. Int. J. Econ. Financ. Stud. **14**(03), 249–263 (2022)

Asiri, M.S., Al-Hanawi, M.A.S., Al-Badidi, H.S., Al Suwayd, A.A.A., Al-Mazni, M.F.: The impact of using artificial intelligence on the quality and automation of audit procedures: a field study on audit firms in the Kingdom of Saudi Arabia (2020)

Atiyah, A.G.: Impact of knowledge workers characteristics in promoting organizational creativity: an applied study in a sample of Smart organizations. PalArch's J. Archaeol. Egypt/Egyptol. **17**(6), 16626–16637 (2020)

Atiyah, A.G.: Effect of temporal and spatial myopia on managerial performance. J. La Bisecoman **3**(4), 140–150 (2022)

Atiyah, A.G.: Strategic network and psychological contract breach: the mediating effect of role ambiguity. Int. J. Res. Manag. Stud. (IJRMS) **13**(1) (2023)

Atiyah, A.G.: Unveiling the quality perception of productivity from the senses of real-time multisensory social interactions strategies in metaverse. In: Al-Emran, M., Ali, J.H., Valeri, M., Alnoor, A., Hussien, Z.A. (eds.) Beyond Reality: Navigating the Power of Metaverse and Its Applications. IMDC-IST 2024. LNNS, vol. 876, pp. 83–93. Springer, Cham pp. 83–93 (2023). https://doi.org/10.1007/978-3-031-51300-8_6

Atiyah, A.G., Zaidan, R.A.: Barriers to using social commerce. In: Alnoor, A., Wah, K.K., Hassan, A. (eds.) Artificial Neural Networks and Structural Equation Modeling, pp. 115–130. Springer, Singapore (2022). https://doi.org/10.1007/978-981-19-6509-8_7

Atiyah, A.G., Alhasnawi, M., Almasoodi, M.F.: Understanding metaverse adoption strategy from perspective of social presence and support theories: the moderating role of privacy risks. In: Al-Emran, M., Ali, J.H., Valeri, M., Alnoor, A., Hussien, Z.A. (eds.) Beyond Reality: Navigating the Power of Metaverse and Its Applications. IMDC-IST 2024. LNNS, vol. 876, pp. 144–158. Springer, Cham (2023a). https://doi.org/10.1007/978-3-031-51300-8_10

Atiyah, A.G., All, N.D.A., Zaidan, A.S., Bayram, G.E.: Understating the social sustainability of metaverse by integrating adoption properties with users' satisfaction. In: Al-Emran, M., Ali, J.H., Valeri, M., Alnoor, A., Hussien, Z.A. (eds.) Beyond Reality: Navigating the Power of Metaverse and Its Applications. IMDC-IST 2024. LNNS, vol. 895, pp. 95–107. Springer, Cham (2023b). https://doi.org/10.1007/978-3-031-51716-7_7

Atiyah, A. G., Faris, N. N., Rexhepi, G., & Qasim, A. J. (2023, December). Integrating Ideal Characteristics of Chat-GPT Mechanisms into the Metaverse: Knowledge, Transparency, and Ethics. In International Multi-Disciplinary Conference-Integrated Sciences and Technologies (pp. 131–141). Cham: Springer Nature Switzerland

Baghdasaryan, V., Davtyan, H., Sarikyan, A., Navasardyan, Z.: Improving tax audit efficiency using machine learning: the role of taxpayer's network data in fraud detection. Appl. Artif. Intell. **36**(1), 2012002 (2022)

Boer, A., de Beer, L., van Praat, F.: Algorithm Assurance: Auditing Applications of Artificial Intelligence. Adv. Digit. Audit. **149** (2023)

Bose, S., Dey, S.K., Bhattacharjee, S.: Big data, data analytics and artificial intelligence in accounting: an overview. Handbook of Big Data Research Methods 32.9 (2023)

Cazazian, R.: Blockchain technology adoption in artificial intelligence-based digital financial services, accounting information systems, and audit quality control. Rev. Contemp. Philos. (21), 55–71 (2022)

Chew, X., Khaw, K.W., Alnoor, A., Ferasso, M., Al Halbusi, H., Muhsen, Y.R.: Circular economy of medical waste: novel intelligent medical waste management framework based on extension linear Diophantine fuzzy FDOSM and neural network approach. Environ. Sci. Pollut. Res. **30**(21), 60473–60499 (2023)

Dalwai, T.A.R., Madbouly, A., Mohammadi, S.S.: An Investigation of Artificial Intelligence Application in Auditing. In: Alareeni, B., Hamdan, A. (eds.) Artificial Intelligence and COVID Effect on Accounting. Accounting, Finance, Sustainability, Governance and Fraud: Theory and Application, pp. 101–114. Springer, Singapore (2022). https://doi.org/10.1007/978-981-19-1036-4_7

Faúndez-Ugalde, A., Mellado-Silva, R., Aldunate-Lizana, E.: Use of artificial intelligence by tax administrations: an analysis regarding taxpayers' rights in Latin American countries. Comput. Law Secur. Rev. **38**, 105441 (2020)

Fedyk, A., Hodson, J., Khimich, N., Fedyk, T.: Is artificial intelligence improving the audit process? Rev. Acc. Stud. **27**(3), 938–985 (2022)

Flayyih, H.H., Khiari, W.: Empirically measuring the impact of corporate social responsibility on earnings management in listed banks of the Iraqi stock exchange: the mediating role of corporate governance. Ind. Eng. Manag. Syst. **22**(3), 273–286 (2023)

Gašević, D., Siemens, G., Sadiq, S.: Empowering learners for the age of artificial intelligence. Comput. Educ. Artif. Intell. **4**, 100130 (2023)

Goo, Y.J.J., Chi, D.J., Shen, Z.D.: Improving the prediction of going concern of Taiwanese listed companies using a hybrid of LASSO with data mining techniques. SpringerPlus **5**(1) (2016). https://doi.org/10.1186/s40064-016-2186-5

Gultom, J.B., Murwaningsari, E., Umar, H., Mayangsari, S.: Reciprocal use of artificial intelligence in audit assignments. J. Account. Bus. Financ. Res. **11**(1), 9–20 (2021)

Hadi, A.H., Abdulhameed, G.R., Malik, Y.S., Flayyih, H.H.: The influence of information tech-nology (it) on firm profitability and stock returns. Eastern-Eur. J. Enterp. Technol. **124**(13) (2023)

Hasan, A.R.: Artificial Intelligence (AI) in accounting & auditing: a literature review. Open J. Bus. Manag. **10**(1), 440–465 (2021)

Husin, N.A., Abdulsaeed, A.A., Muhsen, Y.R., Zaidan, A.S., Alnoor, A., Al-mawla, Z.R.: Evalu-ation of metaverse tools based on privacy model using fuzzy MCDM approach. In: Al-Emran, M., Ali, J.H., Valeri, M., Alnoor, A., Hussien, Z.A. (eds.) Beyond Reality: Navigating the Power of Metaverse and Its Applications. IMDC-IST 2024. LNNS, vol. 895, pp. 1–20. Springer, Cham (2023). https://doi.org/10.1007/978-3-031-51716-7_1

Huynh-The, T., Pham, Q.V., Pham, X.Q., Nguyen, T.T., Han, Z., Kim, D.S.: Artificial intelligence for the metaverse: a survey. Eng. Appl. Artif. Intell. **117**, 105581 (2023)

Kieslich, K., Keller, B., Starke, C.: Artificial intelligence ethics by design. Evaluating public perception on the importance of ethical design principles of artificial intelligence. Big Data Soc. **9**(1), 20539517221092956 (2022)

Kondratenko, Y.P., Kreinovich, V., Pedrycz, W., Chikrii, A., Gil-Lafuente, A.M. (Eds.): Arti-ficial Intelligence in Control and Decision-making Systems: Dedicated to Professor Janusz Kacprzyk (Vol. 1087). Springer, Cham (2023). https://doi.org/10.1007/978-3-031-25759-9

Krakowski, S., Luger, J., Raisch, S.: Artificial intelligence and the changing sources of competitive advantage. Strategy Manag. J. **44**(6), 1425–1452 (2023)

Lehner, O.M., Ittonen, K., Silvola, H., Ström, E., Wührleitner, A.: Artificial intelligence based decision-making in accounting and auditing: ethical challenges and normative thinking. Account. Audit. Account. J. **35**(9), 109–135 (2022)

Maseer, R.W., Zghair, N.G., Flayyih, H.H.: Relationship between cost reduction and reevaluating customers'desires: the mediating role of sustainable development. Int. J. Econ. Financ. Stud. **14**(4), 330–344 (2022)

Muhsen, Y.R., Husin, N.A., Zolkepli, M.B., Manshor, N.: A systematic literature review of fuzzy-weighted zero-inconsistency and fuzzy-decision-by-opinion-score-methods: assessment of the past to inform the future. J. Intell. Fuzzy Syst. **45**(3), 4617–4638 (2023)

Munoko, I., Brown-Liburd, H.L., Vasarhelyi, M.: The ethical implications of using artificial intel-ligence in auditing. J. Bus. Ethics **167**(2), 209–234 (2020). https://doi.org/10.1007/s10551-019-04407-1

Noordin, N.A., Hussainey, K., Hayek, A.F.: The use of artificial intelligence and audit quality: an analysis from the perspectives of external auditors in the UAE. J. Risk Financ. Manag. **15**(8), 339 (2022)

Oluwasegun, I., Ishola, R., Ayoola, O.: Artificial intelligence and audit quality: implications for practicing accountants. Asian Econ. Financ. Rev. **13**(11), 756–772 (2023)

Pimenov, D.Y., Bustillo, A., Wojciechowski, S., Sharma, V.S., Gupta, M.K., Kuntoğlu, M.: Arti-ficial intelligence systems for tool condition monitoring in machining: Analysis and critical review. J. Intell. Manuf. **34**(5), 2079–2121 (2023)

Qasaimeh, G.M., Jaradeh, H.E.: The impact of artificial intelligence on the effective applying of cyber governance in Jordanian commercial banks. Int. J. Technol. Innov. Manag. (IJTIM) **2**(1) (2022)

Rashwan, A.R.M., Alhelou, E.M.: The impact of using artificial intelligence on the accounting and auditing profession in light of the Corona pandemic. J. Adv. Res. Bus. Manag. Account. (2020). ISSN, 2456, 3544

Rehman, A.: With the mediation of internal audit, can artificial intelligence eliminate and mitigate fraud? In: Handbook of Research on the Significance of Forensic Accounting Techniques in Corporate Governance, pp. 232–257. IGI Global. (2022)

Rodgers, W., Murray, J.M., Stefanidis, A., Degbey, W.Y., Tarba, S.Y.: An artificial intelligence algorithmic approach to ethical decision-making in human resource management processes. Hum. Resour. Manag. Rev. **33**(1), 100925 (2023)

Rodrigues, L., Pereira, J., da Silva, A.F., Ribeiro, H.: The impact of artificial intelligence on audit profession. J. Inf. Syst. Eng. Manag. **8**(1) (2023)

Saad, R.: The role of artificial intelligence techniques in achieving audit quality. Acad. Account. Financ. Stud. J. **26**(5), 1–18 (2022)

Salman, M.D., et al.: The impact of engineering anxiety on students: a comprehensive study in the fields of sport, economics, and teaching methods. Revista iberoamericana de psicología del ejercicio y el deporte **18**(3), 326–329 (2023)

Seethamraju, R., Hecimovic, A.: Adoption of artificial intelligence in auditing: an exploratory study. Aust. J. Manag. **48**(4), 780–800 (2023)

Zaidan, A.S., Alshammary, K.M., Khaw, K.W., Yousif, M., Chew, X.: Investigating behavior of using metaverse by integrating UTAUT2 and self-efficacy. In: Al-Emran, M., Ali, J.H., Valeri, M., Alnoor, A., Hussien, Z.A. (eds.) Beyond Reality: Navigating the Power of Metaverse and Its Applications. IMDC-IST 2024. LNNS, vol. 895, pp. 81–94. Springer, Cham (2023). https://doi.org/10.1007/978-3-031-51716-7_6

Zemankova, A.: Artificial intelligence and blockchain in audit and accounting: Literature review. wseas Trans. Bus. Econ. **16**(1), 568–581. (2019a)

Zemankova, A.: Artificial intelligence in audit and accounting: development, current trends, opportunities and threats-literature review. In: 2019 International Conference on Control, Artificial Intelligence, Robotics & Optimization (ICCAIRO), pp. 148–154. IEEE (2019b)

Zhang, C.A., Cho, S., Vasarhelyi, M.: Explainable artificial intelligence (XAI) in auditing. Int. J. Account. Inf. Syst. **46**, 100572 (2022)

Maximizing the Marketing Capabilities and Digital Sustainability of B2B & B2C Platforms Using Artificial Intelligence

Saad Kathim Khammat[1] and Abbas Gatea Atiyah[2](✉) (iD)

[1] Administration and Economic, University of Thi-Qar – Iraq, Nasiriyah, Iraq
[2] College of Administration and Economic, University of Thi-Qar – Iraq, Nasiriyah, Iraq
`abbas-al-khalidi@utq.edu.iq`

Abstract. The research aims to measure the extent of the possibility of maximizing the marketing capabilities of both the B2B platform (from company to company) and B2C (from company to customer). To achieve this goal, it developed a hypothetical plan for a long period of two hypotheses. Which has been implemented through the Reebok website in the United Arab Emirates. (63) Valid questionnaires were analyzed. Managers and marketing professionals presented the modern arms of statistical technical research, which are structural equation modeling with partial least squares and structural equation modeling with partial least squares (PLS-SEM). Using (Smart-PLS, V.0.4) program. The research revealed several opportunities, the most prominent of which is that artificial intelligence is compatible with the outstanding performance of the marketing capabilities of (B2B & B2C) platforms. The most important thing in the research is that more attention should be paid to artificial intelligence, as it produces marketing capabilities for companies.

Keywords: Artificial Intelligence · Maximizing Marketing Capabilities · (B2B & B2C) Platforms

1 Introduction

Controversy still exists among researchers about the results achieved from the use of artificial intelligence. Some believe that this type of digital tools constitutes a major obstacle to achieving credibility by companies in front of customers on the one hand, and violating the user's privacy on the other hand as a result of obtaining his data, geographical location, etc. elements (Atiyah et al., 2023; Sadaa et al., 2023; Khaw et al., 2023). On the other hand, there are those who believe that artificial intelligence is a very important digital tool that can facilitate a lot for companies, especially in the field of product marketing, by relying on the simulation systems provided by the digital space (Kumar et al., 2024; Alnoor et al., 2024a, b, c; Allal-Chérif et al., 2024). The digital revolution has made it possible for companies to adopt various modern methods in marketing and distributing the product, which means delivering it to its users or retailers (Kyytsönen et al., 2024). In the digital world today, and with the continuation of this massive digital revolution, companies have turned to digital marketing methods for their products, which express a type of communication between the company and the end user on the one hand or between the company and agents on the other hand (Zhang and Prebensen, 2024; Sandberg et al., 2023; Chew et al., 2023). Among the tools that have been adopted in this regard are digital marketing platforms called B2B (from business to business) and B2C (from company to customer). This method works to help individuals and make them aware of the nature of the product or service (Alnoor et al., 2024a, b, c; Atiyah, 2023). On the other hand, artificial intelligence is one of the aspects of the digital revolution in the business world. Its characteristics made it possible to delve into various topics with a cognitive dimension in the product and facilitated the possibility of entering into immersive experiences through the vast digital space (Ali et al., 2024). Therefore, it is may to identify its potential in developing these platforms and raising their level of efficiency (Atiyah et al., 2023). Accordingly, the research addressed two variables of great importance at the theoretical and applied levels at the present time. Regarding the independent variable (artificial intelligence), it is a widespread digital tool in the business world with enormous potential to expand the work of companies (Gignac and Szodorai, 2024). As for the adopted variable (B2B & B2C platforms), companies today are working extensively on owning and developing such platforms (Rėklaitis and Pilielienė, 2019; XinYing et al., 2024; Hesselbarth et al., 2023). The research included the descriptive analytical method, and in light of this, the research produced a set of conclusions.

2 Hypothesis Development

Artificial Intelligence is a technical science that performs human-like tasks, sensing inputs to analyze them to extract solutions with appropriate planning (Zhan et al., 2024). Using a computer, the human mind is simulated (Khan, 2024). Which enables artificial intelligence to perform tasks similar to human actions, such as learning, thinking, and knowledge inputs to solve problems creatively (Kandoth and Shekhar, 2024; Zaidan et al., 2023). On the other hand, the Business To Business platform (B2B) represents an electronic marketing platform concerned with the commercial exchanges that occur between companies (Maduku, 2024; Abbas et al., 2023). In other words, it plays the full role of various commercial operations in terms of buying and selling goods and services (Tzanidis et al., 2024). Which is usually between companies that operate according to the concept of wholesale trade, not retail (Mintz and Lilien, 2024). Through this platform, one company sells its products or services to another company, and this is done through electronic transactions using e-commerce technologies (Shams et al., 2024). Therefore, they are considered effective tools for marketing activities that rely on a variety of technologies that enhance the organization's marketing capabilities (Brown et al., 2024; Alnoor et al., 2023). Using artificial intelligence, companies can collect, analyze and interpret data easily and at a very high speed, which provides them with competitive advantages (Badrinarayanan and Ramachandran, 2024). As well as exploring its potential to quickly transform market research into reality and thus the possibility of communicating with consumers creatively (Onyijen et al., 2024; Alnoor et al., 2024a, b, c). AI marketing tools are cutting-edge software applications that are changing the game for businesses when it comes to marketing (Ravat et al., 2024; Atiyah and Zaidan, 2022). These use machine learning algorithms to simplify and automate marketing tasks, such as examining customer data from various sources such as social media, email, and customer interactions (Verma et al., 2024). Valuable insights generated by AI marketing tools can provide important information about customer behavior, preferences, and trends, enabling companies to make informed decisions and create personalized marketing strategies to connect with their audiences (Kshetri et al., 2023). By integrating AI tools into their operations, companies can enhance their marketing efforts and stay ahead of the competition in the digital landscape (Onyijen et al., 2024; Atiyah et al., 2023). Thus, we assume:

H1: Artificial intelligence has direct and positive influence on business-to-business platforms (B2B).

2.1 Artificial Intelligence and Business to Customers Platforms (B2C)

Artificial intelligence is a highly developed digital tool that contributes to finding solutions and taking measures to confront various circumstances with very high efficiency, speed, and accuracy (Sands et al., 2024). The technical applications used in it have transformed the way companies do business (Zhan et al., 2024). Automation methods and social interactions, based on artificial intelligence, have also helped improve marketing efficiency (Khan, 2024). The Business to Customer (B2C) platform represents an electronic platform for business between the company and the final customer (Treiblmaier and Petrozhitskaya, 2023). This type of platform attracts new participants by lowering barriers to entry, providing greater value to users, and has the potential to significantly change market structures (Feike and Rösch, 2024). It includes e-commerce and transactions between companies, consumers and individuals, as it allows companies to connect customers with merchants in exchange for commissions (Iankova et al., 2019). It maintains legal aspects to protect the consumer's right from violation of his rights after concluding the contract (Hassna et al., 2023). Trust also plays a crucial role in the electronic transaction relationship between companies and consumers, which is reflected in consumer behavior and the number of website visitors (Skare et al., 2023). It is also a business platform that bridges the distance between companies and consumers by processing data for the purpose of retail, thus enhancing the customer experience, and allowing various customers to assign and save an address to use the site to link them to companies (Kwon et al., 2022). Therefore, it is possible to employ the capabilities of artificial intelligence to better activate the work of these platforms. Valuable insights generated by AI marketing tools can provide important information about customer behavior, preferences, and trends, enabling companies to make informed decisions and create personalized marketing strategies to connect with their audiences (Kshetri et al., 2023). By integrating AI tools into their operations, companies can enhance their marketing efforts and stay ahead of the competition (Onyijen et al., 2024). So, we assume:

H2: Artificial intelligence has direct and positive influence on business to customer platforms (B2C).

3   Research Methodology

In this study, the researchers adopted the survey method. Primary data was collected using a questionnaire (Ahmed et al., 2024). The study sample consisted of (63) individuals who were managers in the food industry company, namely (Reebok), where the researchers received (58) responses. The two basic element (AI) represent the independent variables that can affects the dependent variable (B2B & B2C). In this study, well-established standards in solid literature were adopted to measure the research model. (4) items were used to measure the AI process, taken from (Mikalef et al., 2023). While for the (B2B & B2C) this study used scale of Feike and Rösch. (2024). As for the (B2B) adopted (4) items, and for (B2C) we utilized (4) items. A 5-point scale (1 = strongly disagree, 5 = strongly agree) was used to operationalize all concepts.

4   Data Analysis

To verify the validity and reliability of the research scale, the researcher adopted several indicators based on what was indicated by (Hair et al., 2020; Ahmed et al., 2024; Husin et al., 2023), where demonstrated the importance of three main indicators, which are the following:

1. Average Extracted Variance (AVE), whose value must be greater than 0.50.
2. Composite Reliability (CR), which must have a value greater than 0.70.

   External loads or component loads (Outer Loading, Factor Loading, FL), whose value must be greater than or equal to 0.70.

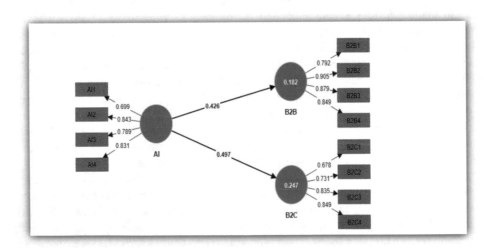

Fig. 1. Measurement of validity and reliability

From the results shown in Fig. 1, we find that the external loadings or factor loadings (FL) for all latent variables in this research range between 0.705 and 0.906, which are

Table 1. Result of measurement model

Variables	Items	FL	CR	AVE
Artificial intelligence	AI1	0.821	0.871	0.693
	AI2	0.819		
	AI3	0.857		
Business to Business	B2B1	0.796	0.917	0.734
	B2B2	0.906		
	B2B3	0.879		
	B2B4	0.843		
Business to Customer	B2C1	0.705	0.863	0.680
	B2C2	0.865		
	B2C3	0.892		

acceptable values. Except for the value of the latent variable (B2C1) and (AI1), they were less than the indicator required in the third item, therefore they were deleted.

After deleting the latent variables and re-analyzing, as shown by the results presented in Table 1 and Fig. 1, the factor loadings for all latent variables (FL) appeared between (0.75–0.906), which is higher than the acceptable value of (0.70), as confirmed by Hair et al. (2010). Also, the average variance extracted (AVE) values for all latent variables in this research range between (0.680–0.734), which are acceptable values. In addition, the composite reliability of the latent variables (CR) ranged between (0.863–0.917), which is also higher than the minimum required value (0.70). These results show that the scale items used in the model are reliable.

Table 2. Result of measurement model

	AI	B2B	B2C
AI	0.833		
B2B	0.446	0.857	
B2C	0.518	0.467	0.825

To confirming the discriminant validity of the scale used as shown in Table 2, the Fornell and Larcker (1981) criterion was used to measure the discriminant validity. It is an expression of the square root value of (AVE), which must be greater than (0.50) for all combinations. This indicates that all variables are clearly different from their counterparts (Hair et al., 2010; Abbas et al., 2023; Muhsen et al., 2023).

To confirming the discriminant validity of the scale used as shown in Table 3, the Fornell and Larcker (1981) criterion was used to measure the discriminant validity. It is an expression of the square root value of (AVE), which must be greater than (0.50)

Table 3. Hypotheses test

Variables	(O)	T statistics	P-values	Result
AI - > B2B	0.446	4.655	0.000	Supported
AI - > B2C	0.518	4.811	0.000	Supported

for all combinations. This indicates that all variables are clearly different from their counterparts (Hair et al., 2010; Abbas et al., 2023; Muhsen et al., 2023). Table 3 displays the results of the structural evaluation of the model. The research model is based on two hypotheses. The results shown in Table 3 and Fig. 1 indicate that there is a direct impact of AI on the business-to-business platforms (B2B) and business to customers platforms (B2C).

5   Discussion and Conclusion

The results show that there is a focused role for artificial intelligence in activating digital marketing platforms H1 and H2. This pivotal role focuses on technology in the sales processes of both platforms, and their ability to provide valuable offers to all consumers. Therefore, it is possible to achieve effective interactions between customers, through daily work related to purchases and sales, communication with customers, access to data, how to access platforms, work force, and harmonized distribution of marketing roles. The results presented by this study support the literature (Kumar et al., 2024; Allal-Chérif et al., 2024), as this literature confirmed that marketing through these two digital platforms can be developed based on artificial intelligence. We notice the extent to which the digital aspects of marketing tools can expand. With the tremendous technological developments, marketing tools can develop and the product can reach the consumer in its natural form without the slightest effort (virtual reality). These technologies can enable the development of more advanced platforms in the future and achieve the sustainability of successful marketing.

References

Abbas, S., et al.: Antecedents of trustworthiness of social commerce platforms: a case of rural communities using multi group SEM & MCDM methods. Electron. Commer. Res. Appl. **62**, 101322 (2023)

Adjei, M.T., Clark, M.N.: Relationship marketing in A B2C context: the moderating role of personality traits. J. Retail. Consum. Serv. **17**(1), 73–79 (2010)

Ahmed, A.D., Salih, M.M., Muhsen, Y.R.: Opinion weight criteria method (OWCM): a new method for weighting criteria with zero inconsistency. IEEE Access (2024)

Ali, J., Hussain, K.N., Alnoor, A., Muhsen, Y.R., Atiyah, A.G.: Benchmarking methodology of banks based on financial sustainability using CRITIC and RAFSI techniques. Decis. Mak. Appl. Manag. Eng. **7**(1), 315–341 (2024)

Allal-Chérif, O., Puertas, R., Carracedo, P.: Intelligent influencer marketing: how AI-powered virtual influencers outperform human influencers. Technol. Forecast. Soc. Chang. **200**, 123113 (2024)

Alnoor, A., Atiyah, A.G., Abbas, S.: Toward digitalization strategic perspective in the European food industry: non-linear nexuses analysis. Asia-Pac. J. Bus. Adm. (2023)

Alnoor, A., Atiyah, A.G., Abbas, S.: Unveiling the determinants of digital strategy from the perspective of entrepreneurial orientation theory: a two-stage SEM-ANN approach. Glob. J. Flex. Syst. Manag. 1–18 (2024a)

Alnoor, A., Chew, X., Khaw, K.W., Muhsen, Y.R., Sadaa, A.M.: Benchmarking of circular economy behaviors for Iraqi energy companies based on engagement modes with green technology and environmental, social, and governance rating. Environ. Sci. Pollut. Res. **31**(4), 5762–5783 (2024b)

Alnoor, A., et al.: How positive and negative electronic word of mouth (eWOM) affects customers' intention to use social commerce? A dual-stage multi group-SEM and ANN analysis. Int. J. Hum.-Comput. Interact. **40**(3), 808–837 (2024c)

Atiyah, A.G.: Unveiling the quality perception of productivity from the senses of real-time multisensory social interactions strategies in metaverse. In: Al-Emran, M., Ali, J.H., Valeri, M., Alnoor, A., Hussien, Z.A. (eds.) Beyond Reality: Navigating the Power of Metaverse and Its Applications. IMDC-IST 2024. LNNS, vol. 876, pp. 83–93. Springer, Cham (2023). https://doi.org/10.1007/978-3-031-51300-8_6

Atiyah, A.G., Zaidan, R.A.: Barriers to using social commerce. In: Alnoor, A., Wah, K.K., Hassan, A. (eds.) Artificial Neural Networks and Structural Equation Modeling, pp. 115–130. Springer, Singapore (2022). https://doi.org/10.1007/978-981-19-6509-8_7

Atiyah, A.G., All, N.D.A., Zaidan, A.S., Bayram, G.E.: Understating the social sustainability of metaverse by integrating adoption properties with users' satisfaction. In: Al-Emran, M., Ali, J.H., Valeri, M., Alnoor, A., Hussien, Z.A. (eds.) Beyond Reality: Navigating the Power of Metaverse and Its Applications. IMDC-IST 2024. LNNS, vol. 895, pp. 95–107. Springer, Cham (2023a). https://doi.org/10.1007/978-3-031-51716-7_7

Atiyah, A.G., Faris, N.N., Rexhepi, G., Qasim, A.J.: Integrating ideal characteristics of chat-GPT mechanisms into the metaverse: knowledge, transparency, and ethics. In: Al-Emran, M., Ali, J.H., Valeri, M., Alnoor, A., Hussien, Z.A. (eds.) Beyond Reality: Navigating the Power of Metaverse and Its Applications. IMDC-IST 2024. LNNS, vol. 895, pp. 131–141. Springer, Cham (2023b). https://doi.org/10.1007/978-3-031-51716-7_9

Badrinarayanan, V., Ramachandran, I.: Relational exchanges in the sales domain: A review and research agenda through the lens of commitment-trust theory of relationship marketing. J. Bus. Res. **177**, 114644 (2024)

Brown, D.M., et al.: Sustainability starts from within: a critical analysis of internal marketing in supporting sustainable value co-creation in B2B organisations. Ind. Mark. Manag. **117**, 14–27 (2024)

Cao, G., Weerawardena, J.: Strategic use of social media in marketing and financial performance: the B2B SME context. Ind. Mark. Manag. **111**, 41–54 (2023)

Carloman, A.N., Bermudo, U.V.W., Estilloso, E.M., Llantos, O.E.: Bundle AI: an application of multiple constraint knapsack problem (MCKP) through genetic algorithm (GA). Procedia Comput. Sci. **231**, 24–31 (2024)

Casidy, R., Mohan, M., Nyadzayo, M.: Integrating B2B and B2C research to explain industrial buyer behavior. Ind. Mark. Manag. **106**, 267–269 (2022)

Chatterjee, S., Chaudhuri, R., Ferraris, A., Sakka, G., Chaudhuri, S.: Implications of dynamic capabilities on triple bottom line performance after the COVID-19 pandemic: an empirical insight from B2B marketing perspective. Ind. Mark. Manag. **115**, 240–252 (2023)

Chew, X., Khaw, K.W., Alnoor, A., Ferasso, M., Al Halbusi, H., Muhsen, Y.R.: Circular economy of medical waste: novel intelligent medical waste management framework based on extension linear Diophantine fuzzy FDOSM and neural network approach. Environ. Sci. Pollut. Res. **30**(21), 60473–60499 (2023)

Deitz, G.D., Hansen, J.D., Fox, J.D., Morgan, R.M.: Masterful scholarly competences and the development of impactful B2B marketing theory: contributions of Dr Shelby D. Hunt. Ind. Mark. Manag. **115**, 368–377 (2023)

Feike, M., Rösch, J.: Nuanced but important: a literature-based comparison between B2B and B2C platforms. Decis. Anal. J. **10**, 100383 (2024)

Gaczek, P., Leszczyński, G., Mouakher, A.: Collaboration with machines in B2B marketing: Overcoming managers' aversion to AI-CRM with explainability. Ind. Mark. Manag. **115**, 127–142 (2023)

Gignac, G.E., Szodorai, E.T.: Defining intelligence: bridging the gap between human and artificial perspectives. Intelligence **104**, 101832 (2024)

Hair (2020) Multivariate Data Analysis. 7th Edition, Pearson, New York

Hair, J.F., Black, W.C., Babin, B.J., Anderson, R.E.: Multivariate Data Analysis, 7th edn. Pearson, New York (2010)

Hair, J.F., Black, W.C., Babin, B.J., Anderson, R.E.: Multivariate Data Analysis, 7th edn. Pearson, New York (2017)

Hassna, G., Rouibah, K., Lowry, P.B., Paliszkiewicz, J., Mądra-Sawicka, M.: The roles of user interface design and uncertainty avoidance in B2C ecommerce success: using evidence from three national cultures. Electron. Commer. Res. Appl. **61**, 101297 (2023)

He, J., Zhang, S.: How digitalized interactive platforms create new value for customers by integrating B2B and B2C models? An empirical study in China. J. Bus. Res. **142**, 694–706 (2022)

Hesselbarth, I., Alnoor, A., Tiberius, V.: Behavioral strategy: a systematic literature review and research framework. Manag. Decis. **61**(9), 2740–2756 (2023)

Husin, N.A., Abdulsaeed, A.A., Muhsen, Y.R., Zaidan, A.S., Alnoor, A., Al-mawla, Z.R.: Evaluation of metaverse tools based on privacy model using fuzzy MCDM approach. In: Al-Emran, M., Ali, J.H., Valeri, M., Alnoor, A., Hussien, Z.A. (eds.) Beyond Reality: Navigating the Power of Metaverse and Its Applications. IMDC-IST 2024. LNNS, vol. 895, pp. 1–20. Springer, Cham. https://doi.org/10.1007/978-3-031-51716-7_1

Iankova, S., Davies, I., Archer-Brown, C., Marder, B., Yau, A.: A comparison of social media marketing between B2B, B2C and mixed business models. Ind. Mark. Manag. **81**, 169–179 (2019)

Kandoth, S., Shekhar, S.K.: Scientometric visualization of data on artificial intelligence and marketing: analysis of trends and themes. Science Talks (2024)

Khan, M.R.: Customer inspiration and artificial intelligence: a paradigm shift in marketing (2024)

Khaw, K.W., Alnoor, A., Al-Abrow, H., Tiberius, V., Ganesan, Y., Atshan, N.A.: Reactions towards organizational change: a systematic literature review. Curr. Psychol. **42**(22), 19137–19160 (2023)

Kshetri, N., Dwivedi, Y.K., Davenport, T.H., Panteli, N.: Generative artificial intelligence in marketing: applications, opportunities, challenges, and research agenda. Int. J. Inf. Manag 102716 (2023)

Kumar, V., Ashraf, A.R., Nadeem, W.: AI-powered marketing: what, where, and how? Int. J. Inf. Manag. 102783 (2024)

Kwon, J., Chan, K.W., Gu, W., Septianto, F.: The role of cool versus warm colors in B2B versus B2C firm-generated content for boosting positive eWOM. Ind. Mark. Manag. **104**, 212–225 (2022)

Kyytsönen, M., Vehko, T., Jylhä, V., Kinnunen, U.M.: Privacy concerns among the users of a national patient portal: a cross-sectional population survey study. Int. J. Med. Inform. **183**, 105336 (2024)

Maduku, D.K.: Social media marketing assimilation in B2B firms: an integrative framework of antecedents and consequences. Ind. Mark. Manag. **119**, 27–42 (2024)

Mero, J., Vanninen, H., Keränen, J.: B2B influencer marketing: conceptualization and four managerial strategies. Ind. Mark. Manag. **108**, 79–93 (2023)

Mikalef, P., Islam, N., Parida, V., Singh, H., Altwaijry, N.: Artificial intelligence (AI) competencies for organizational performance: a B2B marketing capabilities perspective. J. Bus. Res. **164**, 113998 (2023)

Mintz, O., Lilien, G.L.: Should B2B start-ups invest in marketing? Ind. Mark. Manag. **117**, 220–237 (2024)

Moradi, M., Dass, M.: Applications of artificial intelligence in B2B marketing: challenges and future directions. Ind. Mark. Manag. **107**, 300–314 (2022)

Muhsen, Y.R., Husin, N.A., Zolkepli, M.B., Manshor, N.: A systematic literature review of fuzzy-weighted zero-inconsistency and fuzzy-decision-by-opinion-score-methods: assessment of the past to inform the future. J. Intell. Fuzzy Syst. **45**(3), 4617–4638 (2023)

Onyijen, O.H., Oyelola, S., Ogieriakhi, O.J.: Food manufacturing, processing, storage, and marketing using artificial intelligence. In A Biologist s Guide to Artificial Intelligence, pp. 183–200. Academic Press (2024)

Ravat, L., Hemonnet-Goujot, A., Hollet-Haudebert, S.: Exploring how to develop data-driven innovation capability of marketing within B2B firms: toward a capability model and process-oriented approach. Ind. Mark. Manag. **118**, 110–125 (2024)

Rėklaitis, K., Pilelienė, L.: Principle differences between B2B and B2C marketing communication processes. Organizacijø Vadyba: Sisteminiai Tyrimai **81**, 73–86 (2019)

Sadaa, A.M., Ganesan, Y., Yet, C.E., Alkhazaleh, Q., Alnoor, A.: Corporate governance as antecedents and financial distress as a consequence of credit risk. Evidence from Iraqi banks. J. Open Innov. Technol. Mark. Complex. **9**(2), 100051 (2023)

Salonen, A., Mero, J., Munnukka, J., Zimmer, M., Karjaluoto, H.: Digital content marketing on social media along the B2B customer journey: the effect of timely content delivery on customer engagement. Ind. Mark. Manag. **118**, 12–26 (2024)

Sandberg, H., Alnoor, A., Tiberius, V.: Environmental, social, and governance ratings and financial performance: evidence from the European food industry. Bus. Strategy Environ. **32**(4), 2471–2489 (2023)

Sands, S., Campbell, C., Ferraro, C., Demsar, V., Rosengren, S., Farrell, J.: Principles for advertising responsibly using generative AI. Organ. Dyn. 101042 (2024)

Sekaran, U., Bougie, R.: Research Methods for Business: A Skill-Building Approach, 7th Edition, Wiley & Sons, West Sussex (2016)

Shah, P.: Managing customer reactions to brand deletion in B2B and B2C contexts. J. Retail. Consum. Serv. **57**(C) (2020)

Shams, R., Sohag, K., Islam, M.M., Vrontis, D., Kotabe, M., Kumar, V.: B2B marketing for industrial value addition: how do geopolitical tension and economic policy uncertainty affect sustainable development? Ind. Mark. Manag. **117**, 253–274 (2024)

Skare, M., Gavurova, B., Rigelsky, M.: Innovation activity and the outcomes of B2C, B2B, and B2G E-Commerce in EU countries. J. Bus. Res. **163**, 113874 (2023)

Teo, T.S.: Usage and effectiveness of online marketing tools among Business-to-Consumer (B2C) firms in Singapore. Int. J. Inf. Manag. **25**(3), 203–213 (2005)

Treiblmaier, H., Petrozhitskaya, E.: Is it time for marketing to reappraise B2C relationship management? The emergence of a new loyalty paradigm through blockchain technology. J. Bus. Res. **159**, 113725 (2023)

Tzanidis, T., Magni, D., Scuotto, V., Maalaoui, A.: B2B green marketing strategies for European firms: Implications for people, planet and profit. Ind. Mark. Manag. **117**, 481–492 (2024)

Vangeli, A., Małecka, A., Mitręga, M., Pfajfar, G.: From greenwashing to green B2B marketing: a systematic literature review. Ind. Mark. Manag. **115**, 281–299 (2023)

Verma, S., Tiwari, R.K., Singh, L.: Integrating technology and trust: trailblazing role of AI in reframing pharmaceutical digital outreach. Intell. Pharm. (2024)

XinYing, C., Tiberius, V., Alnoor, A., Camilleri, M., Khaw, K.W.: The dark side of metaverse: a multi-perspective of deviant behaviors from PLS-SEM and fsQCA findings. Int. J. Hum.–Comput. Interact 1–21 (2024)

Zaidan, A.S., Alshammary, K.M., Khaw, K.W., Yousif, M., Chew, X.: Investigating behavior of using metaverse by integrating UTAUT2 and self-efficacy. In: Al-Emran, M., Ali, J.H., Valeri, M., Alnoor, A., Hussien, Z.A. (eds.) Beyond Reality: Navigating the Power of Metaverse and Its Applications. IMDC-IST 2024. LNNS, vol. 895, pp. 81–94. Springer, Cham (2023). https://doi.org/10.1007/978-3-031-51716-7_6

Zhan, Y., Xiong, Y., Han, R., Lam, H.K., Blome, C.: The impact of artificial intelligence adoption for business-to-business marketing on shareholder reaction: a social actor perspective. Int. J. Inf. Manag. 102768 (2024)

Zhang, Y., Prebensen, N.K.: Co-creating with ChatGPT for tourism marketing materials. Ann. Tour. Res. Empir. Insights 5(1), 100124 (2024)

Zheng, B., Wang, H., Golmohammadi, A.M., Goli, A.: Impacts of logistics service quality and energy service of Business to Consumer (B2C) online retailing on customer loyalty in a circular economy. Sustain. Energy Technol. Assess. 52, 102333 (2022)

Zhou, Z., Ding, Y., Feng, W., Ke, N.: Extending B2B brands into the B2C market: whether, when, and how brands should emphasize B2B industry background. J. Bus. Res. 130, 364–375 (2021)

Using LSTM Network Based on Logistic Regression Model for Classifying Solar Radiation Time Series

Zinah Mudher ALbazzaz and Osamah Basheer Shukur[✉]

Department of Statistics and Informatics, Faculty of Computer Sciences and Mathematics, University of Mosul, Mosul, Iraq
{zeenamudhar,drosamahannon}@uomosul.edu.iq

Abstract. An accurate knowledge of the types of solar radiation is important. Some machine learning and deep learning models have been used to classify solar radiation time series. However, applying some methods on a large spatial and temporal scale has only been hardly investigated. Therefore, in this study, two algorithms were used: long-term memory (LSTM) networks and the traditional statistical method represented by the logistic regression (LR) model to classify solar radiation patterns based on some meteorological time series data over a large time range. The results show that LSTM achieves acceptable results with clear superiority of the LSTM method for classification tasks. As a result, it is feasible to deduce that the capability of LSTM networks to learn long-term and determine the relation between time series of solar radiation and meteorological data has been more significant for more sophisticated applications, and hence the study underscores the importance of deep learning models, including LSTM networks, for large-scale applications comparing to LR and other traditional models.

Keywords: Long-short-term memory networks (LSTM) · Logistic regression (LR) model · Solar radiation · Meteorological time Series · Classification

1 Introduction

In this study, the classification of one of the most important weather variables was dealt with, is very important especially for the future events to know the extent of their impact on humans, animals, plants and other living organisms, and planning for the future free from the problems of the negative effects of different weather variables (Atiyah, 2023a, b; Ali et al., 2024; Alnoor et al., 2024a, b, c). The data of the time series variable of the total solar radiation (SLR) will be used and classified based on the best logistic regression (LR) model by diagnosing the mathematical relationship between it and several climatic effects to classify the SLR variable in terms of high or low radiation (Alnoor et al., 2023; Atiyah and Zaidan, 2022).

The most important machine learning and artificial intelligence methods represented by the Long Short Term Memory (LSTM) Network will be used to classify the time series data of the SLR variable based on other climatic time series variables such as

© The Author(s), under exclusive license to Springer Nature Switzerland AG 2024
A. Alnoor et al. (Eds.): AIRDS 2024, LNNS 1033, pp. 375–388, 2024.
https://doi.org/10.1007/978-3-031-63717-9_25

air temperatures in the maximum and minimum states, in addition to the time series of relative humidity in the maximum and minimum states in addition to the evaporation variable (Abbas et al., 2023; Muhsen et al., 2023; Ahmed et al., 2024; Atiyah, 2022). Most of the weather data and air pollutants are of the non-linear type, so the use of some linear methods and models may lead to results with little accuracy, because the weather data in general is one of the types of time series based on the principle of self-regression that contains many seasonal as well as cyclical changes that may negatively affect making this type of data heterogeneous, as well as affecting the prediction and classification results and their accuracy (Chaichan et al., 2020; Sekertekin et al., 2020; AL-Fatlawey et al., 2021).

The data was classified into two types of classifications according to the nature of the weather in Nineveh Governorate. The first group of data was classified into relatively hot months (May, June, July, August, September and October), while the second group of data included relatively cold months (November, December, January, February, March and April). In this study, it will be proposed to use the LSTM method as a modern method to improve the classification results of the SLR variable, depending on many climatic factors (Alnoor et al., 2024a, b, c; Husin et al., 2023; Al-Enzi et al., 2023; Atiyah et al., 2023a, b). The work of the LSTM method combines the principle of regression and classification work in one method in order to obtain a high and distinct classification accuracy (Chew et al., 2023; Atiyah, 2020; Zaidan et al., 2023). Hence, the ability of LSTM to learn long term and determine the relationship between solar radiation and meteorological data over the time series becomes more essential for complex applications, which is why the study draws attention to the significance of deep learning models, including LSTM with respect to large scale applications in comparison with LR and other traditional models (Karevan & Suykens, 2020; Liu et al., 2019).

Gao et al. (2019) proposed one day-ahead time-series classification and forecasting methods based on meteorology data for ideal weather condition by using LSTM networks. K.-S. Kim et al. (2020) predicted the ocean weather worldwide by using a deep learning model represented by convolutional LSTM. Alameen (2022) improved the accuracy of detecting plant diseases by proposing system uses LR model with LSTM neural networks for the soil content classification and forecasting by detecting the plant diseases. Naware and Mitra (2022) proposed a classical LSTM network for a combined day-ahead load and solar classification and forecasting based on weather classification. In this paper, LR model and LSTM neural network are used to classify binary SLR time series variable. Median measurement is used as threshold to determine the binary classes (high and low SLR). For satisfying the homogeneity, the studied datasets divided into two sets (hot and cold seasons). Every set of data will be also divided into other two time periods (for training and testing).

2   Methods and Materials

In this section, the basic theoretical concepts of the methods used in this research will be detailed, which include the basics and components of the LSTM network and the LR model, and the most important methods, tools and measurements that will be used.

The multiple linear regression (MLR) model contains a continuous dependent variable, while the dependent variable is binary in the logistic regression (LR) model. The independent variables can be quantitative, categorical variables, or both (Marill, 2004). MLR model supposed that the independent and dependent variables had a linear relationship to get good accurate results, but for logistic regression these variables can be include nonlinear relationships. LR is commonly used as valuable approach for modeling and classifying data based on a response binary variable (Hosmer et al., 1997; Midi et al., 2010; Pohlman & Leitner, 2003; Atiyah et al., 2023a, b; Alnoor et al., 2024a, b, c). LR associated with the probabilities of the observations of dependent (target) variable cases. We suppose that y denotes the target variable and will be take 1 or 0 values. LR is based on a linear model of the log of probability (Dayton, 1992).

The probability will be depended to classify the cases. If this probability will be more than 0.5, the case should classify the current observation as 1, while when the probability will be lower than 0.5, the current observation will be classified as 0 (Santner & Duffy, 1986). LSTM network can be employed to deal with time series datasets and sequences for regression and classification as machine learning and intelligent approach. LSTM is a type of recurrent neural network (RNN) as one of deep learning techniques that can learn long-term periods among time steps of time series datasets. LSTM can insert sequence or time series data into the input layer of network, and make classification and forecasting more accurately based on the univariate time steps related to time series data. Figure 1 below explains the flow of a time series x for C variables of length S of LSTM layer.

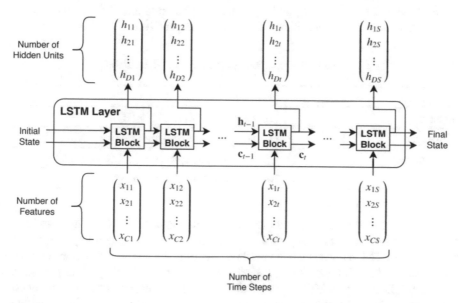

Fig. 1. The flow of a time series x with C variables (features) and length S in LSTM layer.

h_t and c_t represent the output or hidden and cell states respectively in the next step at time step t. The first block of LSTM network in the left side is used as the initial state at the first-time step of the time series or sequence to calculate the first output observation and the updated ct.

At the block of time step t, the current state of (ct − 1, ht − 1) are used and the block of next time step of the time series to calculate output and updated ct. The cell state ct includes the information that learned from cell state ct-1 at the previous time step. Figure 2 and Table 1 below explains the flow of data at time step t.

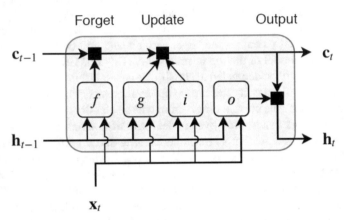

Fig. 2. The flow of data at time step t in LSTM layer.

In the Fig. 2 above, i, f, g, and o gates are highlighted and they denote the input, forget, Cell candidate (updates), and output gates respectively.

Table 1. The components at time step t description

Component	Formula
Input gate	$i_t = \sigma_g(W_i x_t + R_i h_{t-1} + b_i)(1)$
Forget gate	$f_t = \sigma_g(W_f x_t + R_f h_{t-1} + b_f)(2)$
Cell candidate	$g_t = \sigma_c(W_g x_t + R_g h_{t-1} + b_g)(3)$
Output gate	$o_t = \sigma_g(W_o x_t + R_o h_{t-1} + b_o)(4)$

Table 1 shows the state transfer (activation) function of LSTM layer that can be used by default is the hyperbolic tangent function (tanh). While the gate transfer (activation) function of LSTM layer that can be used by default is the sigmoid function formulated (Hochreiter & Schmidhuber, 1997; J. Kim et al., 2016; Li et al., 2017; Özdoğan-Sarıkoç et al., 2023; B. Xu et al., 2023; Y. Xu et al., 2014). Some measures are used to measure the accuracy of classification performance, including the confusion matrix (Luque et al., 2019), whose structure is shown in Table 2.

One of the measures used in the classification performance is Accuracy measurement, in which the ratio of expected cases matching actual cases to the total number of all expected and actual observations.

Table 2. The confusion matrix structure

Yes: positive No: negative		Actual	
		Yes	No
Predicted	Yes	True positive (TP)	False positive (FP)
	No	False negative (FN)	True negative (TN)

3 Results and Discussions

In this section, the practical results of the methods (LSTM network and the LR model) used in this research will be detailed, in addition to the most important discussions. In this study, the LSTM network will be used by writing a program using (MATLAB) software. The data used in this study included the use of six types of climate variables. The first variable is the total solar radiation (SLR) variable which is assumed as the dependent variable (target variable). As for the input variables, they are four variables, which are the two variables of air temperature (AT) in the maximum and minimum cases, and the two variables of relative humidity (RH) in the maximum and minimum cases, in addition to the variable of the amounts of evaporation (ET). To achieve the required homogeneity, which achieves results with high classification accuracy, the data has been aligned depending on the nature of the atmospheric temperature. Through time stratified approach, two homogeneous sets of data were obtained. The first group relates to the hot season for the months (May, June, July, August, September, and October), while the second group of data includes the relatively cold months (November, December, January, February, March, and April) for the years 2018, 2019, and 2020. In the second stage, the data for each season will divided into two groups, the first for training and includes most of the data for approximately 70%, and the second group for testing and includes approximately 30%, associating with the beginnings and ends of the months. The hot season includes 250 observations, of which 170 are for training and 80 are for the probationary period. As for the cold season, it included 266 observations, with 181 observations during the training period and 85 observations during the testing period. The target variable was converted from its originally continuous type into a categorical (binary) by using the median measurement as threshold for each period and season.

The general framework for implementing the LR model can be summarized in several sequential steps, such as follows.

Determining the positive and negative binary categories of the target variable.

Dividing the time series observations into two groups, the first for training and the second for testing, corresponding for all input variables as well as the target variable.

Inserting the inputs (explanatory) and dependent variables into Minitab worksheet according to sets detailed in step 1 and step 2. Table 3 determining the best models which include significant parameters with better classification accuracies. After modelling the studied time series datasets, four models are concluded as the best models according to parameters status and classification accuracy where x1: is maximum AT, x2: is minimum AT, x3: is maximum RH, x4: is minimum RH, and x5: is ET. The significance of the parameters and other details of model 1 are such as in Table 3.

Table 3. The significance and other parameters details of model 1.

Term	Parameters	Z-Value	P-Value	VIF
Constant	−4.8400	−2.37	0.018	*
x_2	0.4480	5.29	0.000	1.99
x_4	−0.1313	−2.04	0.041	1.60
x_5	−0.9780	−3.57	0.000	2.50

The significance of the parameters and other details of model 1 are such as in Table 4.

Table 4. The significance and other parameters details of model 2.

Term	Parameters	Z-Value	P-Value	VIF
Constant	−13.7700	−4.99	0.000	*
x_1	0.2474	2.66	0.008	2.67
x_2	0.3682	4.05	0.000	2.31
x_5	−1.1030	−3.88	0.000	2.76

The significance of the parameters and other details of model 1 are such as in Table 5.

Table 5. The significance and other parameters details of model 3.

Term	Parameters	Z-Value	P-Value	VIF
Constant	−4.1700	−1.97	0.048	*
x_1	0.5080	4.64	0.000	3.95
x_2	−0.4180	−4.16	0.000	3.54
x_4	−0.0489	−2.42	0.015	1.23

The significance of the parameters and other details of model 1 are such as in Table 6. For all models above, all parameters are significant because their p-values are less than the level of significance $\alpha = 0.05$. Variance inflation factor (VIF) is also acceptable that means there is no multicollinearity problem among explanatory variables. The classification accuracy for model 1, model 2, model 3, and model 4 are 66.47%, 75.29%, 69.01%, 68.45% respectively for training, while the accuracy are 50.00%, 50.00%, 61.76%, 67.06% respectively for testing.

Table 6. The significance and other parameters details of model 4.

Term	Parameters	Z-Value	P-Value	VIF
x_1	0.5910	6.58	0.000	3.99
x_2	−0.5141	−5.33	0.000	2.10
x_3	−0.0811	−6.44	0.000	1.39

The general framework for implementing the LSTM network algorithm is summarized in several sequential steps, such as follows.

1. Determining the positive and negative binary categories of the target variable.
2. Dividing the time series observations into two groups, the first for training and the second for testing, corresponding for all input variables as well as the target variable.
3. Determining the structure of the LSTM neural network (network structure) by defining the learning and transfer functions, gates, weights, bias and according to the nature of the data. The proposed transfer function is (tanh) explained through the theoretical side because it is suitable for the nature of the data.
4. Combining the bias value with the additive function of the weighted input variables.
5. Training after determining the learning rate to determine the interruption of the training process depending on the desired learning rate.
6. Computing the classification accuracy of the LSTM network model.

The classification accuracy for training and testing periods are 70.5882%, 51.25% respectively. The training process details above can be shown clearly by following the training diagram such as follows in Fig. 3.

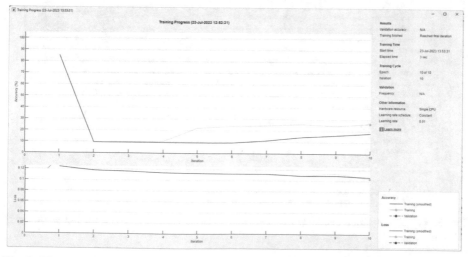

Fig. 3. The classification accuracy and loss scales during training process for hot season data using model 1.

The structure of the confusion matrices was used to express the classification accuracy in its details, and the results were as shown in the Fig. 4.

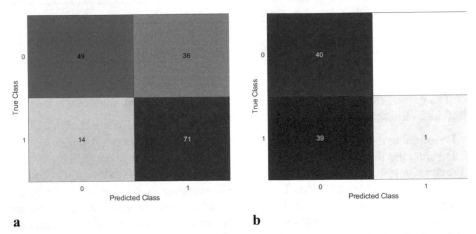

a   **b**

Fig. 4. Classification Confusion Matrix Chart for hot season data using model 1. a: for the training period, b: for the test period.

The classification accuracy for training and testing periods are 88.8235%, 50% respectively. The training process details above can be shown clearly by following the training diagram such as follows in Fig. 5.

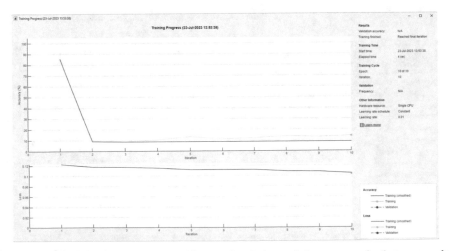

Fig. 5. The classification accuracy and loss scales during training process for hot season data using model 2.

The structure of the confusion matrices was used to express the classification accuracy in its details, and the results were as shown in the Fig. 6.

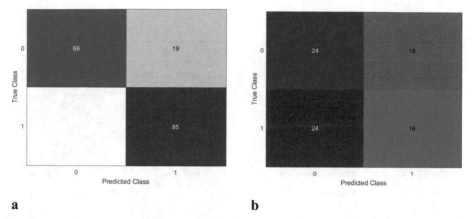

a   b

Fig. 6. Classification Confusion Matrix Chart for hot season data using model 2. a: for the training period, b: for the test period.

The classification accuracy for training and testing periods are 75.6906%, 76.4706% respectively. The training process details above can be shown clearly by following the training diagram such as follows in Fig. 7.

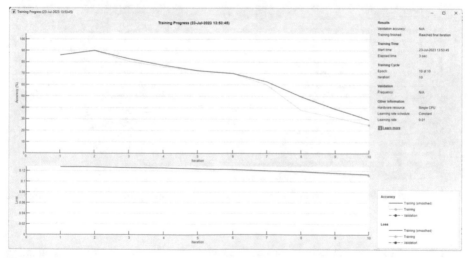

Fig. 7. The classification accuracy and loss scales during training process for cold season data using model 3.

The structure of the confusion matrices was used to express the classification accuracy in its details, and the results were as shown in the Fig. 8.

The classification accuracy for training and testing periods are 76.2431%, 77.6471% respectively. The training process details above can be shown clearly by following the training diagram such as follows in Fig. 9.

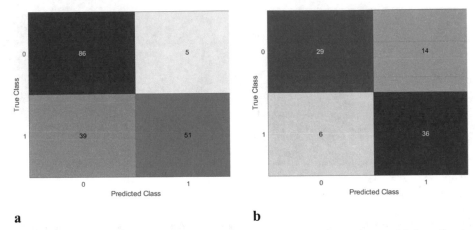

Fig. 8. Classification Confusion Matrix Chart for cold season data using model 3. a: for the training period, b: for the test period.

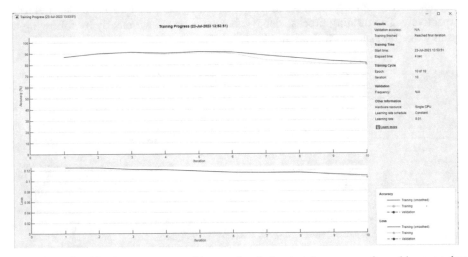

Fig. 9. The classification accuracy and loss scales during training process for cold season data using model 4.

The structure of the confusion matrices was used to express the classification accuracy in its details, and the results were as shown in the Fig. 10.

In this study and from all results in the figures and tables above, it clear that LSTM network outperform LR model for SLR classification. There are some problems in mode 3 and model 4 because of their components. The dispersions of explanatory variables have variety ranges. The variables of RH in maximum and minimum cases have high dispersions and the variables of AT in maximum and minimum cases have low dispersions, while the best variable is ET which has a minimal dispersions and better homogeneity. All of these reasons lead to some problems in models 3 and 4. The classification accuracy

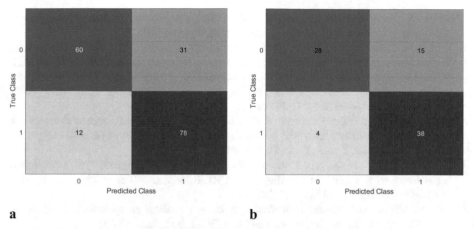

Fig. 10. Classification Confusion Matrix Chart for cold season data using model 4. a: for the training period, b: for the test period.

of testing period for models 3 and 4 are outperform the accuracy of training periods. Over fitting problem is appeared in models 3 and 4 because of that. But mostly, LSTM network performed high accuracy when it used with this type of time series datasets.

4   Conclusions

Through the mentioned results and discussions in the previous section, it is possible to conclude that it is possible to obtain a very accurate classification, as was noted in the results. LSTM is one of RNN techniques and deep learning approach. It has an ability of LSTM networks to learn long-term and determine the relationship between the time series of SLR and its relationship and several meteorological time series datasets. Therefore, it is possible to propose modern method represented by LSTM network as an intelligent and machine learning technique based on LR model as a traditional method to obtain high classification accuracy Accurate algorithms ensure efficient problem solving in a variety of domains and provide many advantages. First, they guarantee accuracy of decision therefore, consistent results and errors minimization. Furthermore, precise algorithms improve resource utilization through time and computational power optimization, which makes the process efficient and cost-effective. Moreover, such projects allow innovation by becoming basis for more research and development, hence, pushing the technology and scientific knowledge upon.

References

Abbas, S., et al.: Antecedents of trustworthiness of social commerce platforms: a case of rural communities using multi group SEM & MCDM methods. Electron. Commer. Res. Appl. **62**, 101322 (2023)

Ahmed, A.D., Salih, M.M., Muhsen, Y.R.: Opinion weight criteria method (OWCM): a new method for weighting criteria with zero inconsistency. IEEE Access (2024)

Alameen, A.: Improving the accuracy of multi-valued datasets in agriculture using logistic regression and LSTM-RNN method. TEM J. **11**(1), 454–462 (2022)

Al-Enzi, S.H.Z., Abbas, S., Abbood, A.A., Muhsen, Y.R., Al-Hchaimi, A.A.J., Almosawi, Z.: Exploring research trends of metaverse: a bibliometric analysis. In: Al-Emran, M., Ali, J.H., Valeri, M., Alnoor, A., Hussien, Z.A. (eds.) Beyond Reality: Navigating the Power of Metaverse and Its Applications. IMDC-IST 2024. LNNS, vol. 895, pp. 21–34. Springer, Cham (2023). https://doi.org/10.1007/978-3-031-51716-7_2

AL-Fatlawey, M.H., Brias, A.K., Atiyah, A.G.: The role of strategic behavior in achievement the organizational excellence analytical research of the manager's views of Ur state company at Thi-Qar governorate. J. Adm. Econ. **10**(37) (2021)

Ali, J., Hussain, K.N., Alnoor, A., Muhsen, Y.R., Atiyah, A.G.: Benchmarking methodology of banks based on financial sustainability using CRITIC and RAFSI techniques. Decis. Mak. Appl. Manag. Eng. **7**(1), 315–341 (2024)

Alnoor, A., Atiyah, A.G., Abbas, S.: Toward digitalization strategic perspective in the European food industry: non-linear nexuses analysis. Asia-Pac. J. Bus. Adm. (2023)

Alnoor, A., Atiyah, A.G., Abbas, S.: Unveiling the determinants of digital strategy from the perspective of entrepreneurial orientation theory: a two-stage SEM-ANN approach. Glob. J. Flex. Syst. Manag. 1–18 (2024a)

Alnoor, A., Chew, X., Khaw, K.W., Muhsen, Y.R., Sadaa, A.M.: Benchmarking of circular economy behaviors for Iraqi energy companies based on engagement modes with green technology and environmental, social, and governance rating. Environ. Sci. Pollut. Res. **31**(4), 5762–5783 (2024b)

Alnoor, A., et al.: How positive and negative electronic word of mouth (eWOM) affects customers' intention to use social commerce? A dual-stage multi group-SEM and ANN analysis. Int. J. Hum.-Comput. Interact. **40**(3), 808–837 (2024c)

Atiyah, A.G.: Impact of knowledge workers characteristics in promoting organizational creativity: an applied study in a sample of smart organizations. PalArch's J. Archaeol. Egypt/Egyptol. **17**(6), 16626–16637 (2020)

Atiyah, A.G.: Effect of temporal and spatial myopia on managerial performance. J. La Bisecoman **3**(4), 140–150 (2022)

Atiyah, A.G.: Strategic network and psychological contract breach: the mediating effect of role ambiguity. Int. J. Res. Manag. Stud. (IJRMS) **13**(1) (2023a)

Atiyah, A.G.: Unveiling the quality perception of productivity from the senses of real-time multisensory social interactions strategies in metaverse. In: Al-Emran, M., Ali, J.H., Valeri, M., Alnoor, A., Hussien, Z.A. (eds.) Beyond Reality: Navigating the Power of Metaverse and Its Applications. IMDC-IST 2024. LNNS, vol. 876, pp. 83–93. Springer, Cham (2023b). https://doi.org/10.1007/978-3-031-51300-8_6

Atiyah, A.G., Zaidan, R.A.: Barriers to using social commerce. In: Alnoor, A., Wah, K.K., Hassan, A. (eds.) Artificial Neural Networks and Structural Equation Modeling, pp. 115–130. Springer, Singapore (2022). https://doi.org/10.1007/978-981-19-6509-8_7

Atiyah, A.G., Alhasnawi, M., Almasoodi, M.F.: Understanding metaverse adoption strategy from perspective of social presence and support theories: the moderating role of privacy risks. In: Al-Emran, M., Ali, J.H., Valeri, M., Alnoor, A., Hussien, Z.A. (eds.) Beyond Reality: Navigating the Power of Metaverse and Its Applications. IMDC-IST 2024. LNNS, vol. 876, pp. 144–158. Springer, Cham (2023). https://doi.org/10.1007/978-3-031-51300-8_10

Atiyah, A.G., Faris, N.N., Rexhepi, G., Qasim, A.J.: Integrating ideal characteristics of chat-GPT mechanisms into the metaverse: knowledge, transparency, and ethics. In: Al-Emran, M., Ali, J.H., Valeri, M., Alnoor, A., Hussien, Z.A. (eds.) Beyond Reality: Navigating the Power of Metaverse and Its Applications. IMDC-IST 2024. LNNS, vol. 895, pp. 131–141. Springer, Cham (2023). https://doi.org/10.1007/978-3-031-51716-7_9

Chaichan, M.T., Kazem, H.A., Al-Waeli, A.H., Sopian, K.: The effect of dust components and contaminants on the performance of photovoltaic for the four regions in Iraq: a practical study. Renew. Energy Environ. Sustain. **5**, 3 (2020)

Chew, X., Khaw, K.W., Alnoor, A., Ferasso, M., Al Halbusi, H., Muhsen, Y.R.: Circular economy of medical waste: novel intelligent medical waste management framework based on extension linear Diophantine fuzzy FDOSM and neural network approach. Environ. Sci. Pollut. Res. **30**(21), 60473–60499 (2023)

Dayton, C.M.: Logistic regression analysis. Stat 474–574 (1992)

Gao, M., Li, J., Hong, F., Long, D.: Day-ahead power forecasting in a large-scale photovoltaic plant based on weather classification using LSTM. Energy **187**, 115838 (2019)

Hochreiter, S., Schmidhuber, J.: Long short-term memory. Neural Comput. **9**(8), 1735–1780 (1997)

Hosmer, D.W., Hosmer, T., Le Cessie, S., Lemeshow, S.: A comparison of goodness-of-fit tests for the logistic regression model. Stat. Med. **16**(9), 965–980 (1997)

Husin, N.A., Abdulsaeed, A.A., Muhsen, Y.R., Zaidan, A.S., Alnoor, A., Al-mawla, Z.R.: Evaluation of metaverse tools based on privacy model using fuzzy MCDM approach. In: Al-Emran, M., Ali, J.H., Valeri, M., Alnoor, A., Hussien, Z.A. (eds.) Beyond Reality: Navigating the Power of Metaverse and Its Applications. IMDC-IST 2024. LNNS, vol. 895, pp. 1–20. Springer, Cham (2023). https://doi.org/10.1007/978-3-031-51716-7_1

Karevan, Z., Suykens, J.A.: Transductive LSTM for time-series prediction: an application to weather forecasting. Neural Netw. **125**, 1–9 (2020)

Kim, J., Kim, J., Thu, H.L.T., Kim, H.: Long short term memory recurrent neural network classifier for intrusion detection. Paper presented at the 2016 international conference on platform technology and service (PlatCon) (2016)

Kim, K.-S., Lee, J.-B., Roh, M.-I., Han, K.-M., Lee, G.-H.: Prediction of ocean weather based on denoising autoencoder and convolutional LSTM. J. Mar. Sci. Eng. **8**(10), 805 (2020)

Li, X., et al.: Long short-term memory neural network for air pollutant concentration predictions: method development and evaluation. Environ. Pollut. **231**, 997–1004 (2017)

Liu, Y., Su, Z., Li, H., Zhang, Y.: An LSTM based classification method for time series trend forecasting. Paper presented at the 2019 14th IEEE Conference on Industrial Electronics and Applications (ICIEA) (2019)

Luque, A., Carrasco, A., Martín, A., de Las Heras, A.: The impact of class imbalance in classification performance metrics based on the binary confusion matrix. Pattern Recognit. **91**, 216–231 (2019)

Marill, K.A.: Advanced statistics: linear regression, part II: multiple linear regression. Acad. Emerg. Med. **11**(1), 94–102 (2004)

Midi, H., Sarkar, S.K., Rana, S.: Collinearity diagnostics of binary logistic regression model. J. Interdiscip. Math. **13**(3), 253–267 (2010)

Muhsen, Y.R., Husin, N.A., Zolkepli, M.B., Manshor, N.: A systematic literature review of fuzzy-weighted zero-inconsistency and fuzzy-decision-by-opinion-score-methods: assessment of the past to inform the future. J. Intell. Fuzzy Syst. **45**(3), 4617–4638 (2023)

Naware, D., Mitra, A.: Weather classification-based load and solar insolation forecasting for residential applications with LSTM neural networks. Electr. Eng. **104**(1), 347–361 (2022)

Özdoğan-Sarıkoç, G., Sarıkoç, M., Celik, M., Dadaser-Celik, F.: Reservoir volume forecasting using artificial intelligence-based models: artificial neural networks, support vector regression, and long short-term memory. J. Hydrol. **616**, 128766 (2023)

Pohlman, J.T., Leitner, D.W.: A comparison of ordinary least squares and logistic regression (2003)

Santner, T.J., Duffy, D.E.: A note on A. Albert and JA Anderson's conditions for the existence of maximum likelihood estimates in logistic regression models. Biometrika **73**(3), 755–758 (1986)

Sekertekin, A., Arslan, N., Bilgili, M.: Modeling diurnal land surface temperature on a local scale of an arid environment using artificial neural network (ANN) and time series of Landsat-8 derived spectral indexes. J. Atmos. Solar Terr. Phys. **206**, 105328 (2020)

Xu, B., Pooi, C.K., Tan, K.M., Huang, S., Shi, X., Ng, H.Y.: A novel long short-term memory artificial neural network (LSTM)-based soft-sensor to monitor and forecast wastewater treatment performance. J. Water Process Eng. **54**, 104041 (2023)

Xu, Y., Du, J., Dai, L.-R., Lee, C.-H.: A regression approach to speech enhancement based on deep neural networks. IEEE/ACM Trans. Audio Speech Lang. Process. **23**(1), 7–19 (2014)

Zaidan, A.S., Alshammary, K.M., Khaw, K.W., Yousif, M., Chew, X.: Investigating behavior of using metaverse by integrating UTAUT2 and self-efficacy. In: Al-Emran, M., Ali, J.H., Valeri, M., Alnoor, A., Hussien, Z.A. (eds.) Beyond Reality: Navigating the Power of Metaverse and Its Applications. IMDC-IST 2024. LNNS, vol. 895, pp. 81–94. Springer, Cham (2023). https://doi.org/10.1007/978-3-031-51716-7_6

Author Index

A. Alnoor et al. (Eds.): AIRDS 2024, LNNS 1033, pp. 389–390, 2024.
https://doi.org/10.1007/978-3-031-63717-9

Printed in the United States
by Baker & Taylor Publisher Services